Lecture Notes in Civil Engineering

Volume 570

Lecture Notes in Civil Engineering (LNCE) publishes the latest developments in Civil Engineering—quickly, informally and in top quality. Though original research reported in proceedings and post-proceedings represents the core of LNCE, edited volumes of exceptionally high quality and interest may also be considered for publication. Volumes published in LNCE embrace all aspects and subfields of, as well as new challenges in, Civil Engineering. Topics in the series include:

- Construction and Structural Mechanics
- Building Materials
- Concrete, Steel and Timber Structures
- Geotechnical Engineering
- Earthquake Engineering
- Coastal Engineering
- Ocean and Offshore Engineering; Ships and Floating Structures
- Hydraulics, Hydrology and Water Resources Engineering
- Environmental Engineering and Sustainability
- Structural Health and Monitoring
- Surveying and Geographical Information Systems
- Indoor Environments
- Transportation and Traffic
- Risk Analysis
- Safety and Security

To submit a proposal or request further information, please contact the appropriate Springer Editor:

- Pierpaolo Riva at pierpaolo.riva@springer.com (Europe and Americas);
- Swati Meherishi at swati.meherishi@springer.com (Asia—except China, Australia, and New Zealand);
- Wayne Hu at wayne.hu@springer.com (China).

All books in the series now indexed by Scopus and EI Compendex database!

Y. K. Guruprasad · G. R. Dodagoudar ·
K. S. Nanjunda Rao · P. C. Pandey · H. B. Nagaraj
Editors

Recent Advances in Infrastructure Development and Management—Volume 1

Select Proceedings of MTCEIDM 2023

Springer

Editors
Y. K. Guruprasad
Department of Civil Engineering
BMS College of Engineering
Bangalore, Karnataka, India

G. R. Dodagoudar
Department of Civil Engineering
Indian Institute of Technology Madras
Chennai, Tamil Nadu, India

K. S. Nanjunda Rao
Department of Civil Engineering
Indian Institute of Science Bangalore
Bangalore, Karnataka, India

P. C. Pandey
Department of Civil Engineering
Indian Institute of Science
Bangalore, Karnataka, India

H. B. Nagaraj
Department of Civil Engineering
BMS College of Engineering
Bangalore, Karnataka, India

ISSN 2366-2557 ISSN 2366-2565 (electronic)
Lecture Notes in Civil Engineering
ISBN 978-981-96-1553-7 ISBN 978-981-96-1554-4 (eBook)
https://doi.org/10.1007/978-981-96-1554-4

This Springer imprint is published by the registered company Springer Nature Singapore Pte Ltd.
The registered company address is: 152 Beach Road, #21-01/04 Gateway East, Singapore 189721, Singapore

Preface

Advancement and modernization of construction practices in the field of civil engineering has led into rapid development in infrastructure and in construction management. The first volume of this book aims to serve as a comprehensive resource for academicians and professionals. The chapters in this volume cover a wide range of topics, and each chapter presents research contributions that addresses current challenges and throws light on future developments in the respective areas.

This volume contains select proceedings of the International Conference on 'Modern Trends in Civil Engineering Infrastructure Development and Management', held during 8th to 10th December 2023.This gathering brings together leading researchers, practitioners, and innovators to discuss and share advancements in civil engineering that shape the future of infrastructure. Civil engineering has always been at the forefront of transforming society, and today, it faces a unique challenge: to develop sustainable, resilient, and technologically advanced solutions that cater to the growing needs of our world.

The themes of this conference cover a broad range of pressing topics and advancements in civil engineering. **IoT applications in monitoring of civil engineering Infrastructure** highlights the role of the Internet of things (IoT) in enhancing infrastructure safety and efficiency through real-time monitoring and predictive maintenance. **Induction of recycled waste in construction** emphasizes sustainable construction practices that reduce environmental impact by incorporating recycled materials and fostering a more circular economy in the construction industry. Parallel to this, **futuristic building materials** explores newer materials that promise greater durability, flexibility, and sustainability in modern structures.

The theme of **innovation in foundation design and soil stabilization** is crucial as engineers face the complexities of diverse soil types and challenging terrains. Novel approaches in foundation design and soil stabilization ensure stability and safety, meeting the demands of various environments. Additionally, **modalities in design and behaviour of mega structures** delves into the unique challenges associated with large-scale structures, which require sophisticated engineering approaches to withstand extreme loads and environmental stresses. **Effective construction management practices** addresses the importance of efficient project management, emphasizing strategies to minimize delays, reduce costs, and maximize resource utilization. Infrastructure for **water resources and transportation engineering** is essential to support growing populations, and this theme explores sustainable water resource management and advanced transportation solutions that are crucial for economic and social development. Finally, **GIS & remote sensing applications to civil engineering** sheds light on how spatial data technologies are revolutionizing planning, design, and infrastructure management by enabling more precise and informed decision-making.

Each chapter in this volume has been selected after a thorough check and peer-review to ensure the quality and academic rigour is maintained. The editors appreciate the contributions made by the authors and thank the reviewers for their invaluable feedback. This volume would not have been possible without their dedication and expertise. The variety of topics covered in this volume shows the multifaceted aspects in civil engineering. From theoretical research to practical applications, each chapter contributes to a deeper understanding of the individual domain and addresses towards the current challenges. The interdisciplinary nature of the research also underscores the importance of collaboration across different domains of expertise, fostering innovation.

We hope that this volume serves as a valuable reference for researchers, engineers, and students, providing them with the inspiration leading them towards application of this knowledge into practice.

Bengaluru, India Dr. Y. K. Guruprasad, Ph.D. (IISc)

Contents

About the Editors

Dr. Y. K. Guruprasad PhD(IISc) is working as an Associate Professor in the Department of Civil Engineering at BMS College of Engineering, Bengaluru. He received his Ph.D. from the Indian Institute of Science (IISc), Bengaluru. He carried out part of his Ph.D. research at the University of Edinburgh, Scotland, United Kingdom, in 2009 as part of the UK-India Education and Research Initiative (UKIERI) program on repair of structures adopting fibre-reinforced polymer (FRP) and the behaviour of structures exposed to fire. Dr. Guruprasad has also completed his post-doctoral research at IISc. He completed M.Tech. in Structural Engineering from M. S. Ramaiah Institute of Technology (MSRIT), Bengaluru, and his B.E. in Civil Engineering from BMS College of Engineering, Bengaluru. His areas of research include but are not limited to repair and retrofit of structures, retrofit of structures adopting FRP composites, adoption of IoT in structural health monitoring, non-destructive testing evaluation, earthquake engineering and seismic-resistant analysis and design of structures, structural dynamics and vibration analysis, fracture mechanics, and microanalysis of concrete exposed to high temperature. Dr. Guruprasad has multiple publications in refereed international and national journals, conferences, and book chapters. He has guided seven M.Tech. students in Structural Engineering. He has also guided many final-year B.E. projects, which have been recognized and funded by the Karnataka State Council for Science and Technology (KSCST) at the IISc Campus. He has also carried out consultancy work in non-destructive evaluation of structures and structural analysis and design of structures. Dr. Guruprasad is a reviewer for international journals. He is a life member of the Institution of Engineers (India), Chartered Engineer (CE) and Institution of Engineers (India), Indian Concrete Institute(ICI), Indian Society of Nondestructive Testing (ISNT), Indian Association for Computational Mechanics (IndACM), and Institute for Research Development and Training Of Construction Trades & Management (INSTRUCT). He is also a fellow member of the Association of Consulting Civil Engineers (ACCE).

Dr. G. R. Dodagoudar PhD(IITB) is presently working as a Professor in the Department of Civil Engineering, Indian Institute of Technology (IIT) Madras, India. He obtained his Ph.D. degree from IIT Bombay. His areas of research interest include computational geomechanics, contaminant transport modelling, reliability analysis of geotechnical systems, soil dynamics and earthquake engineering, and seismic hazard and risk assessment. Dr. Dodagoudar has guided 21 Ph.D. and seven M.S. (research) thesis and many M.Tech. dissertations. He has extensive publications in national and international refereed journals. Dr. Dodagoudar has been actively involved in many continuing education programmes. He is an active member of the New Delhi Standing Appellate Committee of the All India Council for Technical Education (AICTE). Dr. Dodagoudar has completed many prestigious sponsored research and consultancy projects.

Dr. K. S. Nanjunda Rao PhD(IISc) is currently the Chief Research Scientist in the Department of Civil Engineering at IISc, Bangalore. He obtained his B.E. (Civil Engineering) and M.E. from BMS College of Engineering, Bangalore, and Ph.D. (Structural Engineering) from IISc Bangalore. His research interests include mechanics of composite plate and shell structures, finite element structural analysis, structural masonry, reinforced and prestressed concrete, earthquake-resistant design of structures, and alternative building materials. Dr. Dodagoudar is actively involved in experimental facilities like shock table testing and strong floor development for structural testing. The book co-authored by him titled *Alternative Building Materials and Technologies* received the ACCE—NAGADI AWARD in 2008 for Best Publication in Civil Engineering. He has supervised 10 Ph.D. and M.Sc. (Engineering) theses. He has published over 60 papers in refereed journals and conferences. Dr. Dodagoudar has successfully completed several sponsored research projects. He has served as the chair and as a member of several academic and administrative committees.

Dr. P. C. Pandey PhD(UK) retired as a Professor of Structural Engineering from IISc, Bengaluru, and is currently serving as the professor emeritus in Civil Engineering at REVA University, Bangalore. Dr. Pandey has obtained his B.E. from BIT Mesra, ME from IISc, Bengaluru, and Ph.D. from the University of Liverpool, United Kingdom (UK). He has previously served as a faculty at Nanyang Technological University, Singapore, and at BITS Pilani. He completed his post-doctoral fellowship at the University of Wales, Swansea, and later served there as an Erasmus Mundus visiting professor. His broad areas of research interest are computational mechanics, finite element analysis, and damage mechanics. Dr. Pandey has supervised six Ph.D. students and has published over 120 papers in international journals and conferences.

Dr. H. B. Nagaraj PhD(IISc) completed his bachelor's in civil engineering from SJCE, Mysore, in 1986, and obtained his post-graduation in geotechnical engineering from the National Institute of Technology (NIT), Surathkal. He completed his doctoral degree in soil mechanics from IISc, Bengaluru. He has been serving as a faculty in the Civil Engineering Department at BMS College of Engineering since 1997 and is currently heading the department. He has keen research interests in the engineering behaviour of fine-grained soils and alternate building materials. He has published more than 40 technical papers in international journals and conferences. Dr. Nagaraj is an active reviewer for most of the leading international journals in the field of geotechnical engineering and building materials. He has been a recipient of many international awards as a reviewer from 'Publons' and has also received the Outstanding Reviewer Award from two international journals the 'Journal of Cleaner Production' and 'Applied Clay Science'. He has recently been successful in implementing his laboratory-scale research findings in the construction of a rammed earth structure in one of the residential building projects in Bengaluru, taken up in collaboration with Made in Earth-Sustainable Architecture and Building Practices, Bengaluru. In 2016, he was invited to review an international project proposal (seeking funding) submitted by a leading materials manufacturing and research organization to the United States Department of Agriculture, National Institute of Food and Agriculture under the grant programme Small Business Innovation Research (SBIR), USA. Dr. Nagaraj has supervised two doctoral dissertations along with four master of science and forty master's degree theses. He has filed several patents, a few of which have been published. He has handled various externally funded projects, including those of the Department of Science & Technology (DST) and AICTE. He has also reviewed the BIS Code: Review of Geotechnical Provisions in Indian Seismic Code IS 1893 (Part 1): 2002 as a part of the Gujarat State Disaster Management Authority, Gandhinagar (GSDMA)—IIT Kanpur project. He has delivered more than 20 lectures for e-learning programmes on geotechnical engineering. He has handled more than 50 major consultancy projects in the fields of geotechnical investigation and forensic geotechnics and has recently managed two major industry-sponsored research projects to use iron ore tailings from Rai Bahadur Seth Shreeram Narasingdas (RBSSN) Pvt. Ltd., Karnataka, to produce building blocks using iron ore tailings. He is presently supervising the practical implementation of the production units.

Utilisation of Industrial Fines in the Manufacturing of Lime-Stabilised Compressed Earth Bricks for Construction

R. K. Preethi and B. V. Venkatarama Reddy

Abstract This study addresses the use of blends of lime and ground granulated blast furnace slag (GGBS) in stabilising the natural soil. Tests were conducted to obtain the optimum mix ratio for producing the lime—GGBS bricks. The results of the lime—GGBS-stabilised compressed specimens are presented. The investigations showed that the optimum clay content of the soil mix was 20% which resulted in yielding maximum compressive strength of the soil compacts. Tests were conducted to determine the characteristics of the bricks such as strength, absorption, durability, and stress–strain characteristics considering two different lime contents (6 and 10%) and 10% GGBS content. Considerable amount of difference among the wet and dry compressive strengths of the bricks was noticed.

Keywords Lime · Pozzolanic activity · Compressive strength · Absorption

1 Introduction

Mud bricks such as adobe are the earliest masonry units used in the construction industry, that is because of the ease in casting, low cost, good thermal properties, and at the end of the buildings life they can be reused by simply breaking them into powder. To enhance the brick properties, bricks have been subjected to high temperature burning up to 1100 °C to increase the strength and durability characteristics [1, 2]. Later, stabilisation came into practice to improve the properties of the naturally occurring soil to avoid the burning of clay bricks. Stabilisation is the method of altering the properties of the naturally occurring soil to obtain the satisfactory engineering properties required for the construction [3–5]. Utilisation of industrial

R. K. Preethi (✉)
Department of Civil Engineering, Global Academy of Technology, Bengaluru, India
e-mail: drpreethirk@gat.ac.in

B. V. Venkatarama Reddy
Department of Civil Engineering, Indian Institute of Science, Bengaluru, India
e-mail: venkat@iisc.ac.in

Y. K. Guruprasad et al. (eds.), *Recent Advances in Infrastructure Development and Management—Volume 1*, Lecture Notes in Civil Engineering 570,
https://doi.org/10.1007/978-981-96-1554-4_1

fines in stabilising the natural soil is in practise because of its pozzolanic property. To reduce the consumption of cement in the manufacturing process of compressed earth bricks (CEBs), lime pozzolana-based binder was used as an alternative to the cement binder. Ground granulated blast furnace slag (GGBS) is one of the most commonly used pozzolanic materials in the combination with lime to improve the properties of the stabilised soil.

Utilisation of lime-activated GGBS in stabilising the soils to manufacture the unfired masonry units was another application [6–12]. In these studies, the effect of lime—GGBS binder in producing the unfired clay bricks was studied. A series of trial mixes were attempted for the purpose of casting bricks in large scale by varying the clay fraction, GGBS, water, and lime content (% by mass). From the trial mixes, two mix proportions having clay fraction, GGBS, lime, and water content of 87, 5.5, 1.5, and 6%, respectively, were considered, and both quick lime and hydraulic lime were used in casting the bricks. The bricks were moist cured for 7, 28, 56, and 90 days. The wet compressive strength of the bricks was determined at the different ages of curing. The results of lime-stabilised bricks were compared with the cement—GGBS-stabilised bricks. The 28-day compressive strength of the lime-stabilised bricks was around 3.5 MPa, whereas the 90-day strength of the bricks ranged from 6 to 7.5 MPa. The 90-day compressive strength of the PC-stabilised bricks was around 5.5 MPa. The test results demonstrated that the lime—GGBS-stabilised bricks achieved higher strength when compared with the bricks cast using cement—GGBS blends. The rate of moisture absorption of the lime—GGBS and PC—GGBS bricks during the moist curing periods (7, 28, 56, and 90 days) was determined following the guidelines of the code [13]. The water absorption of the lime—GGBS bricks was in the range of 17–20%, whereas the water absorption of cement-stabilised bricks was 20–22%. These studies conclude that the engineering properties of the lime—GGBS-stabilised bricks were within the limits for the satisfactory performance.

The concept of stabilisation of natural soils using lime is in practice in the past 5000 years ago. Greater part of the studies on lime stabilisation of soils were focused on the construction of roads, pavements, embankments, and foundations. From the literature available on lime—GGBS compressed bricks, the experimental investigations resulted in promising results in producing the lime—GGBS-stabilised bricks for the construction purpose. There are inadequate studies on manufacturing the lime—GGBS-stabilised compressed earthen bricks and investigating the properties. The current study is mainly focused on optimising the mix proportions to produce the compacted lime—GGBS bricks and exploring the engineering properties.

2 Materials Used

Materials used in the study include a natural soil, river sand, ground granulated blast furnace slag, and calcium hydroxide (lime).

Table 1 Properties of natural and reconstituted soil

Physical properties	Natural soil	SC20
Specific gravity	2.68	2.67
Liquid limit (%)	31	17.76
Plastic limit (%)	19.48	11.11
Shrinkage limit (%)	15.99	14.85

River sand was used in casting the lime-stabilised soil compacts. Natural river sand of grain size < 4.75 mm was used. The sand used in the investigations does not contain silt or clay size fractions. The river sand used in the investigation follows the Zone II grading as per IS 383 (2016) code classifications. Fresh hydrated lime (calcium hydroxide) an assay of 95.5% was used in the study. Natural soil used in the study was taken from Bengaluru, Karnataka. The soil has a clay fraction of 41% (< 2 μm). The soil contains predominantly kaolin clay mineral and traces of montmorillonite and calcite. The elemental components of natural soil obtained by energy dispersive X-ray analysis (EDAX) are silica (15.68%), alumina (24.56%), calcium (0.32%), iron (9.28%), potassium (1.17%), and sodium (0.21%). The natural soil was reconstituted by mixing with the river sand to obtain the desired clay content in the mix. The reconstituted soil with 20% clay content is designated as SC20. The Atterberg's limits and the specific gravity of the SC20 were determined following the code guidelines [14] and [15]. The properties of the natural soil and SC20 examined are given in Table 1.

Ground granulated blast furnace slag supplied by a steel plant was used throughout the research work. The chemical composition of GGBS is Ca (21.67%), Si (18.31%), Al (10.87%), and Mg (4.41%). The lime reactivity of GGBS tested as per the guidelines of the code [16] was 9.74MPa.

3 Experimental Programme

Compacted cylindrical specimens were used in examining the strength of the stabilised soil compacts. The current study involves the experimental work and testing of large number of specimens considering various mix ratios. The trial mix ratios considered in the study are given in Table 2. The Table gives an information about the variations in clay content in the soil, lime content (3, 6, 10, and 12%), and GGBS content (3, 7, 10, and 15%). Use of compressed bricks for such parametric study results in handling large quantity materials, and therefore, small cylindrical specimens of dimensions 38 mm diameter and 76 mm height were considered in examining the compressive strength. From the trial mix ratios attempted, the bricks of dimensions 230 × 110 × 70 mm manufactured with 6% lime are designated as LG6, the bricks with 10% lime are designated as LG10, and the GGBS content of 10% was kept constant in both the types of the bricks. The bricks were examined for the compressive strength and absorption characteristics.

4 Casting and Testing Procedures

The procedures involved in casting and testing cylindrical specimens and the brick specimens are discussed in the following sections.

4.1 Casting and Curing the Specimens

4.1.1 Cylindrical Specimens

The casting procedure adopted for cylindrical specimens is as follows.

(a) The soil was ground in a small ball mill for 10 min to break soil lumps and then manually mixed dry to ensure uniformity. Water was then added to the mixture.
(b) The water-soil mixture was mixed using a mechanical mixer for seven minutes to ensure uniformity. The dry density was maintained at 1.8 g/cc while casting the cylindrical specimens.
(c) The partially saturated mix was fed into cylindrical moulds, compacted from both ends in a screw-jack setup, and then immediately extruded and placed on a platform for 24 h in the laboratory.
(d) After 24 h, one set of specimens was cured under wet burlap, while the other set was steam-cured in a water bath at 80°C for 24 h.

4.1.2 Lime—GGBS Bricks

Details of casting procedure adopted for compressed LG bricks are as follows.

(a) Ingredients were mixed in a pan mixer to achieve a uniform dry mix.
(b) Water was slowly added, maintaining a dry density of 1.8 g/cc while casting the bricks.
(c) The wet mix was fed into a manually operated brick-making machine, where it was compacted by a piston.
(d) After compaction, the bricks were extruded and placed on a platform for 24 h. They were then cured under wet burlap for 28 days.

4.2 Testing Procedures

Cured specimens (both cylinders and bricks) were dried in air, inside the laboratory (RH: 41%), and later specimens were tested both in wet and dry conditions. Testing procedures followed to determine the compressive strength and durability characteristics of the stabilised soil compacts are detailed in the following sections.

4.2.1 Cylindrical Specimens

The compressive strength and the X-ray diffraction tests were performed to determine the mechanical and chemical properties of the cylindrical specimens. The compression testing machine was employed, and test was performed by fixing the piston movement at the rate of 0.625 mm/min. The specimens after cured and air dried were tested by soaking them in water for 48 h prior to the test, and the strength was designated as wet compressive strength. Six specimens were tested in each type.

The X-ray diffraction (XRD) analysis was performed on the powdered samples of the cylindrical specimens at different ages after curing to understand the mineralogical changes in the naturally occurring clay minerals present in the soil due to lime—GGBS stabilisation. The X-ray diffraction (XRD) patterns of the lime-stabilised samples were taken using Bruker D8 Advance diffractometer with Cu Kα ($\lambda = 1.5406$ Å) and the using a step size of 0.02°(2θ).

4.2.2 Lime—GGBS Bricks

Compressive Strength

The compressive strength of LG bricks was determined following the procedure outlined in IS 3495—Part 1 [17] code. The dimensions of the specimens were measured, and the mass of the specimens was noted at the time of testing. The specimens were tested in a compression testing machine applying the concentric load at the piston displacement rate of 0.625 mm/min. The moisture content at the time of testing was assessed by drying the sample collected from the failed specimen in an oven at 110 °C for 24 h. Totally six specimens were tested in each category.

Absorption Characteristics

The water absorption of bricks was determined following the procedure given in the IS 3495 [17]code guidelines. The bricks were dried in an oven (at 110 ± 5 °C) until the constant mass was obtained. The percentage of water absorbed by the brick was calculated with respect to the initial dry mass. The dried bricks were then soaked in water, and the weights were recorded at different time intervals. The percentage of water absorbed by the bricks was determined with respect to the dry mass. The initial rate of absorption (IRA) test for GCEB's was conducted following the procedure outlined in American standard [18]. The amount of water absorbed by the brick in one minute was calculated, and the IRA was determined.

5 Results and Discussions

5.1 *Compacted Cylindrical Specimens*

5.1.1 Influence of Clay Content on the Strength of the Lime-Stabilised Soil Compacts

The natural soil mixed with river sand was used to cast the lime-stabilised soil compacts. The natural soil with 41% clay fraction was reconstituted by adding the river sand to obtain a mix with 15, 20, and 30% clay fraction. The cylindrical compacts were cast using lime varying between 3 and 12%. The lime pozzolana compacts were cured using two types of curing conditions: 24 h steam curing at 80 °C and 28 days wet burlap curing. The relationships between wet compressive strength and the lime content in the mix with varying clay fractions are shown in Fig. 1. Figure 1a and b shows the variation in strength of the steam-cured and burlap-cured specimens, respectively. The results shown in the figure represent the mean of three values.

Wet compressive strength of the compacts increased with the increase in the lime content in the mix irrespective of the curing conditions and clay content. There is a considerable difference in compressive strength of the compacts as the lime content varied between 3 and 12% for both the types of curing conditions. Increase in strength with the increase in lime content can be attributed to the higher percentage of pozzolanic activity taking place resulting in the formation of more cementitious gels. The cementitious gel mainly consists of hydrates of silica and alumina formed due to pozzolanic reactions between lime and clay minerals. Considerably higher strength values were noticed for steam-cured specimens when compared with the burlap-cured specimens. Steam curing accelerates the pozzolanic reactions and results in higher strength values [19–21]. Increase in clay content marginally increased the

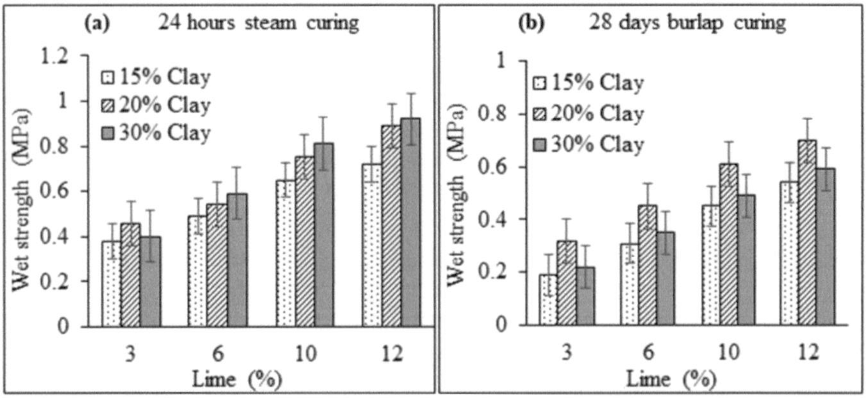

Fig. 1 Variation in strength of the compacts with the varying lime content, **a** 24 h steam curing and **b** 28 days burlap curing

strength of the steam-cured compacts (Fig. 1a). Whereas, strength of the wet burlap-cured compacts increases with increase in clay content (15 to 20%), reaches the peak value, and then drops down when clay content increased from 20 to 30% in the mix (Fig. 1b). The figure clearly depicts that the optimum clay content yielding maximum strength was 20% for wet burlap curing condition. Excess clay fraction in the mix beyond the optimum limit can remain unreacted and remain as a filler leading to the reduction in strength. The maximum wet compressive strength in steam curing process was 0.92 MPa for specimens with 30% clay and 12% lime content and that in wet burlap curing process was 0.7 MPa for specimens with 20% clay and 12% lime content in the mix.

5.1.2 Influence of GGBS Content on the Strength of the Lime-Stabilised Compacts

The results of lime-stabilised soil compacts (discussed in the earlier section) showed maximum strength at 20% clay. Hence, the soil with 20% clay content was further explored to examine influence of GGBS on the lime-stabilised soil compacts. The GGBS content in the mix was varied from 3 to 15% and the lime 3–12%. The 28-day burlap curing and the 24 h steam curing was attempted. The variation in the wet compressive strength of the compacts with varying GGBS and lime content is shown in Fig. 2 for both the curing conditions. The values in the figure represent the mean of three values. From the strength results shown in the figures the following points emerge:

- The wet compressive strength of the lime-stabilised compacts increased with the increase in the GGBS content irrespective of the type of curing. The strength varied nearly linear with increase in GGBS content in the mix.

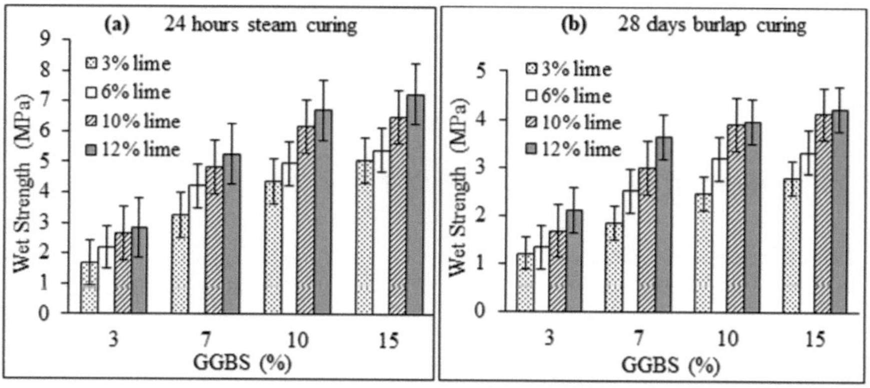

Fig. 2 Variation in strength of the compacts with varying GGBS content (**a**) 24 h steam curing and **b** 28 days burlap curing

- The lime content in the mix has significant influence on the strength of the compacts at the lower dosages of GGBS. At 3% GGBS content the strength of the compacts increased by about 74% as the lime content increased from 3 to 12%, whereas, at 15% GGBS content the strength increased by about 50% irrespective of the curing conditions.
- The maximum wet compressive strength of 7.26 and 4.21 MPa was achieved for steam-cured and burlap-cured specimens, respectively.
- Increase in GGBS content from 3 to 15%, increased the strength of the compacts by about 150 and 100% in the steam-cured and wet burlap-cured specimens, respectively, irrespective of the lime content in the mix.
- The lime-stabilised soil compacts resulted in higher strength with the addition of GGBS, this can be attributed to the fact that the pozzolanic reactions between lime, GGBS, and clay minerals resulted in the formation of additional cementitious gels.

5.1.3 Variation in the Strength of the Lime—GGBS Compacts with Time

Lime-stabilised soil compacts with 20% clay fraction, 10% GGBS, and lime content were cast to determine the strength development of the compacts with time. The cylinders compacted with 6% lime are designated as LG6 and that with 10% lime are designated as LG10. Compacted cylinders were cured under wet burlap for 28 days, after curing the cylinders were exposed to air under laboratory conditions until the time of testing. The strength development of the compacts was determined at different ages of testing such as 7, 14, 28, 60, 180, and 240 days. Figure 3 shows the variation in the strength of the compacts with the time of testing. Each point in the figure represents the mean of three values. The figure shows the relationship for wet compressive strength and the time of testing (days) of the compacts for two lime contents. The compressive strength of the compacts increased steeply with the increase in the time of testing up to 28 days irrespective of the lime content. The increase in strength between 7 and 28 days was about 135%. There was hardly any variation in the strength values beyond 28 days of testing.

5.2 Compacted Lime—GGBS Bricks

5.2.1 Compressive Strength

The compressive strength of the LG bricks was determined in both wet and dry conditions. Indian standard code [17] protocol was followed in determining the compressive strength. The wet compressive strength was determined by testing the bricks soaked in water for 48 h. The dry strength was determined by testing the cured and air-dried specimens. The 28-day wet and dry compressive strengths of the LG6 bricks were 6.51 and 14.22 MPa, respectively, and that of LG10 bricks were 8.64 and 18.42

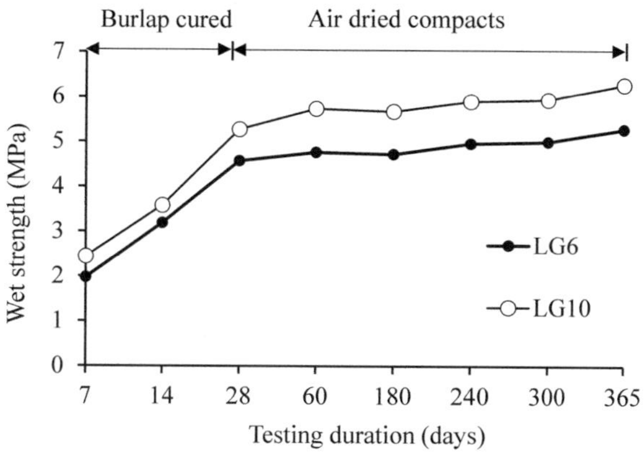

Fig. 3 Compressive strength development with respect to the age of the compacts

MPa, respectively. Studies by Oti et al., reveal that the 28-day compressive strength of the lime—GGBS-stabilised bricks was around 3.5 MPa [9]. The compressive strength of the burnt clay bricks commonly used for construction in India ranges from 3 to 11 MPa [22, 23].

5.2.2 Absorption Characteristics

Water Absorption

The water absorption of LG bricks was determined following the procedure outlined in the code [17] guidelines. The water absorption of LG6 and LG10 bricks after 24 h of soaking in water was 10.55 and 6.72%, respectively. Investigations by Oti et al. show that the water absorption of the lime—GGBS bricks ranges from 17 to 20% [9]. The percentage water absorption of locally available bunt clay bricks ranges from 10 to 19% [22, 23].

Rate of Water Absorption

The rate of water absorption of the bricks was determined to understand the amount of water absorbed by the brick with respect to time. Cured bricks were dried in an oven at 50 °C to attain the constant weight. The oven dried bricks were then soaked in water, and the weights were recorded at different time intervals (1, 2, 3, 4, 5, 10, 20, 30, 40, 50, 60, 120, 180, 240, 1440, 2880, and 4320 min). The results in the figure represent the mean of three values. The figure shows the moisture absorption curves of the two types of bricks (LG6 and LG10). The LG10 bricks absorbed less water content when compared to that of the LG6 bricks. The moisture content of LG6 and LG10 bricks was 8.42 and 12.06%, respectively, at the end of 4320 min of soaking duration. The degree of saturation (S) of the LG6 and LG10 bricks at the end of 4320 min of soaking duration was 66.62 and 46.51%, respectively. This clearly construes that the bricks were not saturated even after 4320 min of soaking duration.

Initial Rate of Absorption

Initial rate of absorption (IRA) test on bricks was conducted following the procedure outlined in the code [18] guidelines. The brick was dried in an oven until the constant mass was attained. The dried brick was allowed to cool down to ambient temperature, and the dry mass was measured. The brick was then kept in a tray over the supports such that 3 mm depth of the brick is in contact with the water for one minute. After one minute the weight of the brick was taken. The amount of water absorbed by the brick in one minute was calculated, and IRA was expressed as kg/m^2/minute. The IRA of LG6 and LG10 bricks was 1.53 and 1.36 kg/m^2/min, respectively. The IRA for conventional burnt clay bricks ranges from 1.31 to 3.53 kg/m^2/min [22, 23]. The IRA test conducted on cement stabilised CEBs showed that the values ranged from 1.93 to 4.5 kg/m^2/min [24].

5.2.3 X-Ray Diffraction (XRD) Analysis

The XRD analysis of the compacts at different ages of testing was determined. The X-ray diffraction (XRD) peaks were obtained using Bruker D8 Advance diffractometer with Cu Kα ($\lambda = 1.5406$ Å). The XRD analysis of the compacts determined at the various ages of testing (7, 14, 28, 60, and 180-days) are shown in Fig. 4. The blends of the stabilised soils displayed the peaks of CSH and traces of residual kaolin clay mineral in the X-ray diffractograms. The peaks at ~ 29° 2-Theta ascribed to the CSH gels [25, 26] although it might be overlapped with a calcite peak. This indicates that the main hydration product of all the regardless of the activator used.

In general, CSH gels do not display sharp peaks in the XRD patterns. While there are sharp peaks displayed at around 29° 2-Theta in the diffractograms for the lime—GGBS-stabilised soils. Formation of these sharp peaks was due to the peaks of the CSH gels overlapping the peaks of calcite (29.405° 2-Theta) [27]. The amount of CSH was small in the 7-day analysis compared to the 180-days analysis in both the cases. This variation in the amount of CSH gels attributes to the formation of more CSH gels due to the pozzolanic reactions which takes place gradually in the lime—GGBS mixtures [28].

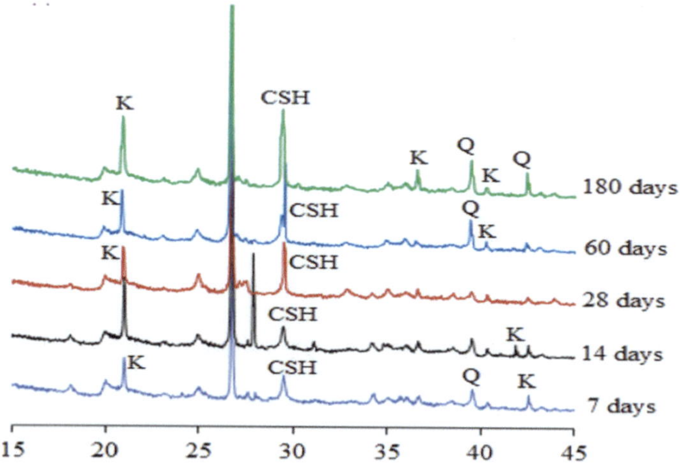

Fig. 4 X-ray diffraction of compacts showing the peaks of K = Kaolinite, Q = Quartz, and CSH = Calcium silicate hydrate

6 Conclusions

The experimental investigations were focused on optimising the mix proportion for producing the lime—GGBS-stabilised bricks and examining the characteristics of bricks such as strength, and absorption characteristics. The following concluding points emerge based on the experimental analysis:

- The optimum clay content yielded maximum strength of the compacts was 20% clay fraction irrespective of the lime content in the mix. Addition of pozzolanic material such as GGBS enhanced the strength of the compacts irrespective of the lime contents.
- The development in the strength of the compacts with the ageing was noticed. The XRD patterns showed the formation of CSH gels.
- The lime—GGBS bricks showed considerable difference among the wet and dry compressive strengths. The dry compressive strength of GCEBs obtained was nearly double that of the wet compressive strength.
- The absorption characteristics of lime—GGBS bricks were well within the acceptable limits.

References

1. Karaman S, Ersahin S, Gunal H (2006) Firing temperature and firing time influence on mechanical and physical properties of clay bricks. J Sci Ind Res 6:153–159
2. Andavan S, Vamsi KP (2020) A study on soil stabilization by addition of fly ash and lime. Mater Today Proc 22:1125–1129. https://doi.org/10.1016/j.matpr.2019.11.323

3. Attoh-Okine NO (1995) Lime treatment of laterite soils and gravels— revisited. Constr Build Mater 9:283–287. https://doi.org/10.1016/0950-0618(95)00030-J
4. Locat J, Tremblay H, Leroueil S (1996) Mechanical and hydraulic behaviour of a soft inorganic clay treated with lime. Can Geotech J 33:654–669
5. Wild S, Kinuthia JM, Robinson RB, Humphreys I (1996) Effects of ground granulated blast furnace of slag (GGBS) on the strength and swelling properties of lime-stabilized kaolinite in the presence of sulphates. Clay Miner 31:423–433
6. Oti JE, Kinuthia JM, Bai J (2008) Developing unfired stabilised building materials in the UK, Proc. Inst. Civ Eng Eng Sustain 161:211–218. https://doi.org/10.1680/ensu.2008.161.4.211
7. Oti JE, Kinuthia JM, Bai J (2008) Using slag for unfired-clay masonry-bricks. Proc Inst Civ Eng Constr Mater 161:147–155. https://doi.org/10.1680/coma.2008.161.4.147
8. Oti JE, Kinuthia JM, Bai J (2009) Compressive strength and microstructural analysis of unfired clay masonry bricks. Eng Geol 109:230–240. https://doi.org/10.1016/j.enggeo.2009.08.010
9. Oti JE, Kinuthia JM, Bai J (2009) Engineering properties of unfired clay masonry bricks. Eng Geol 107:130–139. https://doi.org/10.1016/j.enggeo.2009.05.002
10. Oti JE, Kinuthia JM, Bai J (2009) Unfired clay bricks: from laboratory to industrial production. Proc Inst Civ Eng Eng Sustain 162:229–237. https://doi.org/10.1680/ensu.2009.162.4.229
11. Oti JE, Kinuthia JM, Bai J (2010) Freeze–thaw of stabilised clay brick. Proc Inst Civ Eng Waste Resour Manage 163:129–135. https://doi.org/10.1680/warm.2010.163.3.129
12. Syafiadi RA, Abdullah MMAB, Ahmad R, Nergis DDB, Rahim SZA, Omar MF, Sandu AV, Vizureanu P, Syafwandi (2022) Potential of soil stabilization using ground granulated blast furnace slag (GGBFS) and fly ash via geopolymerization method: a review. Materials (Basel) 15. https://doi.org/10.3390/ma15010375
13. BSI (2003) BS EN 771–1: specification for masonry units—part 1: clay masonry units. Br Stand Inst
14. IS-2720 (Part 5) (1985) Methods of test for soils: Determination of liquid limit and plastic limit. Bur Indian Stand New Delhi India
15. IS-2720 (Part 3) (1980) Indian standard methods of test for soils: determination of specific gravity, Fine grained soils. Bur Indian Stand New Delhi India
16. IS-1727 (2004) Methods of test for pozzolanic materials. Bur Indian Stand New Delhi India
17. IS-3495 (1–4) (1992) Methods of tests for burnt clay building bricks. Bur Indian Stand New Delhi India
18. ASTM C67 (2020) Standard test methods for sampling and testing brick and structural clay tile. Am Soc Test Mater 1–17. https://doi.org/10.1520/C0067
19. Reddy BVV, Hubli SR (2002) Properties of lime stabilised steam-cured blocks for masonry. Mater Struct Constr 34:293–300. https://doi.org/10.1617/13740
20. Reddy BVV, Lokras SS (1998) Steam-cured stabilised soil blocks for masonry construction. Energy Build 29:29–33. https://doi.org/10.1016/s0378-7788(98)00033-4
21. Dhar S, Hussain M (2021) The strength and microstructural behavior of lime stabilized subgrade soil in road construction. Int J Geotech Eng 15:471–483. https://doi.org/10.1080/19386362.2019.1598623
22. Sarangapani G, Venkatarama Reddy BV, Jagadish KS (2005) Brick-mortar bond and masonry compressive strength. J. Mater Civ Eng 17:229–237. https://doi.org/10.1061/(ASCE)0899-1561(2005)17:2(229)
23. Gumaste KS, Venkatarama Reddy BV, Rao KSN, Jagadish KS (2004) Properties of burnt bricks and mortars in India. Mason Int 17:45–52
24. Venkatarama Reddy BV, Richardson L, Rao KSN (2007) Optimum soil grading for the soil-cement blocks. J. Mater. Civ Eng 19:139–148
25. Yi Y, Liska M, Al-Tabbaa A (2014) Properties of two model soils stabilized with different blends and contents of GGBS, MgO, lime, and PC. J Mater Civ Eng 26:267–274. https://doi.org/10.1061/(asce)mt.1943-5533.0000806
26. Yi Y, Liska M, Jin F, Al-Tabbaa A (2016) Mechanism of reactive magnesia—ground granulated blastfurnace slag (GGBS) soil stabilization. Can Geotech J 53:773–782. https://doi.org/10.1139/cgj-2015-0183

27. Bakharev T, Sanjayan JG, Cheng Y (2000) Effect of admixtures on properties of alkali-activated slag concrete. Cem Concr Res 30:1367–1374
28. Nidzam RM, Kinuthia JM (2010) Sustainable soil stabilisation with blastfurnace slag—a review. Proc Inst Civ Eng Constr Mater 163:157–165. https://doi.org/10.1680/coma.2010.163.3.157

R. K. Preethi is Associate Professor in the department of Civil Engineering, Global Academy of Technology, Bangalore, with research interests in strength and durability of Alkali activated-based building materials such as geopolymer compressed earth bricks and lime-stabilised earth bricks utilising the industrial fines as pozzolanic materials.

B. V. Venkatarama Reddy Former Professor, Department of Civil Engineering, Indian Institute of Science, Bangalore, has published extensively on the subject of alternative building materials, sustainability of construction materials, and recycling of materials in concrete and earth-based construction.

Application of Nano-Silica and Polypropylene Fibres to Enhance the Undrained Shear Strength of Black Cotton Soil

Anupkumar G. Ekbote, M. S. Deepak, and Chandrashekharappa Agasnalli

Abstract Due to rampant urbanization and rapid construction industry growth, the structures are being laid upon a weak foundation medium. Thus, it becomes necessary to enhance/modify the strength and serviceability of the weak soil medium. The clayey soil, such as black cotton soil, is weak in strength compared to sandy soil, and it becomes susceptible to cracking when the moisture content is reduced due to volume change. The purpose of this study is to introduce randomly dispersed polypropylene fibres and nano-silica to black cotton soil in order to increase its undrained shear strength. To increase the soil's undrained shear strength, varying percentages of polypropylene fibres, nano-silica, or a mix of the two are considered, depending on the dry weight of the soil. To determine the undrained shear strength, the unconfined compressive strength (UCS) is measured on a total of 80 test samples. The addition of nano-silica is observed to enhance the UCS, due to enhanced bonding, and the same is found to be increased with increase in curing time and imparted brittle failure. Furthermore, by the addition of randomly distributed polypropylene fibre, the brittle nature is reduced. The UCS of the samples is found to be enhanced when the inclusion of polypropylene fibres is made; it increases by 4%, 67%, 94%, 118%, and 213% for 1%, 2%, 3%, 4%, and 5%, respectively. The same increases by 87%, 111%, 187%, 208%, and 231% for 1%, 2%, 3%, 4%, and 5%, respectively, after 14 days of curing with nano-silica addition.

Keywords Soil stabilization · Polypropylene fibres · Nano-silica · Black cotton soil

A. G. Ekbote (✉) · M. S. Deepak · C. Agasnalli
Faculty of Civil Engineering, BMS Institute of Technology and Management, Bengaluru, Karnataka, India
e-mail: anupge@gmail.com

Y. K. Guruprasad et al. (eds.), *Recent Advances in Infrastructure Development and Management—Volume 1*, Lecture Notes in Civil Engineering 570,
https://doi.org/10.1007/978-981-96-1554-4_2

1 Introduction

The soil medium serves as a crucial substance that carries the load transmitted from the forthcoming superstructure through the substructures, so providing support to the foundation. Hence, it is imperative for engineers to assess the soil's suitability for construction purposes. Due to the widespread trend of urbanization, industrialization, and the escalating scale of construction operations, there has been a corresponding rise in the demand for construction sites. The soil present at the building site may not meet the necessary criteria for serviceability due to its inadequate distinctive features. The optimization of soil properties becomes a crucial need when dealing with soil types such as black cotton soil. Soil improvement, also known as ground improvement, has received significant attention from researchers and engineers. The majority of earlier studies on enhancing the properties of soil focused on methods like grouting, stabilizing the soil, and adding admixtures like bitumen, fly ash, lime, and cement. Depending on the particular soil type, these techniques have shown to be somewhat successful in improving the quality of the soil. A substantial quantity of research has been conducted in recent years by several academics worldwide with the aim of examining the engineering features of randomly scattered fibre-reinforced soils and nano-silica. Most of these studies are based on small-scale testing and laboratory research, such as compaction tests, direct shear tests, UCS, and CBR tests.

Using lime and nano-silica to stabilize weak soil, Alireza et al. [1] tested the soil's CBR. In order to find out how these additives affected the characteristics of the soil; they conducted an investigation. According to the study, the addition of nano-silica improved the CBR strength of both the soil and the soil–lime combination. Lime, on the other hand, was shown to have little to no impact on CBR strength. More specifically, the soil's CBR strength increased up to 21 times when nano-silica was added, and up to 7.5 times when the soil–lime combination was added. Choobbasti et al. [2] investigated the effects of cement and nano-silica on compaction and UCS. The investigation involved the combination of four different ratios of nano-silica with three different ratios of cement. The MDD of sand and the cement content was shown to positively correlate. It is noticed that the mechanical properties of cement sand are significantly improved when nano-silica is incorporated in appropriate amounts. The effect of lime particles and ball-milled nano-silica on soil enhancement was examined in this study by Kirithika et al. [3]. When nano-silica was added, the unconfined compression strength increased by a factor of 1.2. The influence of micro- and nano-silica on the permeability coefficient of frost-prone soil was examined in Zięba et al.'s [4] experiment. Samples that were frozen and thawed ten times were the subject of this study. Experiments using 5% micro-silica and nano-silica underwent triaxial testing. It was demonstrated that the use of nano-silica significantly improved the sealing qualities of frost-prone soils. A succinct analysis of soil stabilization methods, such as the use of nanomaterials and nano-silica, was carried out by Meeravali et al. [5]. Materials that have a remarkable propensity for connecting with soils, such as terrasil and nano-silica, were found by the authors. By employing nano-silica as a stabilizing agent, Kannan and Sujatha [6] carried out a thorough investigation that highlighted the material's ability to improve compressibility, hydraulic conductivity, and strength. Nano-silica has been found to boost soil strength, making it a good addition to soils treated with cement, lime, and fibre. Not to be overlooked, though, is the possibility that nano-silica may also lower the soil's hydraulic conductivity and compression index.

Cai et al. [7] conducted a comprehensive series of experiments using polypropylene fibres and lime, which included testing for swelling, shrinkage, unconfined compression, and direct shear. The study highlights the use of fibre, lime concentration, and curing a substantial impact on the engineering parameters of the fibre-lime-treated soil. The clayey soil's cohesion, soil friction, and UCS all amplified when lime was added. However, these traits somewhat declined after that. Tang et al. [8] used UC and direct shear testing in samples that were cured for 7, 14, and 28 days. The axial strain at failure, shear strength, and unconfined compressive strength were significantly increased. Furthermore, there was a decrease in post-peak stiffness and strength. The interaction between the fibre and soil matrix was examined using the SEM technique. Friction at the interface and binding strength are the main parameters governing the reinforcing benefit. Pradhan et al. [9] examined the impact of polypropylene fibre aspect ratio on strength attributes using direct shear testing, unconfined compression tests, and CBR tests. Ali et al.'s [10] investigation looked at a number of topics, including the relationship between moisture and density, unconfined compressive strength and elastic modulus, CBR, and one-dimensional consolidation behaviour. An investigation by Dave et al. [11] included comprehensive CBR and unconfined compression testing. Based on the results, adding polypropylene fibres up to a concentration of 2.25% significantly raised the soil's CBR from 1.54% to 5.75%. Furthermore, the soil's unconsolidated compressive strength showed a notable improvement of 56.9%.

This study examines at the possibilities of adding randomly distributed polypropylene fibres and NS to black cotton soil to increase its undrained shear strength. Due to the varying proportions of NS and polypropylene fibres, the unconfined compression strength and compaction properties are assessed.

2 Materials and Method

For testing purposes, soil samples of black cotton were collected from the land in the Ranibennur Taluk of Karnataka's Haveri District. The soil was gathered at two metres below the surface. To facilitate basic testing, soil samples were treated further after being allowed to air-dry. Table 1 displays the index qualities of the soil used to grow black cotton. The test was carried out in compliance with Indian standards; specifically, the UCS test was certified to IS 2720 part 10 [13] and the compaction test to IS 2720 part 7 [12]. Figure 1 displays the samples ready for UCS testing as well as the UCS configuration.

Table 1 Index characteristics of black cotton soil

Property	Values
Maximum dry density (g/cc)	1.38
Optimum moisture content (%)	22.33
Percentage silt and clay fraction (%)	79
Liquid limit (%)	65.54
Shrinkage limit (%)	8.45
Plastic limit (%)	38.67
Specific gravity	2.73

Fig. 1 Test samples and UCS setup

(a) various samples prepared

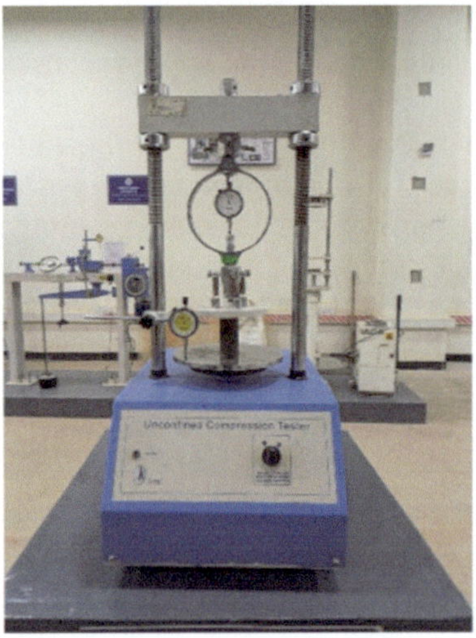

(b) Unconfined Compression Strength setup

2.1 Nano-silica and Polypropylene Fibres

The nano-silica utilized in the current investigation was a colourless, non-crystalline variant of silicon dioxide (SiO_2) with a specific gravity of 2.32. It exhibited a purity level of 98.9% and a density of 0.5 g/cc. Furthermore, the nano-silica's particle sizes varied from 255 to 295 nm. The characteristics of the nano-silica are reliable with the results of Kalhor et al. [15] and Changizi and Haddad [14]. The polypropylene fibre's specific gravity and density are both 0.91 g/cc and 0.91, respectively. The dimensions of the polypropylene fibre were 6 mm long and 18 μm in diameter. Figure 2 displays the samples of the nano-silica and polypropylene fibre that were used in the study.

Fig. 2 Pictures of the materials used in the study

(a) Nano-Silica

(b) polypropylene fibre

3 Discussion on Test Results

Compaction characteristics of soil enhanced with nano-silica and polypropylene fibres generated from black cotton have been noted. The relationship of dry density with moisture content for different ratios of polypropylene and nano-silica fibres is shown in Fig. 3. Interestingly, adding nano-silica always rises the MDD values and reduces the equivalent OMC. However, there seems to be a tendency in the other direction for both variables. The variation seen in OMC and MDD is consistent with the conclusions made by Dave et al. [11] and Al-Shami et al. [16]. In addition to having a greater specific surface area, the nano-silica also causes tighter packing density through pore space-filling and subsequent particle bonding. Thus, the MDD increased from 1.38 to 1.41 g/cc with the addition of nano-silica in the range of 1–4%. However, the MDD decreased from 1.38 to 1.33 g/cc when polypropylene fibres were added in the range of 1–5%.

The specimens were prepared at appropriate OMC with MDD as determined by the compaction test in preparation for the unconfined compression testing. The

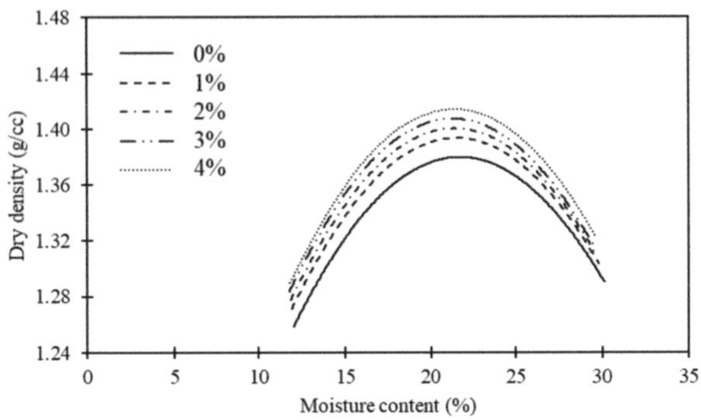

(a) Dry density variation with moisture content for different percentage of Nano-Silica

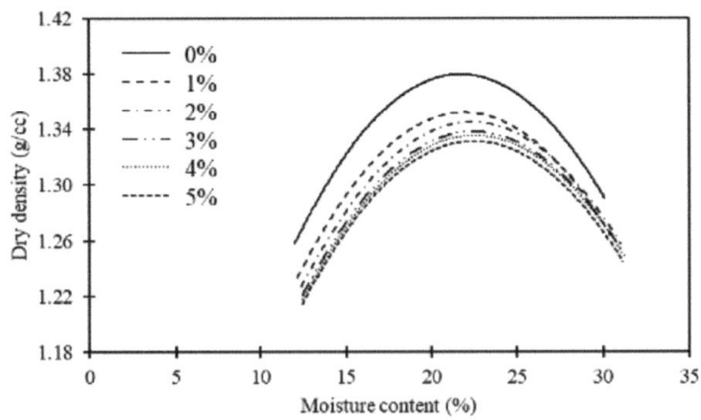

(b) Dry density variation with moisture content for different percentage of polypropylene fibre

Fig. 3 Variation of dry density with moisture content

standard cylindrical specimens were made with different ratios of nano-silica and polypropylene fibres. After adding nano-silica to the samples, they were allowed to cure for 7, 14, and 28 days. The prepared specimens were tested under unconfined compressive loads. The results of UCS for varied quantum of nano-silica are shown in Table 2, whereas the corresponding UCS values for varying percentages of polypropylene fibre are shown in Table 3. Additionally, the blending of polypropylene and nano-silica fibres has been done, and Fig. 4 displays the UCS obtained for various samples. Because there are no weaker planes in the soil as a result of the addition of randomly dispersed fibres, the strength isotropy of the soil is preserved, and the UCS rises as the proportion of fibre increases [17]. The enhancement in the UCS due to nano-silica may attribute due to the pozzolanic reaction causing the flocculated structure. Further microstructural study is required to confirm the same.

Table 2 UCS of black cotton soil with varying percentages of nano-silica

Nano-silica	UCS at 7 days (N/cm^2)	UCS at 14 days (N/cm^2)	UCS at 28 days (N/cm^2)
0%	16.33	16.33	16.33
1%	29.92	30.52	36.15
2%	32.05	34.50	29.80
3%	41.73	47.04	23.79
4%	44.03	50.40	23.02
5%	52.13	54.06	18.83

Table 3 UCS of black cotton soil with varying percentages of polypropylene fibres

Polypropylene fibres	UCS (N/cm^2)
0%	16.33
1%	16.98
2%	27.42
3%	31.85
4%	35.66
5%	51.14

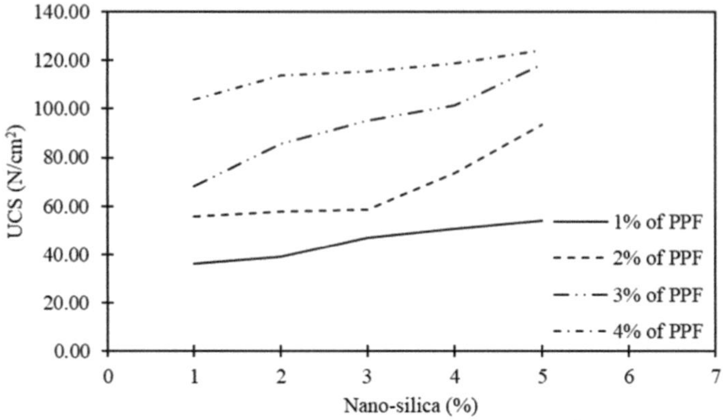

(a) 7 days

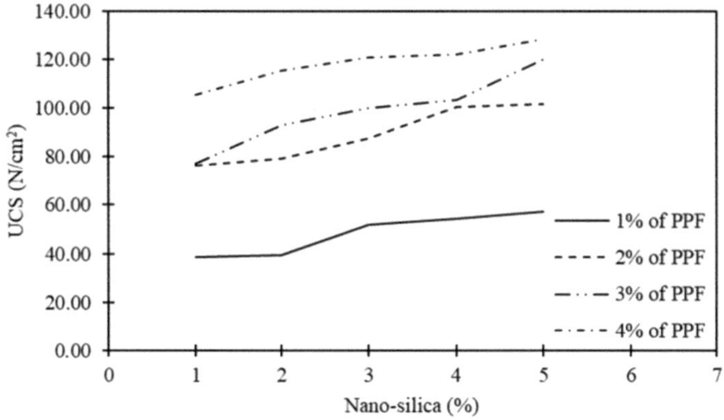

(b) 14 days

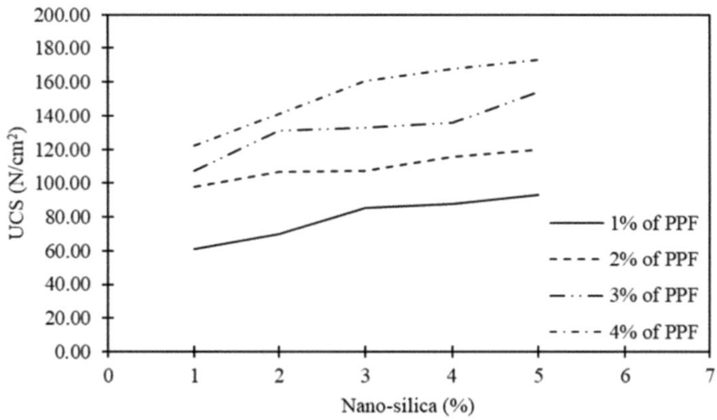

(c) 28 days of curing

Fig. 4 Variation of UCS with different percentage of nano-silica

4 Conclusion

The purpose of the study is to determine whether adding nano-silica and polypropylene fibres to black cotton soil may increase its UCS. Experiments were conducted using a variety of nano-silica and polypropylene fibre percentages to find the OMC and MDD. The observation's outcomes demonstrate that adding nano-silica causes the MDD to rise while adding polypropylene fibres causes it to fall. It has been discovered that the UCS rises with an increase in the quantity of nano-silica and polypropylene fibres. The addition of polypropylene fibres, nano-silica, and a combination of the two has been evaluated in relation to the UCS of the soil. The test results indicate that the UCS increases by 4%, 67%, 94%, 118%, and 213%, respectively, when polypropylene fibres are added at concentrations of 1%, 2%, 3%, 4%, and 5%. Following a 28-day curing, the corresponding samples exhibit sequential increases of 121%, 82%, 46%, 41%, and 58%.

References

1. Alireza SGS, Mohammad MS, Hasan BM (2013) Application of nanomaterial to stabilize a weak soil. In: Seventh international conference on case histories in geotechnical engineering 2013. Chicago, Illinois, pp 1–8
2. Choobbasti AJ, Vafaei A, Kutanaei SS (2015) Mechanical properties of sandy soil improved with cement and nanosilica. Open Eng 5(1):111–116
3. Kirithika M, Prabu T, Stalin V (2015) Influence of nanosized silica and lime particles on the behaviour of soil. In: Proceedings of the 50th Indian geotechnical conference, 17th–19th Dec 2015
4. Zięba Z, Witek K, Mońka J (2018) The effect of micro-and nanosilica on the soil permeability coefficient under cyclic freezing and thawing conditions. In: E3S Web of conferences, vol 66, pp 02004. EDP Sciences
5. Meeravali K, Ruben N, Indira M (2020) Soil stabilization with nanomaterials and extraction of nanosilica: a review. Advances in lightweight materials and structures: In: Proceedings of ICALMS 2020, pp 293–300
6. Kannan G, Sujatha ER (2022) A review on the choice of Nano-Silica as soil stabilizer. SILICON 14(12):6477–6492
7. Cai Y, Shi B, Ng CW, Tang CS (2006) Effect of polypropylene fibre and lime admixture on engineering properties of clayey soil. Eng Geol 87(3–4):230–240
8. Tang C, Shi B, Gao W, Chen F, Cai Y (2007) Strength and mechanical behavior of short polypropylene fibre reinforced and cement stabilized clayey soil. Geotext Geomembr 25(3):194–202
9. Pradhan PK, Kar RK, Naik A (2012) Effect of random inclusion of polypropylene fibres on strength characteristics of cohesive soil. Geotech Geol Eng 30:15–25
10. Ali M, Aziz M, Hamza M, Madni MF (2020) Engineering properties of expansive soil treated with polypropylene fibres. Geomech Eng 22(3):227–236
11. Dave TN, Patel D, Saiyad G, Patolia N (2020) Use of polypropylene fibres for cohesive soil stabilization. In: Advances in computer methods and geomechanics: IACMAG symposium 2019, vol 2, pp 409–417
12. IS 2720-7 (1980) Methods of test for soils, Part 7
13. IS 2720-10 (1991) Methods of test for soils, Part 10

14. Changizi F, Haddad A (2016) Effect of nano-SiO_2 on the geotechnical properties of cohesive soil. Geotech Geol Eng 34:725–733
15. Kalhor A, Ghazavi M, Roustaei M, Mirhosseini SM (2019) Influence of nano-SiO_2 on geotechnical properties of fine soils subjected to freeze-thaw cycles. Cold Reg Sci Technol 161:129–136
16. Alshami AW, Ismael BH, Aswad MF, Majdi A, Alshijlawi M, Aljumaily MM, Hameed MM (2022) Compaction curves and strength of clayey soil modified with micro and nano silica. Materials 15(20):7148
17. Maher MH, Gray DH (1990) Static response of sands reinforced with randomly distributed fibres. J Geotech Eng 116(11):1661–1677

Numerical Investigation of Strip Footing Interference on Saturated Foundation Medium

Anupkumar G. Ekbote, M. S. Deepak, and Chandrashekharappa Agasnalli

Abstract Most early studies on estimating shallow foundations' bearing capacity and settlement characteristics are for dry or low near-to-the-ground moisture-content soils. Practically, the state of saturation becomes the worst-case scenario to estimate the bearing capacity of the foundations. Furthermore, the challenge of dealing with the interference between two adjacent footings has been mostly attributed to the scarcity of appropriate construction sites. Consequently, engineers frequently encounter the need to position footings in close proximity, leading to alterations in the ultimate bearing capacity (UBC) and settlement, as influenced by the chosen spacing. The primary objective of this study is to assess the bearing capacity of strip footings that are put in close proximity on a saturated foundation media. This will be achieved by utilizing a numerical technique, specifically the finite element method, with the use of a commercially available software program known as ABAQUS. The soil is described using the Mohr–Coulomb material model in order to conduct a coupled pore fluid stress analysis, with the assumption that the water table is located at the ground level. The observed interference effect seen in the dry soil has a notable similarity to that reported in the existing literature. Additionally, the saturation level exerts a notable influence on the interference experienced by the bearing capacity, resulting in a large drop for the spacing of major interference. It is found that the drop in the UBC due to saturation for the closely spaced footings are 60%, 59%, 57%, and 55%, respectively, for the spacing 0.5, 0.75, 1.0, and 1.5 times the width of the footing.

Keywords Interference · Saturated soil · Ultimate bearing capacity

---

A. G. Ekbote (✉) · M. S. Deepak · C. Agasnalli
Department of Civil Engineering, BMS Institute of Technology and Management, Bengaluru, Karnataka, India
e-mail: anupge@gmail.com

Y. K. Guruprasad et al. (eds.), *Recent Advances in Infrastructure Development and Management—Volume 1*, Lecture Notes in Civil Engineering 570,
https://doi.org/10.1007/978-981-96-1554-4_3

1 Introduction

The engineer takes a special interest in comprehending the load transmission mechanism to the underlying soil medium. This is due to the fact that the stability and performance of the superstructure are dependent upon the behaviour of the underlying sub-structure, specifically the foundations. The primary objective of the design is to reduce the potential for foundation failure resulting from either shear stresses or excessive settlements. Therefore, the fundamental factors to consider in design are the bearing capacity and settlement characteristics. However, the existing techniques for assessing the bearing capacity are restricted to individual/isolated footings. Likewise, the techniques utilized for assessing settlement are also limited in scope. The close proximity of foundation footings can be attributed to situations, such as excessive urbanization, limited construction areas, and adherence to architectural requirements. As a result, the characteristics of the isolated footing will diverge from typical behaviour.

There exists a substantial body of research pertaining to the impact of interference on neighbouring footings, with a comprehensive overview presented by Ghazavi and Dehkordi [1]. In a recent work by Melese et al. [2], a three-dimensional investigation was undertaken utilizing the finite element program PLAXIS 3D to analyse the differences in settlement. The researchers took into account the impact of unequal footings and loading on settlement characteristics. They discovered that this led to a considerable degree of tilt and an increase in both the magnitude and unevenness of the weight on the footings. Sha et al. [3] conducted a study whereby they employed conformal transformation methods (CTMs) in the analysis of complex variable functions. The objective of their investigation was to examine the effects of both the clear spacing of the footings and the shear modulus of the soil. The investigation was additionally validated using the utilization of the finite element program analysis systems (ANSYS). The study revealed that there is a positive correlation between the differential settlement and the centre spacing, indicating that a rise in centre spacing leads to an increase in differential settlement. However, it was observed that the differential settlement remains constant regardless of any variations in the shear modulus. In their study, Sarvesh et al. [4] proposed a novel approach for predicting the settlement of square and rectangular footings that are in close proximity to each other. This was achieved by using Steinbrennr's influence factor into the equation used for evaluating settlement. The study investigated the relationship between settlement and embedment for footings with varying spacing and aspect ratios. The confirmation of the results was conducted using 3-dimensional finite element analysis and existing experimental data. In their study, Fattah [5] employed the finite element program, Midas GTS-NX, to examine the impact of various factors on the behaviour of footings. These factors include the spacing between footings, the depth of footings, the undrained shear strength of the soil, and the level of the groundwater table. The study revealed that the ultimate bearing capacity exhibited a 10% increase when the spacing between footings was equal to the width of the footing. Furthermore, this effect was observed to be more pronounced under softer soil conditions. In their study,

Eslami et al. [6] conducted a finite element analysis utilizing PLAXIS 3D software to evaluate the impact of the number of adjacent building floors on the settlement and tilt of an existing building. The inquiry was divided into two groups, namely small-scale and full-scale investigations. The findings indicate a positive correlation between the number of stories and the magnitude of induced settlement. Additionally, a comprehensive assessment of a real-time case study was conducted, which involved the examination of numerous complexities.

In relation to the subject under consideration, the saturation of the soil represents the most severe scenario that can impact the bearing capacity of foundations. Hence, in instances of intense precipitation or flood events, the soil appears to reach a state of saturation. Consequently, this has an impact on the load-bearing capability of the foundations. Thus, current study aims to estimate the ultimate bearing capacity (UBC) of adjacent strip footings on a saturated soil medium.

2 Problem Statement

Two strip footings, each with a width denoted as B, are positioned in close proximity to one another on the surface of a homogeneous, isolated, and semi-infinite sandy medium. The footings experience a uniform load of q, simultaneously. In the context of a saturated condition, it is often considered that the water table is located at the surface of the soil medium. Conversely, in a dry state, it is assumed that there is no water table present. The problem's definition is depicted in Fig. 1. The analysis is conducted by altering the clear spacing between the foundations for both the saturated and dry states of the soil medium. Consequently, the load–settlement behaviour is determined, from which the ultimate bearing capacity (UBC) is calculated for each soil condition and spacing configuration.

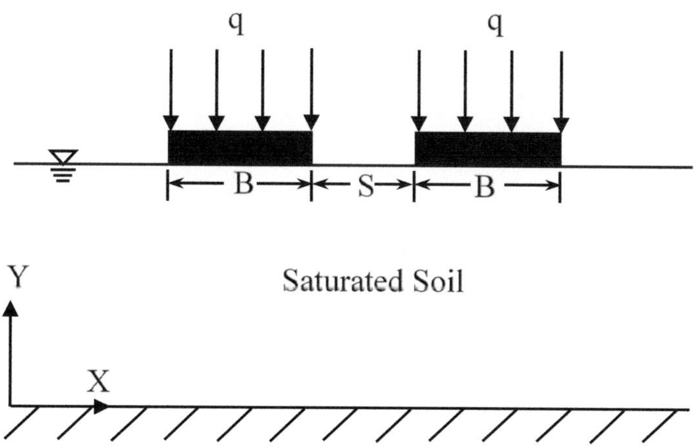

Fig. 1 Definition of the problem

3 Methodology

The analyses were conducted with the finite element program ABAQUS. The soil was characterized by the linear-elastic and perfect-plastic material model, according to the Mohr–Coulomb failure criterion. On the other hand, the footings were described as a linear-elastic material. The unit weight and Young's modulus values utilized in this study were 16 kN/m^3 and 18 kN/m^3, 32 MPa and 12 MPa, correspondingly, for the dry and saturated conditions. The soil cohesion values were considered to be 2 kPa and 0.6 kPa for the dry and saturated (drained) states, respectively. The soil friction angle was considered as 30^0, and it is important to note that this value is for drained condition for saturated soil. The model was discretized using a two-dimensional plane strain approach with quadrilateral elements. The elements were characterized by bilinear pore pressure interpolation and employ reduced integration and hourglass control techniques. Specifically, the continuum element used is CPE8RP. The inclusion of the pore fluid/stress element was implemented to automatically model the impact of pore water pressure inside a geostatic stress environment. The domain sensitivity and mesh convergence analysis has been extensively examined and documented by Ekbote and Nainegali [7–9]. Consequently, the current study has adopted the same dimensions, specifically a size of 15B on both sides of the footings and 12B from the base of the footings. Correspondingly, the roller supports are taken into account in conjunction with the vertical boundaries at the far end in order to confine the horizontal displacement ($u_1 = 0$) while permitting any potential vertical displacement (u_2). Subsequently, the horizontal boundary is regarded as fixed to restrict both horizontal and vertical displacements ($u_1 = u_2 = 0$) [10]. The defined model in the ABAQUS program considering various loads (geostatic stress and uniform load on the footings) and boundary conditions is illustrated in Fig. 2.

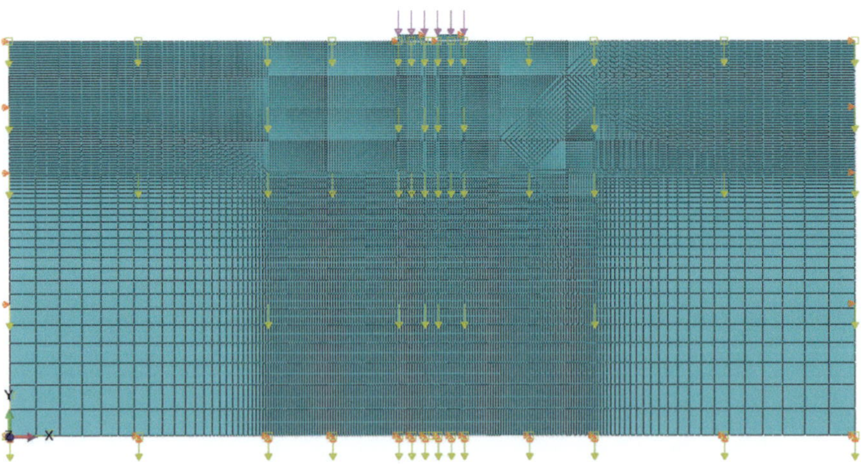

Fig. 2 Model considered in the analysis showing loads and the boundary conditions

4 Results and Discussion

The (UBC) derived from the analysis aligns favourably with the analytical equation proposed by Meyerhof [11], exhibiting a small deviation of 2.2% and 5.6% for dry and saturated conditions, respectively. The load–settlement plot for isolated footing on dry and saturated condition is presented in Fig. 3. The load–settlement characteristics of strip footings situated on both dry and saturated soil, with varying spacing between the footings, have been determined by analysis. These results are depicted in Fig. 4. The double tangent approach, as described by Naderi and Hataf [12] and Saha Roy and Deb [13], is employed for the determination of the ultimate bearing capacity (UBC) based on load–settlement curves (Fig. 3). The load–settlement behaviour as observed for footings in closed spacing (S/B = 0.25, 0.50, 0.75, 1.0, 1.5, 2.0) are all on the right-hand side of the isolated footing for both dry and saturated conditions signifying greater load carrying capacity. The similar observations are made and reported by Ekbote and Nainegali [7].

It is evident that the ultimate loads of footings in close proximity are greater than those of isolated footings in both dry and saturated conditions. Nevertheless, the magnitude of the UBC obtained is lower for saturated conditions, although the trend of variations remains consistent. Moreover, the outcomes obtained for varying spacing have been measured in terms of interference factors (ξ_{dry}/ξ_{sat}), as defined by Eq. (1). Figure 5 displays the relationship between the spacing between the footings and the variation of interference factors, as derived from Eq. (1). The UBC has been shown to exhibit a consistent rise as the spacing between the footings is reduced,

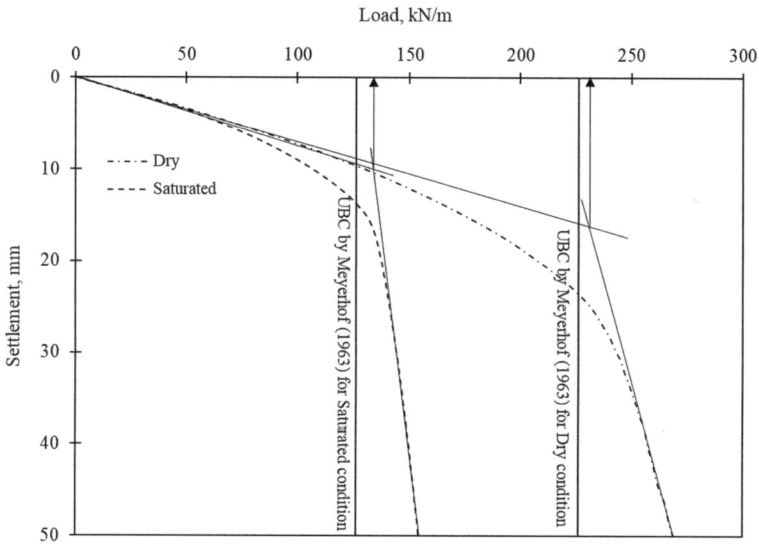

Fig. 3 Load–settlement plot for isolated footing on dry soil and saturated soil

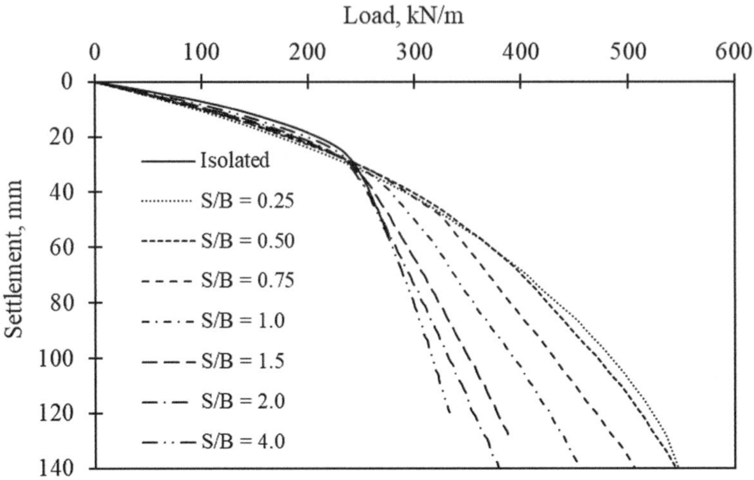

(a)

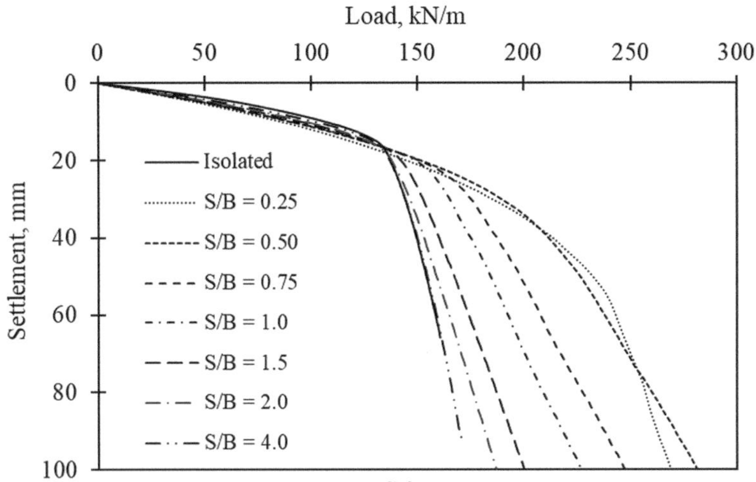

(b)

Fig. 4 Load–settlement plots of the interfering footings on **a** dry soil, **b** saturated soil

regardless of whether the circumstances are dry or saturated. The UBC of an interfering footing under dry conditions exhibits an increasing trend until it reaches its peak at a spacing equal to 0.5 times the width of the footing ($S/B = 0.50$). However, at a spacing of 2.0 times the width of the footing, the effect of interference begins to diminish. The most significant zone of interference for dry conditions is determined to occur at $S/B = 0.50$, while the corresponding region for saturated conditions is identified at $S/B = 0.25$. Furthermore, for saturated conditions, it has been observed that the UBC for an interfering footing is higher in comparison with an isolated footing. The observed increments in UBC for the values of the S/B ratio at 0.25,

0.50, 0.75, and 1.0 are 60.3%, 38%, 21.6%, and 11.6%, respectively. Hence, from the standpoint of UBC, it appears that interference seems to be favourable under saturated conditions (Table 1).

$$\xi_{dry}/\xi_{sat} = \frac{\text{UBC of interfering footings on dry or saturated condition}}{\text{UBC of isolated footings on dry or saturated condition}} \tag{1}$$

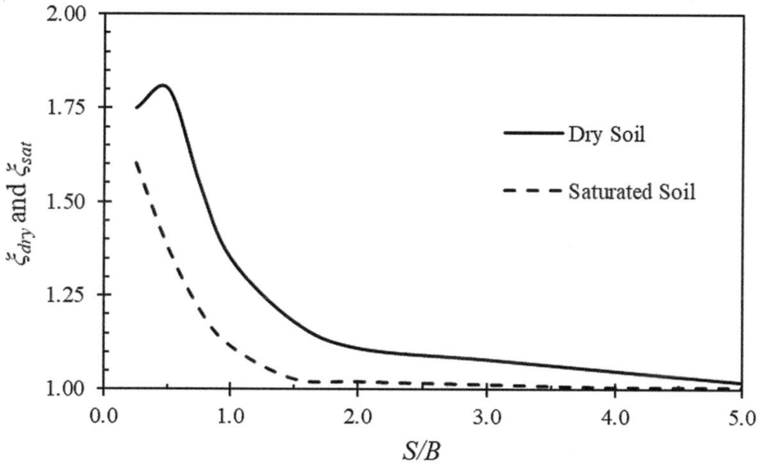

Fig. 5 Variation of interference factors ($\xi_{dry}/\ \xi_{sat}$) with spacing between the footings

Table 1 Percentage decrease in the ultimate bearing capacity for different spacing due to saturated condition of the soil

Spacing between the footings, S/B	UBC, dry (kN/m)	UBC, saturated (kN/m)	(%) decrease
Isolated	190.5	99.4	48
0.25	333.3	159.2	52
0.50	342.8	137.1	60
0.75	295.2	120.8	59
1.0	257.1	110.9	57
1.5	224.7	102.2	55
2.0	211.4	101.5	52
3.0	205.8	100.8	51
4.0	200.1	100.1	50
5.0	194.3	100.1	48

5 Conclusion

The investigation was conducted on two adjacent surface strip footings situated on a uniform soil medium considering dry and saturated conditions. The ultimate bearing capacity (UBC) is subject to a notable impact from the interference effect. Additionally, a drop in the UBC is observed due to the saturated state of the soil medium. The isolated footing experiences a 48% loss in the UBC, whereas for the footings in the major interference zone, result in a 60% decrease in the UBC. However, it has been shown that the UBC of the interfering footings is higher than that of isolated footings when considering a saturated soil medium. The S/B ratio exhibits percentage increases of 60.3%, 38%, 21.6%, and 11.6% at values of 0.25, 0.50, 0.75, and 1.0, respectively. Therefore, the close proximity of footings on saturated soil can be seen as advantageous solely from the perspective of the UBC in comparison with isolated footings.

References

1. Ghazavi M, Dehkordi PF (2021) Interference influence on behavior of shallow footings constructed on soils, past studies to future forecast: a state-of-the-art review. Transp Geotech 27:100502
2. Melese D, Genemo D, Boru Y, Desisa A, Mezmur T, Tamru M (2023) Interference effect of closely spaced foundation footing on settlement variation. Adv Civ Eng 1–11
3. Sha X, Lu A, Zhang N (2023) Complex variable solution of interference effect of two nearby strip footings on layered soil. Int J Geomech 23(10):04023161
4. Sarvesh R, Srinivasan V, Patel A (2023) Elastic settlements of identical angular footings in close proximity. Geomech Eng 32(2):193
5. Fattah MY (2023) Finite element analysis of two nearby interfering strip footings embedded in saturated cohesive soils. Civ Eng J 9(3):752–769
6. Eslami A, Afshar D, Moghadasi H, Akbarimehr D (2023) Numerical and experimental investigations of interference effect of adjacent buildings on sand and fill deposits. Int J Civ Eng 1–16
7. Ekbote AG, Nainegali L (2019) Interference of two closely spaced footings embedded in unreinforced and reinforced soil medium: a finite element approach using ABAQUS. Arab J Geosci 12(683):1–21
8. Ekbote AG, Nainegali L (2019) Interference of two closely spaced strip footings embedded in cohesionless fibre-reinforced foundation soil bed. In: Meehan CL, Kumar S, Miguel A, Coe JT (eds) Eighth international conference on case histories in geotechnical engineering 2019. American Society of Civil Engineers, Reston, VA, pp 454–464
9. Ekbote AG, Nainegali L (2021) Finite element analysis of two nearby interfering asymmetric footings embedded in cohesionless foundation medium. Geomech Geoengin 16(4):263–276
10. Ekbote AG, Nainegali L, Rajhans P, Deepak MS (2022) Behavioural assessment on influence of adjacently placed strip footings at different embedment level. Arch Civ Eng Environ 15(4):93–103

11. Meyerhof GG (1963) Some recent research on the bearing capacity of foundations. Can Geotech J 1(1):16–26
12. Naderi E, Hataf N (2014) Model testing and numerical investigation of interference effect of closely spaced ring and circular footings on reinforced sand. Geotext Geomembr 42(3):191–200
13. Roy SS, Deb K (2018) Closely spaced rectangular footings on sand over soft clay with geogrid at the interface. Geosynth Int 25(4):412–426

Stability Analysis of Reinforced Soil Embankments with Alccofine, Glass Fibres, Metal Mesh, and Geogrid

B. R. Vinod and J. Sumalatha

Abstract In the recent years, the slope stability analysis and stabilization of soil slopes have gained importance since there is increase in the occurrences of landslides and land subsidences. This present study reports the results of stability of embankment soil reinforced with Alccofine, geogrid, glass fibres, and metal mesh. The optimum percentages of Alccofine and glass fibres found from the unconfined compression tests were 4% and 2%, respectively. The embankment models of various slope angles were prepared by casting the moulds with slope angles 60°, 70°, 80°, and 90°. The reinforcement with metal mesh and geogrid was placed with two options, i.e. single layer and double layer of reinforcement. The compression tests were conducted on these embankment models, and the results were analysed by numerical simulations using Plaxis and GeoStudio software tools. The factors of safety of the slopes were estimated using software tools and Taylor's method, and the variations in the factors of safety with respect to each method are analysed. It was found that the factors of safety obtained with Alccofine amended soil are higher than remaining materials, and this is due to the improvement in shear strength properties of the soil. The load carrying capacity was increased with the inclusion of metal mesh and geogrid, and it is further increased with double layers of reinforcements.

Keywords Alccofine · Plaxis · GeoStudio · Geogrid · Glass fibers · Reinforcement

---

B. R. Vinod
Department of Civil Engineering, B M S Institute of Technology and Management, Bengaluru, India

J. Sumalatha (✉)
Department of Civil Engineering, Ramaiah Institute of Technology, Bengaluru, India
e-mail: sumalatha@msrit.edu

Y. K. Guruprasad et al. (eds.), *Recent Advances in Infrastructure Development and Management—Volume 1*, Lecture Notes in Civil Engineering 570,
https://doi.org/10.1007/978-981-96-1554-4_4

1 Introduction

Since there is an increase in the occurrence of natural disasters like landslides which are occurring due to rainfall and resulting in immense loss of life and property, it is necessary to provide a suitable method for stabilization of slopes. Analysing the soil stability deals with finding and comparing the shear stress and shear strength developed along the failure potential surface with the strength of the soil. While analysing, great attention has to be paid to surface drainage conditions, groundwater, geology of the place, shear strength of the soils, etc. To ensure the stability of slopes, factor of safety must be estimated, and it should be greater than the allowable value.

There are many methods to find out the slope stability, one of the most well-known methods is limit equilibrium method (LEM), and it is extensively used in many areas under geotechnical engineering [1–3]. The GeoStudio software is effective for slope stability analysis of different soils and water table conditions [4]. There are variety of materials that can be used for soil reinforcement such as mild steel, glass fibres, natural fibres, various types of wood, various forms of rubber, alloys of aluminium, and thermoplastics. Reinforcement can be used in various forms like strips, grids, anchors, sheets, planks, rope, vegetation, and combinations of these or other material forms.

Alccofine is a cement-based product much finer than cement, fly ash, rice husk ash, etc. The use of Alccofine with other additives to stabilize the expansive soil was studied by [5–10]. They found that Alccofine addition causes decrease in plasticity characteristics of soil, increase in California bearing ratio (CBR), and increase in unconfined compressive strength of soil. The utilization of glass fibres as reinforcement material was studied by several researchers [11–17], and the material was found effective in improving the soil properties.

The slope stability analysis can be conducted using various methods and software tools [18–25]. As part of this study, the red soil was collected from Doddaballapura highway near Bangalore, India and basic properties of the soil were found out using laboratory procedures. The stability of soil slopes was studied by amending red soil with Alccofine, glass fibres, geogrid, and metallic mesh. Various proportions of these materials were tested using unconfined compression tests to determine the optimal dosage. Subsequently, Plaxis software, GeoStudio software, and Taylor's method were used for slope stability analyses. The results indicated that adding Alccofine and glass fibres to the red soil significantly increases the factor of safety of the slopes. The double layer reinforcement provided to the soil slope contributed higher values of factors of safety than with single layer reinforcement.

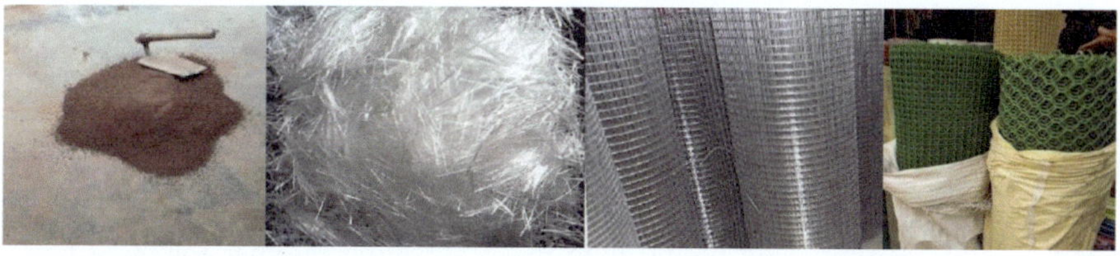

Fig. 1 Materials used in the study (red soil, glass fibres, metal mesh, and geogrid)

Table 1 Shear strength parameters of soil samples

Soil	Unit Weight (kN/m^3)	Cohesion (kPa)	Angle of internal Friction
Soil	15.70	51.0	39°
Soil + 4% Alccofine	17.66	179.8	36°
Soil + 2% Glass Fibre	17.66	40.0	20°

2 Materials and Methods

2.1 Red Soil and Amended Red Soil

The red soil for this study was sourced from the Doddaballapura highway near Bangalore, India (Fig. 1). It has a specific gravity of 2.6, liquid limit of 70%, and plastic limit of 40%. As per Indian Standard classification system, it is classified as low plasticity clay (CL). The maximum dry density and the optimum moisture content were estimated as 1.63 g/cc and 15%, respectively. The soil's cohesion and angle of shearing resistance are 51 kPa and 39°, respectively. To determine the optimal dosage of amendments, the soil was mixed with 2%, 3%, 4%, and 5% Alccofine. Based on unconfined compression tests and compaction tests, the optimal dosages for Alccofine and glass fibres were identified as 4% and 2%, respectively. The shear strength properties of the soil determined from triaxial tests are presented in Table 1.

2.2 Alccofine, Glass Fibres, Metal Mesh, and Geogrid

In this study, Alccofine 1203, a slag-based material with ultra-fine particles and optimized particle size distribution, was used. The glass fibres used in this research belong to the soda-lime-silicate type, which are highly recyclable. The locally available metal mesh, depicted in Fig. 1, was employed in this study. The geogrid used for this study is shown in Fig. 1.

2.3 *Preparation of Embankment Models and Compression Tests*

The embankment models of different slope angles were prepared by casting cubical moulds of size 15 cm (Fig. 2). The red soil passing through 4.75 mm sieve is compacted in the fabricated mould with the compacting rammer for three layers of equal proportions. The three layers of soil are compacted at equal heights of 50 mm each. For the samples reinforced with metal mesh or geogrid, the reinforcement material is placed after compaction of two layers, i.e. at a height of 100 mm (Fig. 2) from the bottom, and then the third layer of soil is placed and compacted. Then the mould is demoulded, and the soil sample is tested in the compression testing machine (Fig. 3). The total load carrying capacity was estimated, and the horizontal and vertical deformations were found by fixing dial gauges within the soil embankment.

Fig. 2 Embankment models of different slopes prepared with fabricated moulds and placement of reinforcement in the embankment model

Fig. 3 Compression tests on embankment models

2.4 Slope Stability Analyses

GeoStudio and Plaxis 2D software tools were used to estimate the factors of safety of amended soil slopes. By using experimental methods and software, the amount of reinforcement required, and its placement can be determined for different slope angles. In Plaxis software, the embankment model was drawn by defining the various dimensions and structural elements. The general soil properties such as permeability, cohesion, angle of internal friction, seepage velocity, and modulus of elasticity were added to the model along with the boundary conditions. The initial flow conditions, pore water pressure, and other stresses were defined, and after running the analysis, the factor of safety was obtained along with other results and figures.

3 Results and Discussions

The compression tests were conducted, and the displacements were noted through dial gauges. The horizontal and vertical displacements of embankment models of different slope angles on embankment models prepared with soil alone are given in Table 2. The horizontal displacement is comparatively lesser than the vertical displacement as the load is applied in the vertical direction. The compression test results of embankment models with different combinations of Alccofine, glass fibre, metal mesh, and geogrid are given in Tables 3, 4, 5, 6, 7 and 8.

The load carrying capacity for slope angle of 80° was maximum for double layer of metal mesh reinforcement with capacity of 4.2 kN, and single layer reinforcement carried a load of 3.7 kN (Table 5). The load carrying capacity of the embankment was enhanced by incorporating reinforcements at uniform intervals from the top and bottom, surpassing the load-bearing capability of unreinforced soil embankments. The Alccofine amended soil slope of 60° carried more load of 3.3 kN without reinforcement (Table 3). At 90°, single layer of reinforcement carried a load of 6 kN, and double layer carried a load of 6.5 kN (Table 7). At 80°, the glass fibre amended soil embankment carried a load of 3.5 kN without reinforcement (Table 4). At 80°, single layer of reinforcement carried a load of 3.5 kN, and double layer of reinforcement carried a load of 7 kN (Table 8).

3.1 Factors of Safety with Taylor's Method

The factors of safety were estimated using Taylor's Formula for different slope angles are shown in Table 9. The results indicate that the maximum factor of safety corresponds to a slope angle of 60°. The factors of safety with Alccofine amendment higher than those with Glass fibre amendment.

Table 2 Compression test results of embankment models with soil alone

Type of embankment	Section of embankment (Dimensions are in cm)	Load (kN)	Dry density (g/cc)	Displacements (mm)	
				Vertical	Horizontal
E1 (90° slope)	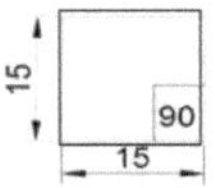	4.6	1.6	8.15	2.66
E2 (80° slope)	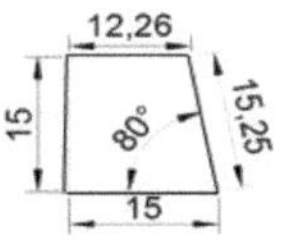	4	1.6	3	5.1
E3 (70° slope)	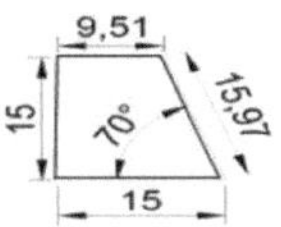	3	1.6	6.2	1.51
E4 (60° slope)	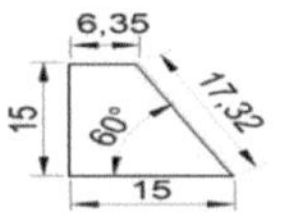	2.5	1.6	3.2	0.95

Table 3 Compression test results of soil with 4% Alccofine

Type of embankment	Load (kN)	Dry density (g/cc)	Displacements (mm)	
			Vertical	Vertical
E1 (90° slope)	2.5	1.8	2.8	3.3
E2 (80° slope)	2	1.8	6.1	3.8
E3 (70° slope)	2.5	1.8	5	1.14
E4 (60° slope)	3.3	1.8	10.3	3.8

Table 4 Compression test results of soil with 2% glass fibre

Type of embankment	Load (kN)	Dry density (g/cc)	Displacements (mm)	
			Vertical	Vertical
E1 (90° slope)	3.4	1.8	9.9	1.55
E2 (80° slope)	3.5	1.8	9.3	2.46
E3 (70° slope)	2	1.8	9.8	4.1
E4 (60° slope)	1.7	1.8	9.9	2.19

Table 5 Compression test results of soil with metal mesh reinforcement

Type of embankment	Soil + Single layer of reinforcement (Chicken Mesh)			Soil + Double layer of reinforcement (Chicken Mesh)		
	Load (kN)	Displacements (mm)		Load (kN)	Displacements (mm)	
		Vertical	Horizontal		Vertical	Horizontal
E1	3.6	10.45	6.85	4.1	2.86	5.3
E2	3.7	9.75	2.21	4.2	2.62	1.1
E3	3.2	9.42	2.55	3	2.44	3.4
E4	2.5	12.55	2.74	2	11	1.7

Table 6 Compression test results of soil with geogrid reinforcement

Type of embankment	Soil + Single layer of geogrid reinforcement			Soil + Double layer of geogrid reinforcement		
	Load (kN)	Displacements (mm)		Load (kN)	Displacements (mm)	
		Vertical	Horizontal		Vertical	Horizontal
E1	7.7	7.60	1.48	10.7	8.65	2.51
E2	7.0	6.80	0.81	8.7	5.70	2.00
E3	3.5	8.13	1.96	6.2	7.40	1.66
E4	3.3	8.50	2.25	3.0	2.25	0.45

Table 7 Compression test results of soil with Alccofine and metal mesh reinforcement

Type of embankment	Soil + 4% Alccofine + Single layer of reinforcement (Chicken Mesh)			Soil + 4% Alccofine + Double layer of reinforcement (Chicken Mesh)		
	Load (kN)	Displacements (mm)		Load (kN)	Displacements (mm)	
		Vertical	Horizontal		Vertical	Horizontal
E1	6	5.3	3.5	6.5	3.6	1.6
E2	5.5	3.2	1.12	6	5.2	2.6
E3	4.5	4	2.3	5.5	6.3	3.6
E4	4	7.2	3.5	5	3.8	5.8

Table 8 Compression test results of soil with 2% glass fibre and metal mesh reinforcement

Type of embankment	Soil + 2% Glass fibre + Single layer of reinforcement (Chicken Mesh)			Soil + 2% Glass Fibre + Double layer of reinforcement (Chicken Mesh)		
	Load (kN)	Displacements (mm)		Load (kN)	Displacements (mm)	
		Vertical	Horizontal		Vertical	Horizontal
E1	3.5	8.85	1.9	6.5	18	1.6
E2	3.5	10	2.74	7	16	9.39
E3	3.5	9.7	1.5	5	11.8	3.58
E4	3	6.2	2.11	4	6.5	0.35

Table 9 Factors of safety (FoS) of soil with Taylor's method

Slope angle	Embankment with soil		Embankment with soil + 4% Alccofine		Embankment with soil + 2% Glass fibre	
	Stability number	Factor of safety	Stability number	Factor of safety	Stability number	Factor of safety
90°	5.6	1.08	5.6	3.78	5.6	1.77
80°	6.8	1.31	6.8	4.59	6.8	2.15
70°	8.2	1.58	8.2	5.54	8.2	2.60
60°	8.7	1.68	8.7	5.88	8.7	2.76

3.2 *Factors of Safety from Plaxis Software Results*

The experimental results of the embankment models were taken as input to the Plaxis software, and the slope stability analysis was performed with respect to each material and slope angle. The factors of safety obtained with respect to each model are presented in Table 10.

3.3 *GeoStudio Results*

The test data of various embankment models corresponding to different slope angles and material combinations were analysed using GeoStudio software tool. The estimated values of factors of safety with respect to each model are presented in Table 10.

4 Conclusions

The experimental and numerical analyses were conducted on embankment slopes of different angles (90°, 80°, 70°, and 60°) and different amendments (Alccofine, glass fibres, and metal mesh). By conducting the unconfined compression tests, the optimum dosage of Alccofine and glass fibre were estimated as 4% and 2%, respectively. Load carrying capacity for soil embankment with reinforcement like metal mesh was more when compared with that of only soil embankment, and double layer of reinforcement carried much more load which shows that in the design of soil embankment a minimum amount of reinforcement is required to increase the slope stability. The soil embankment for slope angle of 60° was carrying more load and less deformations when compared with the other slope angels. The factors of safety were found using software tools like Plaxis and GeoStudio and Taylor's stability method. It was found that the factors of safety with Taylor's method were higher and Plaxis gave higher values than GeoStudio software. The reinforcements like metallic

Table 10 Comparison of factors of safety (FoS) with different methods and materials

Soil embankment			
Embankment slope angle	FoS by Taylor's stability method	FoS by Plaxis software	FoS by GeoStudio software
90°	1.07	1.89	1.70
80°	1.30	1.62	1.53
70°	1.57	1.94	1.84
60°	1.67	2.74	2.54
Soil + 4% Alccofine			
90°	3.78	3.97	3.85
80°	4.59	4.35	4.15
70°	5.54	5.32	5.17
60°	5.58	7.01	6.88
Soil + 2% Glass fibre			
90°	1.77	1.12	0.95
80°	2.15	1.45	1.06
70°	2.60	1.44	1.23
60°	2.76	1.82	1.73
Soil + Single layer metallic mesh			
90°	1.92	1.95	1.66
80°	1.84	1.88	1.59
70°	2.73	1.54	1.85
60°	3.28	2.57	2.38
Soil + Double layer metallic mesh			
90°	1.78	1.92	1.88
80°	2.46	2.35	2.16
70°	3.15	2.93	2.62
60°	3.87	3.67	3.45
Soil + Single layer geogrid			
90°	2.11	1.96	1.71
80°	1.98	1.81	1.79
70°	2.37	2.03	1.85
60°	3.06	2.84	2.55
Soil + Double layer geogrid			
90°	2.32	2.13	1.94
80°	2.18	2.26	2.11
70°	3.01	2.94	2.83
60°	3.83	3.72	3.65

mesh and geogrid improved the slope stability and the load carrying capacity further increased with double layered reinforcement.

References

1. Cheng YM, Lansivaara T, Wei WB (2007) Two-dimensional slope stability analysis by limit equilibrium and strength reduction methods. Comput Geotech 34(3):137–150
2. Hungr O (1994) A general limit equilibrium model for three-dimensional slope stability analysis: discussion. Can Geotech J 31(5):793–795
3. Kalatehjari R, Ali NA (2013) Review of three-dimensional slope stability analyses based on limit equilibrium method. Electron J Geotech Eng 18:119–134
4. Jampani H, Bhupathi N (2017) Stability analysis of slope with different soil types and its stabilization techniques. In: Proceeding of Indian geotechnical conference (GeoNEst), pp 1–4
5. Godayal A, Kapoor A, Garg P (2018) Effect of alccofine, lime on geotechnical properties of cohesive soil. Int J Creative Res Thoughts (IJCRT) 6(2)
6. Dev S, Sharma N (2017) Stabilization of expansive soil with marble dust and alccofine. Int J Adv Res Sci Eng 6:1212–1219
7. Mir SH, Sharma N (2019) Amendment of geotechnical properties of clayey soil using brick kiln dust and Alccofine, vol 1101
8. Rather SA, Sharma N, Najar IA (2019) Effects of rice husk ash (RHA) and alccofine-1101 on stabilization of clay soil. Int Res J Eng Technol 6:474–478
9. Sambyal LS, Sharma N (2018) Utilizing fly ash and alccofine for efficient soil stabilization. Int J Sci Eng Res 9(3)
10. Soni MK, Singh S (2019) Statistical interpretation of marble dust and alccofine for soil stabilization. Int J Innov Technol Explor Eng 8(7):1609–1613
11. Abdeldjouad L, Asadi A, Ball RJ, Nahazanan H, Huat BB (2019) Application of alkali- activated palm oil fuel ash reinforced with glass fibers in soil stabilization. Soils Found 59(5):1552–1561
12. Ateş A (2016) Mechanical properties of sandy soils reinforced with cement and randomly distributed glass fibers (GRC). Compos B Eng 96:295–304
13. Ayininuola GM, Balogun LO (2018) Investigation of glass fiber potential in soil stabilization. Int J Eng Adv Technol 7(5):113–117
14. Baruah H (2015) Effect of glass fibers on red soil. Int J Adv Technol Eng Sci 3(1):217–223
15. Gul N, Mir BA (2022) Influence of glass fiber and cement kiln dust on physicochemical and geomechanical properties of fine-grained soil. Innov Infrastruct Solut 7(6):344
16. Rabab'ah S, Al Hattamleh O, Aldeeky H, Alfoul BA (2021) Effect of glass fiber on the properties of expansive soil and its utilization as subgrade reinforcement in pavement applications. Case Stud Constr Mater 14:e00485
17. Valipour M, Shourijeh PT, Mohammadinia A (2021) Application of recycled tire polymer fibers and glass fibers for clay reinforcement. Transp Geotech 27:100474
18. Umrao RK, Singh R, Ahmad M, TN S (2011) Stability analysis of cut slopes using continuous slope mass rating and kinematic analysis in Rudraprayag district, Uttarakhand. Geomaterials, 2011
19. Kumar V, Parkash V (2015) A model study of slope stability in mines situated in south India. Adv Appl Sci Res 6(8):82–90
20. Mahdi Rasouli M, Mohammad M, Kambiz M (2011) Design of overall slope angle and analysis of rock slope stability of Chadormalu mine using empirical and numerical methods. Engineering, 2011
21. Jacob A, Thomas AA, Nath AG, MP A (2018) Slope stability analysis using Plaxis 2D. Int Res J Eng Technol (IRJET) 5(4):3666–3668
22. Bidisha C, Shivananda P (2017) Two-dimensional slope stability analysis by plaxis-2d. Int J Res Appl Sci Eng Technol 5(9):871–877

23. Sumalatha J (2021) Bearing capacity and settlement analysis of black cotton soil amended with rubber powder using the GEO5 software tool. In: Biswas S, Metya S, Kumar S, Samui P (eds) Advances in sustainable construction materials. Lecture notes in civil engineering, vol 124. Springer, Singapore. https://doi.org/10.1007/978-981-33-4590-4_27
24. Sumalatha J (2022) Slope stability analysis using in situ ground reinforcement techniques for a landslide prone area at Madikere, India. In: Geohazard mitigation. Springer, Singapore, pp 173–183
25. Sumalatha J (2022) Stability analysis of slopes at a landslide prone area: a case study on the landslide at Madikere, India. In: Stability of slopes and underground excavations. Springer, Singapore, pp 105–114

Citizens' Perception of Smart City Projects in Salem

A. Meenachi, R. Malathy, V. Kavinthanjali, and K. Arun Kumar

Abstract This abstract provides an overview of a comprehensive study conducted to explore the citizens' perception of smart city projects in Salem, a city in Tamil Nadu, India. The aim of the research was to understand how the residents of Salem perceive the implementation and impact of various smart city initiatives in their day-to-day lives. The study employed a mixed-methods approach, combining quantitative surveys and qualitative interviews to gather data and analyse the citizens' perspectives. The research sample consisted of a diverse group of participants, including residents from different socioeconomic backgrounds, age groups, and professions. The survey questionnaire was designed to gather demographic information, assess awareness and utilization of smart city services, and gauge citizens' satisfaction levels. The study also highlighted the impact of socioeconomic factors on the perception of smart city projects, with variations observed among different demographic groups. Furthermore, the research explored the factors influencing citizens' perception, including the accessibility and affordability of smart city services, the level of engagement and communication from the local government, and the perceived benefits in terms of improved quality of life and sustainability. The study shed light on the areas where citizens felt more involvement and improvement were needed, such as citizen participation in decision-making processes and the provision of digital literacy programs to ensure equitable access to smart city services. This case study provides valuable insights for policymakers, city planners, and local authorities involved in smart city initiatives. Understanding citizens' perception is crucial for such projects' successful implementation and long-term sustainability. The findings can contribute to the refinement and enhancement of existing smart city strategies in Salem, while also offering valuable lessons for other cities undertaking similar initiatives.

Keywords Smart city projects · Citizens' perception · Sustainable development · Technology adoption

A. Meenachi (✉) · R. Malathy · V. Kavinthanjali · K. Arun Kumar
Department of Civil Engineering, Sona College of Technology, Salem, Tamil Nadu, India
e-mail: meenachia@sonatech.ac.in

Y. K. Guruprasad et al. (eds.), *Recent Advances in Infrastructure Development and Management—Volume 1*, Lecture Notes in Civil Engineering 570,
https://doi.org/10.1007/978-981-96-1554-4_5

1 Introduction

Developing citizen involvement that enhances quality of life is the main task of the smart city. Infrastructure, cultural capital, and behavioural capital are all part of a smart community. Information and communication technologies (ICT) are used in smart cities and are simple for citizens to access. When the city is improved intelligently, the standard of living of the citizens improves. Salem is the fifth-largest metropolis in the state of Tamil Nadu, which has 11 smart towns. Salem City's urban residents must have total economic, environmental [1], and sustainable growth sustainability. Public adoption, which is intended to promote and alter management, is one of the three components that is crucial for a smart community. The second component is converting some apps' raw data into activities, and the third and final component is offering the residents of Salem City powerful technology. Salem is renowned for its manufacturing foundation and has expanded with small-scale businesses.

In order to facilitate economic growth and projects, Salem Smart City has planned to build an industrial route from Coimbatore to Bangalore that is like Coimbatore-Salem. Streets & traffic, infrastructure, and the river waterway are a few of the major problems with area growth in Salem Smart City is shown in Fig. 1. By utilizing ICT-based solutions, the central city area can be transformed into a thriving commercial business centre. The city has undertaken a range of initiatives that leverage technology to improve governance, transportation, energy, public safety, and other services [2]. In this case study, we will explore the citizens' perceptions of smart city projects in Salem. This study will provide valuable insights into the benefits and challenges of implementing smart city projects, and the role of citizens in shaping the future [3] of their cities. To better understand the citizens' perceptions of these smart city projects, a questionnaire survey was conducted in Salem [4, 5].

A smart city questionnaire survey can provide valuable insights into the needs and preferences of residents and stakeholders, helping to identify areas where smart city initiatives can be focused and tailored to meet specific community needs. The importance of community input and citizens' perception on smart city projects in Salem cannot be overstated [6].

Fig. 1 Salem smart city project

This can help ensure the success of smart city projects in Salem. By involving citizens in the planning and implementation of smart city projects, the solutions can be designed to improve the quality of life for residents. This could include initiatives that address issues such as traffic congestion, air pollution, and access to public services [7]. This can help build trust and confidence among citizens, leading to greater support for these initiatives. It can lead to tailored solutions, improved quality of life, greater transparency and accountability, and foster collaboration and partnerships among different stakeholders [8]. This will provide a comprehensive overview of the citizens' perceptions of smart city projects in Salem, highlighting the successes and challenges of the initiatives and identifying opportunities for further development. Salem, a city in India, has also embraced this trend and implemented several smart city projects aimed at improving the quality of life for its citizens [9, 10]. The Statistical Package for Social Sciences (SPSS) software, a highly effective statistical analysis instrument used to spot developments and patterns in data, was used to assess the survey data. This study contributed to a more thorough grasp of how Salem's residents feel about smart city initiatives [11].

1.1 Questionnaire Design and Data Collection

1. What are the key factors that influence citizens' perceptions of smart city projects in Salem?
2. To what extent do citizens in Salem perceive smart city projects as beneficial to their daily lives?
3. How do demographic factors, such as age, income, and education, influence citizens' perceptions of smart city projects in Salem?
4. What are the main concerns or challenges that citizens in Salem have with smart city projects, and how can they be addressed?
5. What are the differences in perception between citizens who have experienced smart city projects first hand and those who have not?
6. What role do communication and public engagement play in shaping citizens' perceptions of smart city projects in Salem?
7. To what extent do citizens in Salem feel involved in the planning and implementation of smart city projects, and how does this affect their perceptions of these projects?
8. How do citizens in Salem compare their perception of smart city projects to their perception of traditional city projects, such as road maintenance or public transport?
9. How do citizens in Salem perceive the trade-offs between the benefits of smart city projects and potential concerns, such as data privacy or surveillance?
10. How do citizens' perceptions of smart city projects in Salem relate to broader societal and political attitudes towards technology and innovation?

2 Project Outcomes

2.1 Qualitative Studies on Smart Cities

Here are some common findings from qualitative studies on smart cities based on survey responses:

- Improved Quality of Life [12]: Many respondents perceive smart city technologies as beneficial for improving the overall quality of life. They appreciate features such as enhanced public transportation, better healthcare services, improved safety and security measures, and increased access to information and services.
- Environmental [13] Sustainability: Smart city initiatives often focus on promoting sustainability and reducing environmental impacts [14]. Survey respondents commonly express a positive attitude towards eco-friendly practices, such as smart energy management, waste reduction, and sustainable transportation options.
- Enhanced Mobility: Improved transportation systems, including intelligent traffic management, real-time navigation, and smart parking solutions, are frequently mentioned as desirable features. Respondents value the potential for reduced congestion, enhanced connectivity, and convenient mobility options.
- Citizen Engagement: Many studies emphasize the importance of involving citizens in the design and implementation of smart city initiatives. Survey responses often highlight the need for transparent governance, participatory decision-making processes, and platforms for citizen feedback and engagement [15].
- Privacy and Data Security Concerns: Survey respondents frequently express concerns about the collection, storage, and use of personal data in smart cities. They emphasize the importance of privacy protection measures, data security protocols, and clear consent mechanisms to address these concerns.
- Challenges in Implementation: Qualitative studies often highlight challenges faced during the implementation of smart city projects [16]. These challenges may include technological barriers, lack of interoperability between systems, bureaucratic obstacles, and resistance to change from stakeholders. The responses of smart cities, current challenges, adoption of smart cities, and infrastructure layer with its attributes are shown in Figs. 2, 3, 4, and 5, respectively.

2.1.1 Benefits of Smart City Projects

- Improved Infrastructure: Smart city projects can enhance the quality of infrastructure, such as transportation systems, energy grids, and waste management. This can lead to improved efficiency, reduced costs, and better resource allocation (Figs. 6, 7, 8, 9, and 10).
- Enhanced Efficiency and Sustainability [17]: By utilizing data-driven technologies, smart city projects can optimize resource usage, reduce energy consumption,

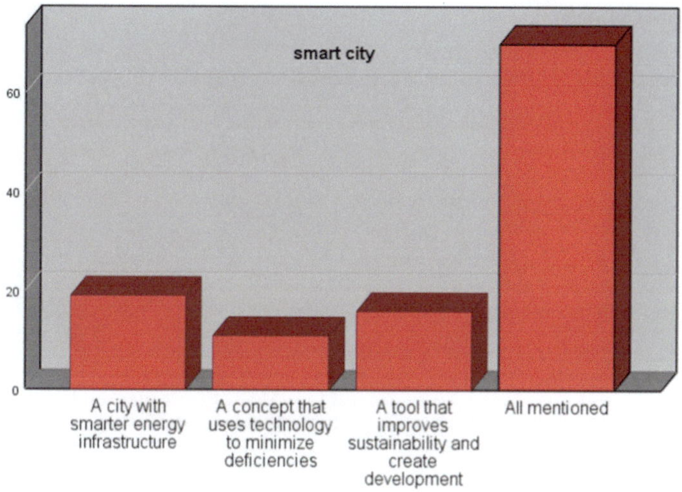

Fig. 2 Smart city

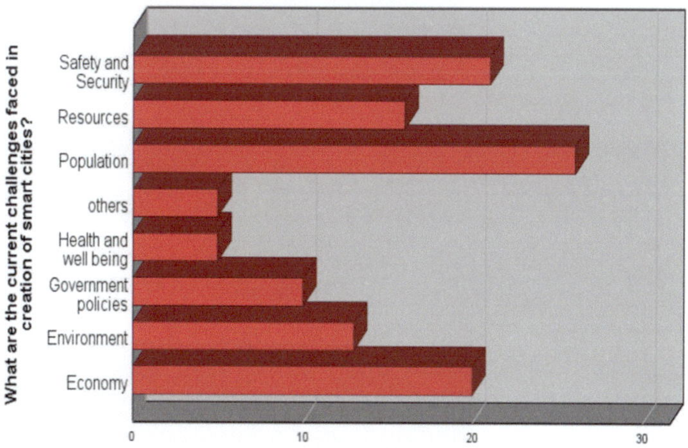

Fig. 3 Current challenges

and promote sustainable practices. This can result in lower emissions, improved air quality, and more sustainable urban development [18].

- Enhanced Public Services: Smart city initiatives often aim to improve public services such as health care, education, safety, and governance. This can include advanced healthcare systems, smart classrooms, better public safety measures, and e-governance solutions, leading to improved quality of life for residents.
- Economic Growth and Innovation: Smart city projects can attract investments, foster innovation, and promote economic growth. By creating an environment that supports entrepreneurship and innovation, cities can attract businesses and

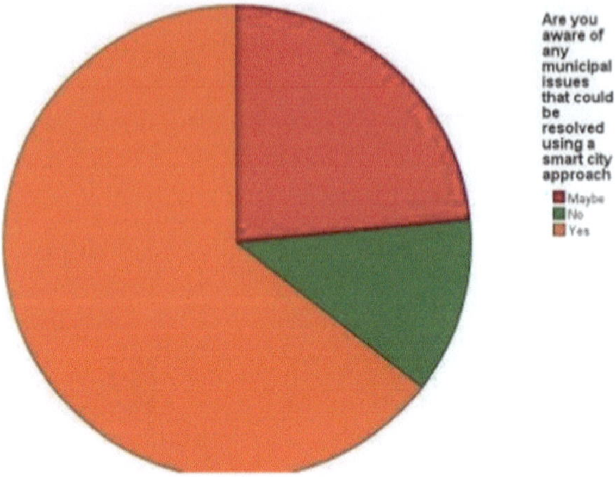

Fig. 4 Adoption of smart cities

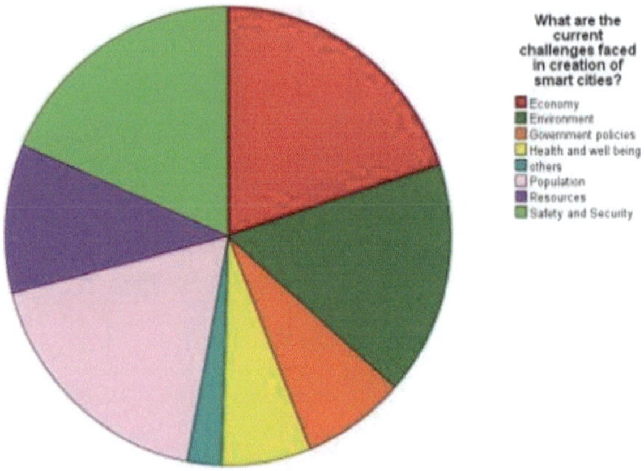

Fig. 5 Infrastructure layer

Fig. 6 Area development-key interventions

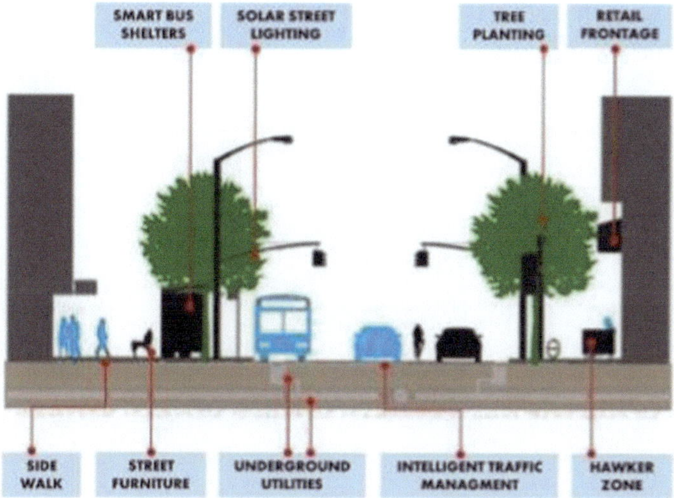

Fig. 7 Smart street designs

Fig. 8 Re-development of old bus stand

talent, leading to job creation and economic prosperity. The responses are shown in Figs. 11, 12, 13, 14, 15 and 16 along with its attributes.

2.1.2 Downsides of Smart City Projects

- Privacy and Data Security: The collection and use of vast amounts of data in smart city projects raise concerns about privacy and data security. There may be risks of data breaches, unauthorized access, and surveillance, which can undermine public trust and raise ethical questions.
- Inequality and Exclusion: The adoption of smart city technologies can exacerbate existing social and economic disparities. Certain populations, such as low-income

Fig. 9 Solid waste disposal

Fig. 10 Metro lite

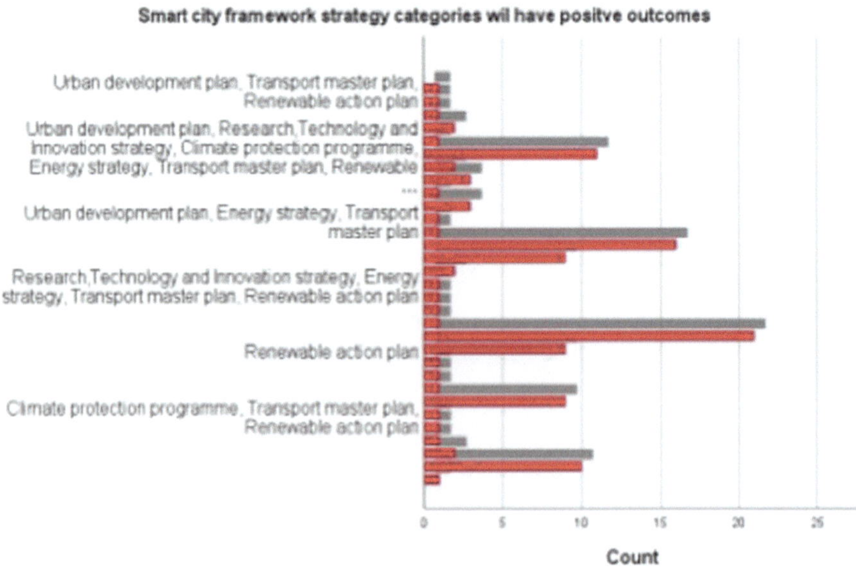

Fig. 11 Positive outcomes on smart cities

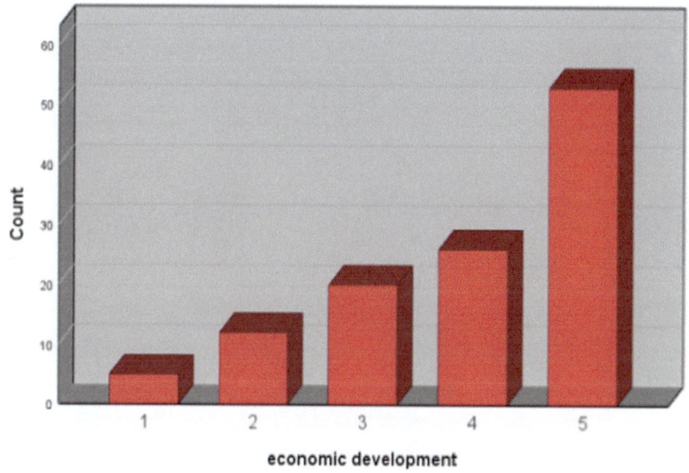

Fig. 12 Economic development

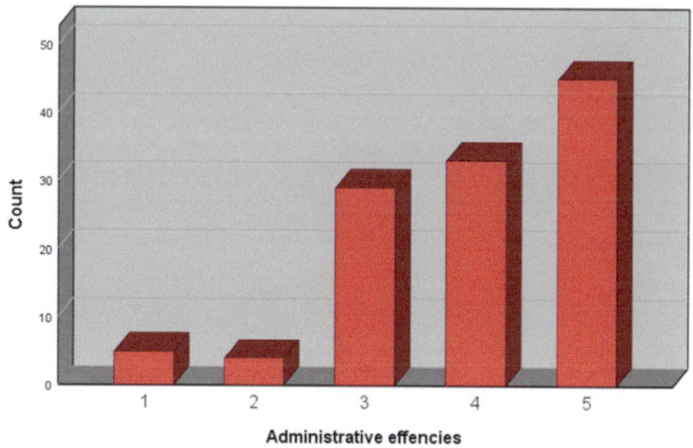

Fig. 13 Administrative effects

communities or elderly residents, may face difficulties in accessing or utilizing these technologies, leading to a digital divide and exclusion from the benefits of the projects.

- Cost and Sustainability: Implementing and maintaining smart city projects can be expensive, requiring significant financial investments. Additionally, ensuring the long-term sustainability of the projects and avoiding technology obsolescence can pose challenges.
- Dependency on Technology: Smart city projects heavily rely on technology infrastructure, connectivity, and data analysis. System failures, technical glitches, or

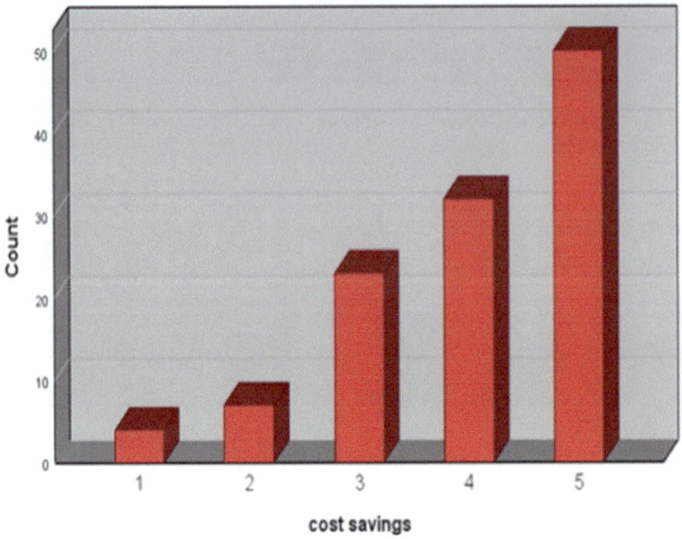

Fig. 14 Cost savings

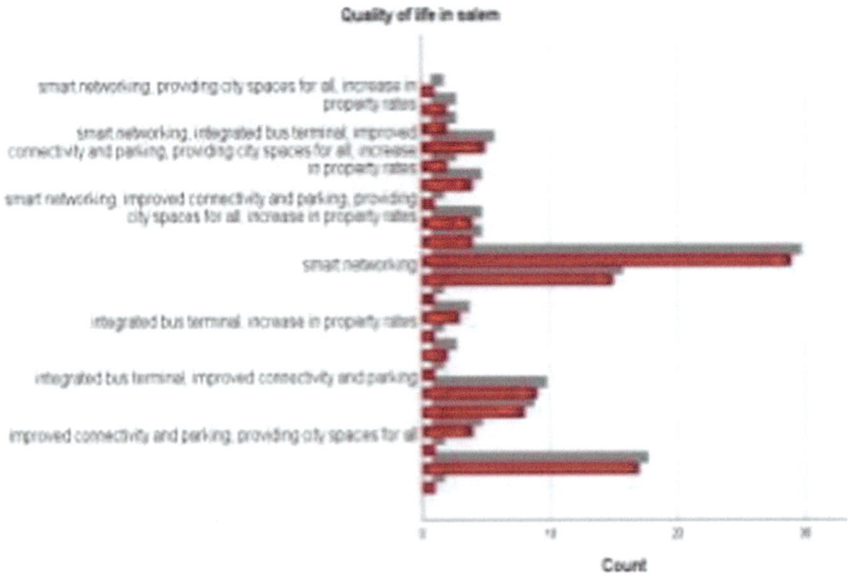

Fig. 15 Quality of life in Salem

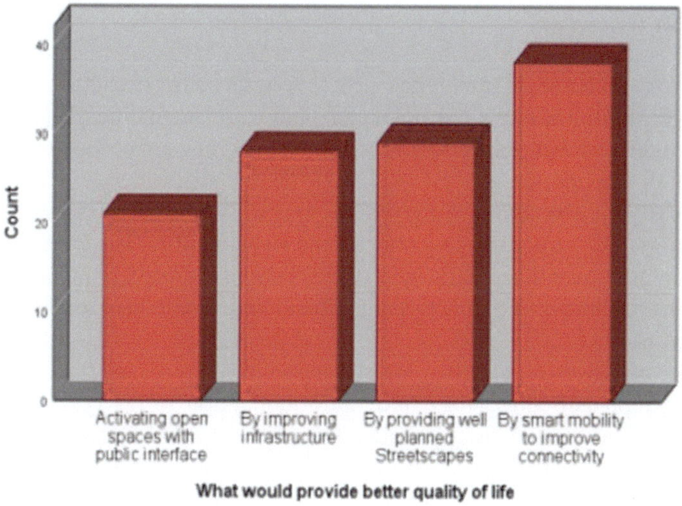

Fig. 16 Better quality of life

cyber-attacks can disrupt essential services and impact the daily lives of residents
.

The smart city projects and initiatives are shown in Figs. 6, 7, 8, 9, and 10.

2.2 *Frequency Data Analysis Using SPSS*

According to responses nearly 60% of people are aware of what a smart city is, and only about 12% are unaware of its significance. 20% of respondents stated that the present issue in creating a smart city with economy, including innovative communication networks, sensors, and data collection systems, is the economics. 18% of respondents stated that smart city will be provided for safety and security and population. 24.5% of the people said that the current challenges are due to the resources, government policies, and health being. This is because some are not known about the resources available [8]. 33% of the respondents say that culture play a significant role in the development of smart cities. While the concept of smart cities focuses on leveraging technology and data to enhance efficiency, sustainability, and quality of life, it is essential to consider the unique identity and cultural aspects of a city or community during the planning and implementation stages [21, 22].

2.2.1 Low Responses

- Lack of Awareness: As mentioned earlier, low response rates may occur if individuals have limited knowledge or understanding of smart cities. If people are not aware of the concept or its potential benefits, they may not perceive the survey as relevant or important.
- Complexity of the Topic: Smart cities encompass various aspects, including technology, infrastructure, governance, and sustainability. The complexity of the subject matter may deter some individuals from participating in the survey, especially if they find it difficult to comprehend or express their opinions on such complex issues.
- Lack of Incentives or Benefits: Without any incentives or perceived benefits for participation, individuals may have little motivation to respond to the survey. Offering rewards, such as gift cards, discounts, or entry into a prize draw, can help increase response rates.
- Time Constraints: Busy schedules and competing priorities can lead to low response rates. If the survey is lengthy or perceived as time-consuming, individuals may be less inclined to allocate the necessary time to complete it.

2.2.2 High Responses

- Strong Interest or Engagement: A high response rate could indicate a strong interest or engagement in the topic of smart cities among the target population. If the survey addresses issues that individuals are passionate about or directly affects their lives, they are more likely to participate.
- Clear Communication and Relevance: If the purpose and relevance of the survey are effectively communicated to potential respondents, they are more likely to understand the importance of their participation [23]. Clear instructions and a well-structured questionnaire that aligns with their experiences and interests can also contribute to higher response rates.
- Incentives or Rewards: Providing incentives or rewards for participation, as mentioned earlier, can significantly increase response rates. Individuals may be more motivated to complete the survey if they perceive tangible benefits or rewards associated with their participation.
- Easy Accessibility: Offering multiple channels or formats for survey completion, such as online and offline options, can make the survey more accessible to a wider range of individuals. Compatibility with different devices and user-friendly interfaces also contribute to higher response rates.
- Clear and Concise Survey Design: A well-designed survey with clear, concise, and focused questions are more likely to elicit responses. A streamlined survey that respects respondents' time and effort can encourage greater participation.
- Effective Communication and Marketing: Promoting the survey through various channels, such as social media, email, or community organizations, can increase its visibility and reach a larger audience. Engaging and persuasive messaging

about the importance and potential impact of the survey can also drive higher response rates.

3 Conclusion

In conclusion, the questionnaire survey conducted on citizens' perception [24] of smart city projects in Salem has provided valuable insights into their attitudes, expectations, and concerns. The findings indicate that overall, there is a positive perception, and a significant portion of the respondents (approximately 75%) were aware of the ongoing smart city projects in Salem. This suggests that there has been effective communication and information dissemination about these initiatives to the citizens.

- Public Services: A considerable portion of respondents (around 40%) expressed apprehension about the collection and use of personal information in smart city projects. This highlights the need for robust privacy frameworks and transparent data governance practices to address these concerns.
- Citizen Participation and Involvement: The analysis also revealed a desire for increased citizen participation and involvement in the planning and decision-making processes of smart city projects.
- Addressing these concerns through robust privacy frameworks and transparent data governance practices will be crucial to maintaining public trust and confidence in the long run.
- In conclusion, the questionnaire survey has provided valuable insights into citizens perceptions of smart city projects in Salem. It is evident that while there is an overall positive perception, there are also concerns that need to be addressed to maximize the benefits of these projects. By considering the feedback and suggestions of citizens, local authorities and stakeholders can continue to enhance the smart city initiatives in Salem, fostering a sustainable and inclusive urban environment for all residents.

References

1. Sinofsky A et al (2017) Citizen participation in smart cities: Evaluation framework proposal. In: 2017 IEEE 19th conference on business informatics (CBI), vol 1. IEEE
2. Pellicer, Soledad, et al (2013) A global perspective of smart cities: a survey. In: 2013 Seventh international conference on innovative mobile and internet services in ubiquitous computing. IEEE
3. LugoSantiago JA, LugoSantiago JA (2020) Is there such a thing as the smart city 1.0, 2.0, or 3.0?. Leadersh Strat Foresight Smart Cities Futur Model 33–42
4. Cocchia A (2014) Smart and digital city: a systematic literature review. Smart city: how to create public and economic value with high technology in urban space, pp 13–43
5. Yin CT et al (2015) A literature survey on smart cities. Sci China Inf Sci 58(10):1–18

6. Madakam S, Ramaswamy R, Date H (2019) Quality of life @ Palava smart city: a case study. Glob Bus Rev 20(3):708–742
7. Allam Z, Jones DS (2020) On the coronavirus (COVID-19) outbreak and the smart city network: universal data sharing standards coupled with artificial intelligence (AI) to benefit urban health monitoring and management. Healthcare 8(1)
8. Tadili J, Fasly H (2020) General smart city experts' perceptions of citizen participation: a questionnaire survey. In: Innovations in smart cities applications edition 3: the proceedings of the 4th international conference on smart city applications 4. Springer International Publishing, pp 3–15
9. Zhang Y et al (2019) Research on smart city evaluation based on hierarchy of needs. Procedia Comput Sci 162:467–474
10. Xydis G et al (2021) SMARTIES project: the survey of needs for municipalities and trainers for smart cities. Challenges 12(1):13
11. Molnár L et al (2021) Miskolc as a "Smart City"–Experiences of a Questionnaire Survey. Theory Methodol Pract Rev Bus Manage 17(Special Issue): 11–21.70
12. Siokas G, Kelaidi V, Tsakanikas A (2022) The smart city as a hub for nourishing public-private partnerships. Sustain Cities Soc 76:103466
13. Lai CMT, Cole A (2022) Levels of public trust as the driver of Citizens' perceptions of smart cities: the case of Hong Kong. Procedia Comput Sci 207:1919–1926
14. Jothimani P, Chenniappan P, Chidambaranathan V (2022) Factors impinge on the development of a smart city: a field study. Environ Sci Pollut Res:1–10
15. Belanche D, Casaló LV, Orús C (2016) City attachment and use of urban services: benefits for smart cities. Cities 50:75–81
16. Zenker S, Rütter N (2014) Is satisfaction the key? The role of citizen satisfaction, place attachment and place brand attitude on positive citizenship behavior. Cities 38:11–17
17. Albino V, Berardi U, Dangelico RM (2015) Smart cities: definitions, dimensions, performance, and initiatives. J Urban Technol 22(1):3–21
18. Georgiadis A, Christodoulou P, Zinonos Z (2021) Perception of smart cities: a case study. Appl Sci 11(6):2517
19. Giffinger R, Gudrun H (2010) Smart cities ranking: an instrument for the positioning of the cities?. ACE Arch City Environ 4(12):7–26
20. Angelidou M (2014) Smart city policies: a spatial approach. Cities 41:S3–S11
21. Castelnovo W (2016) Co-production makes cities smarter: Citizens' participation in smart city initiatives. Co-production in the public sector: experiences and challenges, pp 97–117
22. Schelings C, Elsen C (2017) Smart city concepts: from perception to acceptability. In: 21st conference of the environmental and sustainability management accounting network (EMAN)-sustainability accounting and control for smart cities
23. Julsrud TE, Krogstad JR (2020) Is there enough trust for the smart city? Exploring acceptance for use of mobile phone data in Oslo and tallinn. Technol Forecast Soc Chang 161:120314
24. Kavithanjali V, Meenachi A, Soundararajan M, Gencel O, Balasubramanian M, Palanisamy T (2024) Sustainable Innovations in Construction Management Select Proceedings of ICC-IDEA 2023 Perception of Proposed Bridge Using Augmented Reality of BIM in Educational Institute—Case Study Springer Nature Singapore Singapore 613–623

Solar Panel Technologies for Implementation in Civil Structures to Achieve a Greener Future: A Review

P. Velmurugan, M. N. A. Gulshan Taj, R. Malathy, and S. R. R. Senthil Kumar

Abstract 80% of the world's population will reside in cities by 2030 accounting for more than 70% of carbon emissions. So, governments all around the world are considering reducing their usage of fossil fuels in this decade as they significantly increase air pollution. The use of solar energy in energy-saving techniques has become increasingly prevalent across all major sectors, including residences and industries. Globally, the percentage of electricity produced from solar energy in 2019 was 2.7%. According to projections made by the International Energy Agency, solar energy will produce more than 16% of global electrical power by 2050. Therefore, research is being conducted globally to integrate renewable energy sources, particularly solar energy, into building practices. Recent advancements in solar panel technology have received a great deal of attention in the building industry, particularly from 2020 to 2023. This review article has discussed more than 30 world-class research done throughout the last ten years in this area. This article discusses the diagnosis of solar panel faults using algorithms and provides a thorough description of the many types of solar panel faults. To demonstrate the adaptability of solar cells in buildings, this article also highlights their inclusion in civil constructions, particularly facades and envelopes. To attain a green future, several studies are being conducted globally to use this renewable energy in the construction industry.

Keywords Solar energy · Renewable energy · Construction sector · Photovoltaic cells · Envelope

P. Velmurugan (✉) · M. N. A. Gulshan Taj · R. Malathy · S. R. R. Senthil Kumar
Sona College of Technology, Salem, Tamil Nadu, India
e-mail: veluvelmurugan001@gmail.com

Y. K. Guruprasad et al. (eds.), *Recent Advances in Infrastructure Development and Management—Volume 1*, Lecture Notes in Civil Engineering 570,
https://doi.org/10.1007/978-981-96-1554-4_7

1 Introduction

The incorporation of solar panels into building structures is leading the way in the construction industry's shift towards sustainability. The building architecture is being significantly altered by solar energy, which is also paving the path for a more environmentally and energy-conscious future. This review highlights the benefits of solar panels for creating a sustainable environment while analysing their impact on the building sector [24]. The capacity to convert solar energy into clean power has given rise to the technology of integrating solar panels into building structures. Legislators, architects, and developers anticipate solar energy as a sustainable energy source that reduces greenhouse gas emissions as talks about climate change and environmental degradation gain attraction [7]. The capacity of solar panels to generate power on-site is one of its first benefits for the building sector. Solar panels help buildings meet some of their electricity needs by turning solar energy into electrical energy. This technique reduces the need for traditional grid electricity, lowers utility costs, and increases energy independence [10]. To achieve sustainable energy solutions, building-integrated photovoltaics (BIPVs) and building applied photovoltaics (BAPVs) have demonstrated a noteworthy advancement. The restrictions in the usage of conventional solar panel applications have been solved by the concept of integrating solar technology into building structures and architectural designs. While BAPVs entail retrofitting solar panels into already-existing structures to produce solar energy from the built environment, BIPVs deal with the integration of solar panels into building components [13]. The solar system's power conversion mechanism is depicted in Fig. 1.

This technique increases the energy self-sufficiency of buildings and decentralizes the energy system. The use of solar panel technology in the building sector offers immediate environmental benefits in addition to financial ones. Buildings may efficiently minimize their carbon footprints, lower air pollution, and conserve fossil fuel

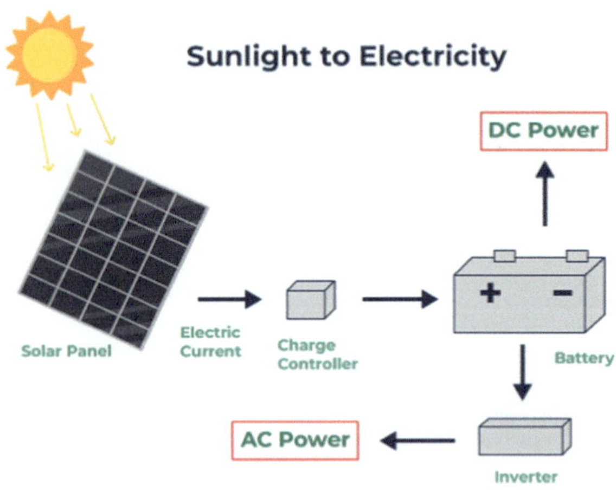

Fig. 1 Solar energy generation process (https://www.elprocus.com/solar-power-energy-with-advantages-and-disadvantages/)

resources by utilizing solar energy. This switch to greener energy sources is consistent with reducing carbon emissions and combating climate change [18]. To be used effectively, solar panels in buildings must overcome certain obstacles. The two biggest barriers to implementing this technology for a building owner are its high cost and installation challenges. Since people's economies grow and technology advances quickly in this century, installing solar panels can be done affordably, attracting greater investment in this area [2].

2 Integration of Solar Panels in Building Designs

To achieve the objective of sustainability in the construction sector, solar panel integration technology is simply amazing. The thermal efficiency and energy observation of the solar thermal collector structure, which was created by utilizing sheet metal half-shells made of extremely inventive incremental sheet forming with an integrated channel inside and a unique absorber coating outside, was proven by Scholz et al. In the two coastal cities of Naples, Italy, and Thessaloniki, Greece, Vassiliades et al. examined the effects of temperature conditions in public areas before and following the installation of photovoltaic facades in buildings. The street-level energy simulation was carried out using Envi-MET software. Italos et al. provided an example of the application of a double-skin façade with integrated photovoltaics in a building located in the southeastern Mediterranean region. Energy-Plus software is used to conduct energy simulations, which leads to a 63% reduction in building energy consumption when compared to standard energy sources. Constructing integrated photovoltaics results in a multifaceted strategy that addresses design architecture advancements and reducing carbon emissions [17, 31, 33]. These strategies include thermal insulation, electricity generation, and shading effect. According to Attoye et al., it is possible to enhance conventional building-integrated photovoltaic facades to create effective products that combine the production of electricity with aesthetically pleasing uses. Moreover, the integration of solar panels on double facades is employed to guarantee airflow over the cavity, which cools the solar panels and results in increased efficiency all provide a general description of the impact of ventilation on the performance of building-integrated photovoltaics. Researchers are examining the double facade geometry that integrates photovoltaics to show how this geometric design affects apartment building energy performance. The photovoltaics system, which is mounted on the building's facade to convert solar energy into electrical energy, is shown in Fig. 2.

Meanwhile, other researchers used this geometrical change to create a concentrating collector on the facades. Peng et al. state that adopting a building-integrated photovoltaic double-skin façade made by semi-transparent solar modules can result in electricity savings of around 50%. A naturally ventilated double-skin façade combined with semi-transparent photovoltaic modules can reduce the amount of solar radiation that enters the structure while providing a satisfactory degree of daylighting. Furthermore, Theokli et al. focus on the climate of the Mediterranean

Fig. 2 Building with solar façade (https://a2-solar.com/en/building-integrated-pv/solar-facades/)

area and effectively show that integrating photovoltaics into a structure may transform a low-energy performance building into one with almost zero energy use. Following the previously stated concepts, several researchers have also studied the air cavity in double façades. Other researchers sought to increase the absorption of solar energy in buildings by combining building-integrated photovoltaics with other systems like unglazed transpired collectors. Yang et al. investigated open-area air-based active double facades in the building. While some researchers are working to combine phase change materials and building-integrated photovoltaics, to observe the characteristics of these systems, some researchers investigated with semi-transparent building-integrated photovoltaics systems, thus improving the building's ability to have natural insolation.

3 Techniques Adopted for Solar Panel Inspection

There are two primary methods for inspecting big solar plants: measuring the intensity using cameras and measuring the voltage and current within the photovoltaic modules. The use of cameras for inspection is becoming more popular due to technical breakthroughs in the specialized fields of robotics and unmanned aerial vehicles (UAVs), which enable quicker defect identification in solar modules. Certain kinds of voltage and current sensors attached to photovoltaic strings rather than a single solar panel are sometimes used for photovoltaic plant monitoring. While there is still room for improvement in machine learning methods for fault categorization, detection, and localization, these sensors perform a far better function in terms of accuracy [8]. Additionally, without resolving the communication aspect of the problem, Ansari and Pereira discussed about the use of smart sensors in defect detection in solar plant monitoring. When installing solar panels, non-electrical failures can lead to poor clamping, which is one type of failure that occurs. About 10% of power loss can be attributed to a variety of issues. The most common flaw to affect all photovoltaic modules is the hotspot defect, which may be found using detection techniques including visible light, thermography, and high-resolution electroluminescence. Camera-based monitoring, also known as contactless monitoring, makes use of a range of 2D camera models. The obvious flaws that quickly reduce the effectiveness of photovoltaic plants such as broken or cracked cells, bird droppings, or delamination—are detected using visible light cameras [20, 27]. Using these cameras, a technical battle between machine learning algorithms and artificial intelligence approaches like VGG19 and AlexNet have been characterized to find and categorize photovoltaic problems. The International Technical Commission is not setting any international standards for these sorts of infrared and electroluminescence camera examinations of solar modules. Although electroluminescence is a useful method for identifying small fractures and failures, it is not able to forecast how such damages would affect power output. The most effective way to identify the impending hotspot defect is made up of both cooled and uncooled varieties in infrared cameras [21]. Figure 3 shows how a drone equipped with a camera inspects the solar panels in a solar farm.

Accurate fault type, magnitude, and position assessment are some of the primary benefits of passive camera-based hotspot detection [12]. This method works better at night than during the day because of the low radiation levels that cause the MWIR

Fig. 3 Inspection of solar panel with camera inbuilt in a drone (https://www.azorobotics.com/Article.aspx?ArticleID=568)

range to change [15]. Numerous studies have been conducted utilizing specific limitations imposed by various image segmentation techniques in order to forecast the impact of additional radiations operating on photovoltaic panels inside the infrared camera's range [22]. Using radiometric thermal cameras makes photovoltaic examination easier by providing precise temperature information for each pixel in the picture. These days, 1 MW solar facilities in South Korea are testing the use of infrared cameras integrated into unmanned aerial vehicles for autonomous photovoltaic plant monitoring [16, 19, 29]. Nippon and Telephone Corporation researchers have developed many algorithms for solar panel monitoring and have demonstrated that single shot detection algorithms yield good results. With the UAV-based thermal camera detection technology, the problem of extra installation labour may be solved for precise power loss detection in big plants. When visible light and infrared cameras are used in tandem, some image processing methods, such Canny edge detection, can identify faults more accurately [14]. Wesii Company has just released a cutting-edge solar array monitoring system on the market. In order to move forward with a routine monitoring procedure, a geotagged picture has been provided by the GPS system. An expert drone pilot is needed to operate the device during a drone-based solar plant inspection. The drone transmits photographs to the ground, where a thermos-diagnostician analyses them. Hotspot defect detection approach was also proven by researchers based on data made in a variety of domains.

4 Case Study

An Education Institution, namely Sona College of Technology, located in Salem, Tamil Nadu, India (Latitude: 11.67810 N, Longitude: 78.12440 E), has been utilized to provide the study's results, which are displayed in Fig. 4.

Fig. 4 Solar panel in the study area

In this case study, the DJI Phantom 4 Pro was used to capture images of the solar panels installed on the institution's building rooftops. Drones are controlled while in flight using a smartphone app called DJI GO 4 PRO. The software used to build three-dimensional models of all the standard drone photos is called Context Capture. Three-dimensional models of any size, ranging from a few centimetres to an entire structure, may be created using this programme. The resolution of the images entered into the context capture programme determines the calibre of the resulting three-dimensional model. Context Capture software on a PC or laptop can process around 20 gigapixels a day. Specifically, the software that turns two-dimensional images into three-dimensional models is called Context Capture Master. The programme runs on the Context Capture Engine, which requires a certain amount of time to finish the conversion process. With the help of the program's add-on Context Capture Viewer, users may inspect and measure the volume inside the three-dimensional model that has been built.

The ecologically beneficial results of the net zero energy building are mostly due to solar electricity. The educational institution has many buildings with upper-level solar panels for usage as a potential future dynamic. Drones are being used to take more than 30 photos of each solar panel, which are then converted into a 3D model using Context Capture Software. Once the process of extracting the 3D model from the high-definition photographs taken with drones is finished, the solar panel's area can be measured. The solar panel's four corners are delineated by a polygon. Manually measuring the solar panel with conventional methods will take time. Figure 5 shows how to use Context Capture Software to rapidly estimate the size of a solar panel installed on a building. The outcome showed that some solar panels had perimeters of 24.56 m and area of 36.54 m, respectively.

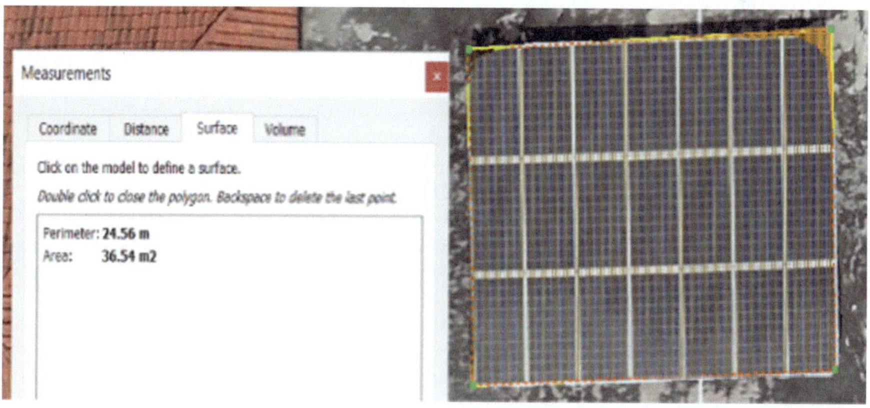

Fig. 5 Perimeter and area calculation of the solar panel from 3D model generated using context capture software

5 Conclusions

It is widely accepted that solar energy is the most significant renewable energy source on the planet. Its tremendous growth across a range of industries, including the building sector, has demonstrated its ability to help fight climate change. There has been no environmental harm as a result of the generation of solar energy, including no noise or chemical emissions. Installing solar panels on buildings leads in a significant increase in job prospects throughout the installation and maintenance operations, as well as a long-term economical solution to reduce energy costs. Because solar energy is becoming more and more affordable every year, along with improving panel efficiency, many houses are adopting it. It provides a method for balanced energy distribution and use, which raises energy efficiency. Discussions are being held about the topic of integrating photovoltaics into building facades, and the findings of different researchers provide positive results. Researchers have experimented with a variety of methods for monitoring solar panels, including the integration of cameras into drones for flaw identification and explanation. Using traditional methods to measure the area of the solar panel takes a lot of time and requires a lot of human labour. However, using digital methodologies, like using drones to take pictures and Context Capture software to process those images, makes the process possible in a matter of seconds, and the case study clearly outlines the results.

References

1. Advanced and automotive homepage. https://a2-solar.com/en/building-integrated-pv/solar-facades/
2. Al-Radhi Y, Roy K, Liang H, Ghosh K, Clifton GC, Lim JBP (2023) Thermal performance of different construction materials used in New Zealand dwellings comparatively to international practice—a systematic literature review. J Build Eng 72:106346
3. Ansari S (2021) A review of monitoring technologies for solar PV systems using data processing modules and transmission protocols: progress, challenges and prospects, pp 1–34
4. Attoye DE, Aoul KAT, Hassan A (2017) A review on building integrated photovoltaic facade customization potentials. Sustainability 9:2287
5. Azo robotics homepage. https://www.azorobotics.com/Article.aspx?ArticleID=568
6. Italos C, Patsias M, Yiangou A, Stavrinou S, Vassiliades C (2022) Use of double skin façade with building integrated solar systems for an energy renovation of an existing building in Limassol, Cyprus: energy performance analysis. Energy Rep 8:15144–15161
7. Dai SF, Liu HJ, Yang JH, Peng HY (2022) Wind loads on roof-mounted isolated solar panels of tall buildings through wind tunnel testing. Sol Energy 231:607–622
8. Daliento S, Chouder A, Guerriero P, Pavan AM, Mellit A, Moeini R, Tricoli P (2017) Monitoring, diagnosis, and power forecasting for photovoltaic fields: a review. Int J Photoenergy
9. Electronic project focus homepage. https://www.elprocus.com/solar-power-energy-with-advantages-and-disadvantages/
10. El-Khawad L, Bartkowiak D, Kümmerer K (2022) Improving the end-of-life management of solar panels in Germany. Renew Sustain Energy Rev 168:112678
11. Favoino F, Goia F, Perino M, Serra V (2016) Experimental analysis of the energy performance of an ACTive, RESponsive and Solar (ACTRESS) façade module. Sol Energy 133:226–248

12. Galfalk M, Olofsson G, Crill P, Bastviken D (2015) Making methane visible. Nat Clim Chang 6(4):426–430
13. Ghosh A (2020) Potential of building integrated and attached/applied photovoltaic (BIPV/BAPV) for adaptive less energy-hungry building's skin: a comprehensive review. J Clean Prod 276:123343
14. Grimaccia F, Leva S, Dolara A, Aghaei M (2017) Survey on PV modules common faults after an O&M flight extensive campaign over different plants in Italy. IEEE J Photovolt 7(3):810–816
15. Henry C, Pudel S, Lee S, Jeong H (2020) Automatic detection system of deteriorated PV modules using drone with thermal camera. Appl Sci 10(11):3802
16. Higuchi Y, Babasaki T (2018) Failure detection of solar panels using thermographic images captured by drone. In: 7th international conference on renewable energy research and applications Paris, France
17. Kang DW, Ryu J, Konagai M (2019) High-performance amorphous silicon thin film solar cells prepared at 100°C: toward flexible building-integrated photovoltaics. Electron Mater Lett 15:623–629
18. Kumar CMS, Singh S, Gupta MK, Nimdeo YM, Raushan R, Deorankar AV, Kumar TA, Rout PK, Chanotiya C, Pakhale VD (2023) Solar energy: a promising renewable source for meeting energy demand in Indian agriculture applications. Sustain Energy Technol Assess 55:102905
19. Leva S, Aghaei M, Grimaccia F (2015) PV power plant inspection by UAS: correlation between altitude and detection of defects on PV modules. In: 2015 IEEE 15th international conference on environment and electrical engineering (EEEIC). IEEE, pp 1921–1926
20. Madeti SR, Singh S (2017) Monitoring system for photovoltaic plants: a review. Renew Sustain Energy Rev 67:1180–1207
21. Miklós A (2001) Application of acoustic resonators in photoacoustic trace gas analysis and metrology. Rev Sci Instrum 72(4):1937–1955
22. Moradi Sizkouhi AM, Esmailifar SM, Aghaei M, de Oliveira AKV, Rüther R (2019) Autonomous path planning by unmanned aerial vehicle (UAV) for precise monitoring of large-scale PV plants. In: 2019 IEEE 46th photovoltaic specialists conference (PVSC). IEEE, pp 1398–1402
23. Peng J, Curcija DC, Lu L, Selkowitz SE, Yang H, Mitchell R (2016) Developing a method and simulation model for evaluating the overall energy performance of a ventilated semi-transparent photovoltaic double skin facade. Prog Photovolt Res Appl 24:781–799
24. Peng HY, Song SS, Liu HJ, Dai SF, Zhang FL (2022) Investigation of wind loading characteristics of roof-mounted solar panels on tall buildings. Sustain Energy Technol Assess 54:102800
25. Pereira R (2019) IoT embedded systems network and sensors signal conditioning applied to decentralized photovoltaic plants. Measure J 142:195–212
26. Scholz P, Weise D, Schmidt L, Dembski M, Stahr A, Dix M, Duminica F, Le Craz S, Koziorek J (2023) Sheet metal design approach for 3D shaped façade elements with integrated solar thermal functionality. MDPI solar 3(2):213–228
27. Sizhoukieet A (2021) A deep convolutional encoder-decoder architecture for autonomous fault detection of PV plants using multi-copters. Sol Energy 223:217–228
28. Theokli C, Elia C, Markou M, Vassiliades C (2021) Energy renovation of an existing building in Nicosia cyprus and investigation of the passive contribution of a BIPV/T double fac¸ ade system: a case-study. Energy Rep 7:8522–8533
29. Tsanakas JA, Ha LD, Al SF (2017) Advanced inspection of photovoltaic installation by aerial triangulation and terrestrial georeferencing of thermal/visual imagery. Renew Energy 102:224–233
30. Vassiliades C, Savvides A, Buonomano A (2022) Building integration of active solar energy systems for facades renovation in the urban fabric: effects on the thermal comfort in outdoor public spaces in Naples and Thessaloniki. Renew Energy 190:20–47
31. Vassiliades C, Michael A, Savvides A, Kalogirou S (2018) Improvement of passive behaviour of existing buildings through the integration of active solar energy systems. Energy 163:1178–1192

32. Yang T, Athienitis AK (2014) A study of design options for a building integrated photovoltaic/thermal (BIPV/T) system with glazed air collector and multiple inlets. Sol Energy 104:82–92
33. Yang J, Jo H, Choi SW, Kang DW, Kwon JD (2018) All p-i-n hydrogenated amorphous silicon oxide thin film solar cells for semi-transparent solar cells. Thin Solid Films 662:97–102

Finite Element Analysis of Doubly Reinforced Concrete Beams Under Flexural Loading

Krishnakanth Chidre and Sumant Kulkarni

Abstract The moment of inertia, cross-sectional area, and modulus of elasticity are crucial concepts in structural analysis. We need to be aware of the members' geometric and material characteristics in indeterminate constructions. The modulus of elasticity of concrete is expressed in terms of the concrete's characteristic strength, per the Indian codal requirements. Reinforced concrete is a composite material used in real-time structures. The current work aims to analyze the modulus of elasticity of a beam cross section with respect to the proportion of compression and tension reinforcement. The dimensions of the beam are 250 × 350 × 2300 mm, and it has two 12 mm diameter tension and compression bars. It is ensured that the doubly reinforced cross sections selected in the study are under reinforced based on limiting moment of resistance and limiting depth of neutral axis in line with codal provision. The provisions for the minimum and maximum reinforcement in both the tension and compression zones, as well as the spacing requirements for stirrups, adhere to the guidelines set forth in IS 456:2000. For a grade of concrete M30 and steel FE415, FE500, FE550, two-point loading has been taken into consideration when performing the FE analysis. The analysis's result is an empirical formula for the modulus of elasticity for the grade of steel and concrete under consideration.

Keywords Doubly reinforced concrete beams · Modulus of elasticity · Grade of concrete and steel

1 Introduction

A thorough grasp of RC structures is necessary for the successful design of a reinforced concrete structural element. When analyzing high-rise buildings with software, the impact of stirrups and reinforcement bars is usually disregarded, and just

K. Chidre (✉) · S. Kulkarni
REVA University, Bangalore, Karnataka, India
e-mail: krishnakanthchidre123@gmail.com

Y. K. Guruprasad et al. (eds.), *Recent Advances in Infrastructure Development and Management—Volume 1*, Lecture Notes in Civil Engineering 570,
https://doi.org/10.1007/978-981-96-1554-4_8

the cross-sectional area of plain concrete is taken into account. According to IS 456-2000, the Young's modulus of beams made of reinforced concrete changes based on the concrete grade. The elasticity of a material is represented by its Young's modulus, which shows how it rebounds from stress and reverts to its initial shape. When examining reinforced concrete structures, it is standard to consider the modulus of elasticity of the concrete. Practically speaking, reinforced concrete has a far higher modulus of elasticity than plain concrete. Since concrete is naturally weak in tension, reinforcement largely gives it tensile strength. A material with increased overall strength is produced when the strong tensile strength of steel and the compressive strength of concrete are combined.

The two primary types of reinforcing steel that structural designers are permitted to use are mild steel and HYSD steel bars, as per IS: 456-2000 [6]. The ductility, strength, durability, and production technology of reinforcing bars have all seen major changes. Simple mild steel reinforcing bars with a typical yield strength of 250 MPa were widely utilized prior to the 1960s. According to IS: 1786, any reinforcement, comprising steel grade Fe-415 and higher, is referred to as high strength deformed steel (HYSD) bars [9]. Varying steel grades might have varying strengths-to-diameter ratios. The yield strength of the steel is divided by the diameter to get the ratio. Higher strength-to-diameter ratios are frequently preferred in the case of reinforced concrete, where steel bars play a key role. This is so because a larger ratio means that, in relation to its size, a steel bar can support heavier loads. This study aims to determine the best way to use the steel grade while maintaining the M30 concrete grade.

Numerous academics have published the results of their studies on estimating the elastic modulus of concrete. According to Satyaprakash et al. [1], the percentage of reinforcement and concrete grade affect the modulus of elasticity of reinforced concrete. According to Dahmani et al. [2], the ANSYS commercial code's concrete failure model can be used to appropriately evaluate the nonlinear response of RC structures. The E_{RCC} equation was provided by Kulkarni et al. [3] and is helpful for RCC designers when designing different RC sections. Additionally, these can provide guidance on how to securely support the highest loads with the least amount of steel, lowering the overall cost of building a structure. It has been confirmed by Slowik [4] that applying elasticity theory to determine the cracking moment in flexural concrete members is inappropriate. For results that accurately represent the behavior of the structure, fracture mechanics-based design procedures should be applied. Following the observation of the changes in the stress–strain curves of the two primary constituents of reinforced concrete, reinforcing steel, and concrete, Chatterjee and Bhanja [5] deduced that the reduction in the yield strain of the reinforcing steel resulted in an increase in the section's neutral axis depth. As a result, when concrete is compressed, its rectangular parabolic stress block changes, increasing the stress block area and, consequently, the concrete's compressive stress. Consequently, the section's moment of resistance rises as well. According to reports, the modulus of elasticity is overestimated by IS:456-2000 [6] and Eurocode-2 [7] compared to BS-8110 [8]. IS:1786 [9]. In order to ensure that the beam's greatest deflection stays within the ACI's maximum allowed deflection, Rasheed and Agha [10] proposed adjusting the span depth ratio of the ACI Code influenced by steel area in tension, which is expressed in terms of the reinforcement indices ratio (ρ / ρ_b), uniform load (w), strength of concrete in compressive (f_c'), and width of the section (b). Ezeagu and Obasi [11] carried out a comparison study to determine how nominal size aggregate affected concrete E. The authors concluded that 20 mm size aggregate has a greater impact on the modulus of elasticity than 30 and 60 mm size aggregate. According to Noguchi and Takafumi [12], the square of the concrete's unit weight is directly correlated with changes in the material's modulus of elasticity (E). Anbuvelan and Subramanian [13] conducted an experimental study to determine the fiber-reinforced concrete's E value. According to Baalbaki [14], it is relatively difficult to establish a relationship between E of concrete with higher grades and strength of concrete under compression. The formula for deflection of beam specimens, which includes the maximum moment in the beam owing to loads and the cracking moment, is specified in Eurocode-2 [7]. IS:456-2000 is quiet on the details pertaining to a formula for beam deflection in both cracked and uncracked conditions.

1.1 Objectives of Study

Based on the literature review, the current study's objectives have been decided upon.

- The performance of reinforced concrete sections before and after cracking.
- How HYSD bar grade affects stiffness and strength?
- Examining sections made of reinforced concrete analytically.
- To determine the theoretical connection between the grade of steel and concrete and the true modulus of elasticity (E) of RCC.

2 Methodology

The study concentrated on performing finite element analysis for flexural loading on RCC beams with M30 grade. The analysis took into account a number of reinforcement combinations, including Fe-415, Fe-500, and Fe-550. According to clause 21, IS: 456-2000 [6] suggests that structural members have a minimum width of 200 mm. As a result, a cross-sectional dimension of 250 $\times$ 350 mm with a 2300 mm overall span was chosen for the model. Clause 23.2 of the Indian codal regulations determined the permissible deflection for beams using the formula (span/350) mm. The model under consideration has two bars of 12 mm tor steel each serving as both compression and tension reinforcement, totaling 0.283%.

A beam section's load-carrying capacity is the amount of applied load it can withstand and maintain without failing or deflecting excessively. It is a crucial metric in structural engineering that is affected by a number of variables, such as the size, material characteristics, and beam's structural design. The moment of resistance for the section is provided by the equation in accordance with clause 38.1 annexure G, per IS 456–2000[6].

Table 1 Limiting depth of neutral axis for varies grades of steel

Grade of steel (f_y)	(x_{umax} / d)
415	0.48
500	0.46
550	0.44

$$M_{\text{ulim}} = 0.36\left(\frac{x_{u,\max}}{d}\right) f_{\text{ck}}\left[1 - 0.42\left(\frac{x_{u,\max}}{d}\right)\right] bd^2 \tag{1}$$

In the above equation,

x_u = sectional depth at null stress,
d = depth up to centroid of reinforcement,
f_y = characteristic strength of reinforcement,
f_{ck} = characteristic compressive strength of concrete,
b = sectional width,
M_{ulim} = permissible sectional moment capacity,
x_u = limiting value of x_u from clause 39.1.

According to IS:456-2000 [6], clause 38.1, the limiting values of the neutral axis depth for various steel grades are displayed in Table 1.

The load-carrying capacity of a beam subjected to two-point symmetric load can be determined by assessing the bending moment at a midspan.

$$W_u = \frac{3M_{\text{ulim}}}{L} \tag{2}$$

In the above Eq. (2),

W_{ulim} = load-carrying capacity of a beam,
M_{ulim} = limiting moment of resistance of a section,
L = effective span of beam.

3 Finite Element Model of Beam Specimen

Modeling an RCC beam in FEM software is challenging due to the brittle nature of concrete and the simultaneous presence of two materials steel and concrete. For three main reasons, it works well as a technique to assess the static performance of buildings. It improves structural safety, cuts down on design time, and saves money during construction. Three steps are included while analyzing a structure with ANSYS.

(a) Determining the environmental elements to be used and defining the finite element model are part of pre-processing.
(b) Finite element model solving is the main goal of the analysis solver.

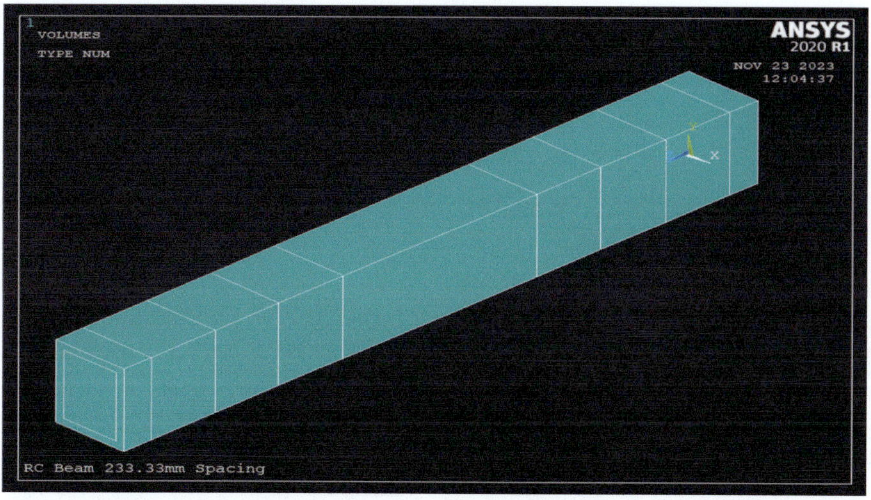

Fig. 1 RC beam

(c) Post-processing include utilizing visualization tools to analyze the data, such as deformation outlines for displacement.

The concrete grade used for this investigation is M30, and the concrete element employed is SOLID65 (see Fig. 1). Eight nodes, each with three degrees of freedom (i.e., translation in the x, y, and z dimensions), define the element. The LINK8 spar element (see Fig. 3) is utilized to simulate the stirrups and primary reinforcement, incorporating three degrees of freedom at each node. Numerous factors, including plasticity, creep, swelling, stress stiffening, and significant deflection, are taken into account by this element. Stirrups and the main reinforcement are modeled using a bi-linear idealized stress–strain curve. The model stiffness in this study is updated using the ANSYS software using Newton–Raphson equilibrium iterations. Force and displacement are the two factors that establish the convergence criteria for solid elements made of reinforced concrete in the study. Below is a display of the concrete solid model and its meshing with a 25 mm element size (See Figs. 1, 2, 3, 4, 5 and 6; Tables 2, 3 and 4).

4 Result and Discussion

4.1 M30 Concrete Grade with Fe-415 Steel Grade (Model 1)

The load-carrying capacity of model (1) mentioned in the study is estimated as 152.20 kN as per Eq. (2). The steel grade used to compute this load is Fe-415. Two-point symmetric loading and simply supported boundary conditions are applied to

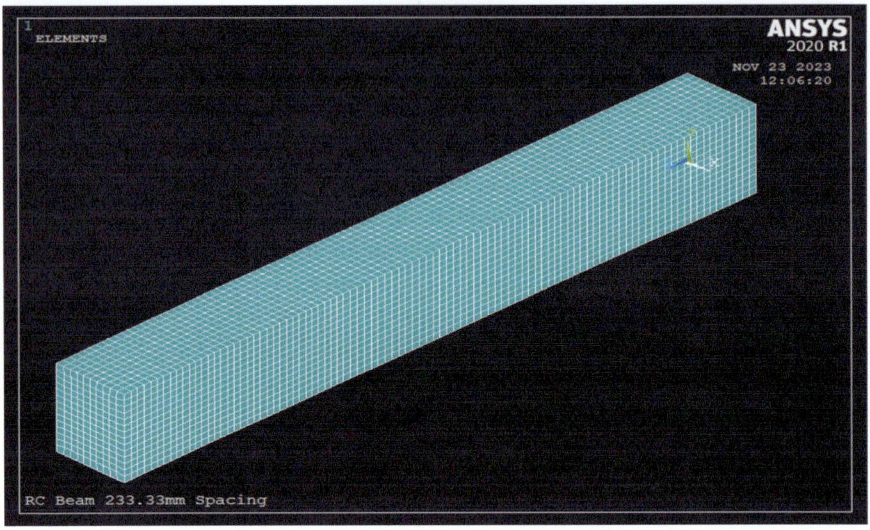

Fig. 2 Concrete beam meshing

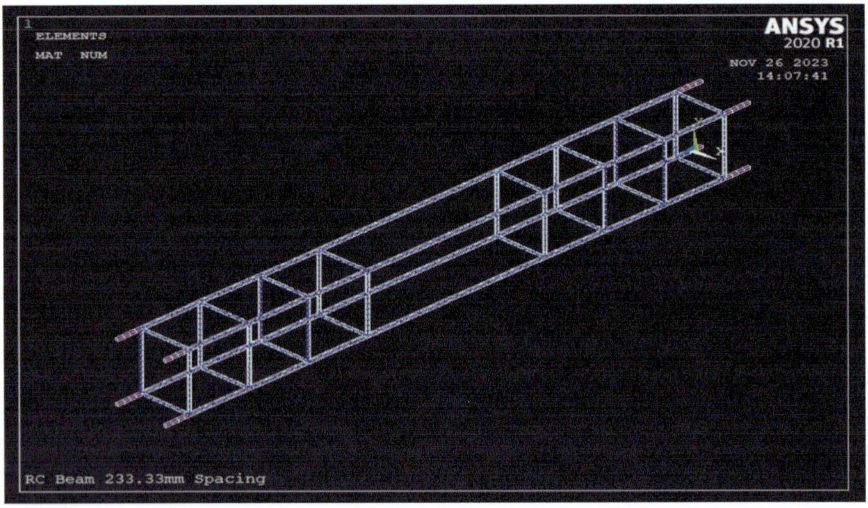

Fig. 3 FE modeling of beams sections

the finite element model. The load is applied in substeps, and as Table 4 illustrates, deflections at three distinct loading phases were recorded. The modulus of elasticity was estimated using the first fracture deflection and the associated load, as indicated by Eq. (3).

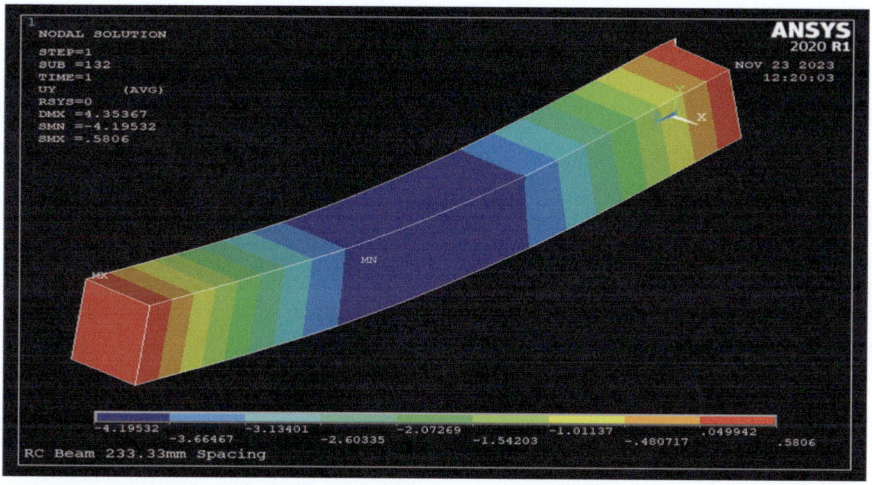

Fig. 4 Deformation for model 1 of M30, Fe-415

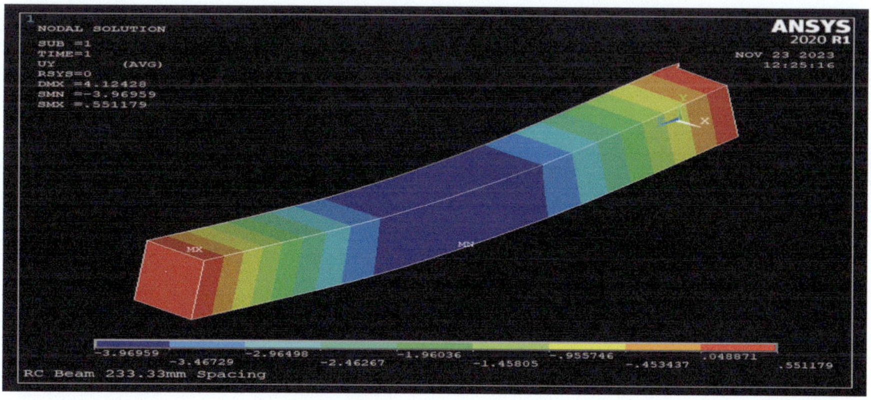

Fig. 5 Deformation for model 2 of M30, Fe-500

$$\delta = \left(\frac{5}{384}\right)\left(\frac{\mathrm{wl}^4}{\mathrm{EI}}\right) + \left(\frac{23}{648}\right)\left(\frac{\mathrm{Wl}^3}{\mathrm{EI}}\right) \tag{3}$$

where, $I = \frac{250 \times 350^3}{12} = 893,229,166.7\ \mathrm{mm}^4$

$W = (30.92 \times 1000) / 2 = 15{,}460\ \mathrm{N}, \mathrm{l} = 2100\ \mathrm{mm}$.

By substituting all the values in the above Eq. (3), we obtain,

$$0.2 = \left(\frac{5}{384}\right)\left(\frac{2.1875 \times 2100^4}{E \times 893229166.7}\right) + \left(\frac{23}{648}\right)\left(\frac{15460 \times 2100^3}{E \times 893229166.7}\right) \tag{4}$$

By solving Eq. (4), we obtain,

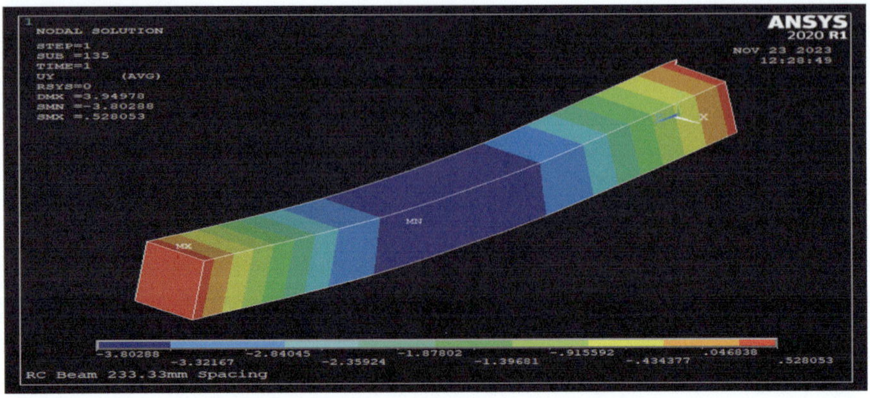

Fig. 6 Deformation for model 3 of M30, Fe-550

Table 2 Elements used for different materials

Material	Element	DoF/node	Properties
Concrete	SOLID65	3	Multi-linear
Main reinforcement	LINK8	3	Bi-linear
Stirrups	LINK8	3	Bi-linear

Table 3 Stress–strain values for M30 grade of concrete

Stress (N/mm^2)	Strain
6.02	0.000222
7.414	0.000275
13.015	0.0005
22.661	0.001
25.275	0.0012

Table 4 Load deflection for every model at various loading stages

Load and deflection (kN, mm)	Model 1	Model 2	Model 3
First crack load Deflection	30.92 0.2	30.70 0.2	30.67 0.2
Yield load Deflection	113.35 3.12	111.74 3.03	110.72 2.98
Ultimate load Deflection	152.20 4.354	147.39 4.12	142.45 3.94

$E = 31{,}554.52$ N/mm^2.

The calculated modulus of elasticity for model (1) is 31554.52 N/mm^2. The value is higher in magnitude owing to tension and compression reinforcement provided in the section.

4.2 M30 Concrete Grade with Fe-500 Steel Grade (Model 2)

The study reports a calculated load-carrying capacity of 147.39 kN for model (2), determined through the application of Eq. (2). This load is calculated for Fe-500 grade of steel. The load-carrying capacity for model (1) is more compared to model (2) which has steel grade of Fe-500. This is due to the limiting values of the depth of neutral axis which is 0.46 for Fe-500 grade of steel, where Fe-415 has 0.48. The load is lower because of the decreased moment-carrying capacity.

The modulus of elasticity computed for model (2) is expressed 31,350.55 N/mm^2. The value is higher in magnitude owing to tension and compression reinforcement provided in the section.

4.3 M30 Concrete Grade with Fe-550 Steel Grade (Model 3)

The study indicates that model (3) has a computed load-carrying capacity of 142.45 kN, as determined by the utilization of Eq. (2). The load, calculated for Fe-550 grade of steel, exhibits lower load-carrying capacity for model (3) in comparison with both model (1) and model (2). The difference in load-carrying capacities can be determined by values for the depth of the neutral axis, 0.44 is for steel of Fe-550 grade. The load is lower because of the decreased moment-carrying capacity.

The calculated modulus of elasticity for model (3) is 31320.77 N/mm^2. The increased magnitude of the value is a result of incorporating reinforcement for both tension and compression within the section.

According to IS:456–2000 [6], modulus of elasticity of concrete is $5000\sqrt{f_{\text{ck}}}$. The effect of compression and tension reinforcement gives rise to the following three equations for each grade of steel in study.

$$E_{\text{RCC}} = 5000\sqrt{f_{\text{ck}}} + 1000p_t + 1000p_c \text{ for Fe415} \tag{5}$$

$$E_{\text{RCC}} = 5000\sqrt{f_{ck}} + 900p_t + 900p_c \text{ for Fe500} \tag{6}$$

$$E_{\text{RCC}} = 5000\sqrt{f_{ck}} + 800p_t + 800p_c \text{ for Fe550} \tag{7}$$

Equations (5), (6), and (7) are empirical equations derived from a regression analysis.

5 Conclusion

The current work uses finite element analysis to investigate reinforce concrete beam models made of M30 grade concrete. The three grades of steel used in the study are Fe-415, Fe-500, and Fe-550. The following results are drawn from the analytical work done on models of reinforced concrete beams.

- The modulus of elasticity of reinforced concrete beam model in study is 13.21, 12.64, and 12.56% more compared to that of plain concrete section for Fe-415, Fe-500, and Fe-550 steel grade.
- Economical design will result from the updated analysis of reinforced concrete beam sections.
- When analyzing reinforced concrete beam sections, the percentage of tension and compression reinforcement in the beams is crucial.

6 Future Scope

It is possible to estimate the modulus of elasticity for the steel grades Fe-415, Fe-500, and Fe-550 by changing the percentage of tension and compression reinforcement for a doubly reinforced cross section from minimum to maximum.

References

1. SatyaPrakash MKK, Imran A, Rajiv B (2016) Study of modulus of elasticity of RC beam under flexural loading using ANSYS. Int J Sci Eng Res 7(3)
2. Dahmani L, Khennane A, Kaci S (2010) Crack identification in reinforced concrete beams using ANSYS software. Strength Mater 42(2):232–240
3. Kulkarni SK, Shiyekar MR, Wagh B (2012) Elastic properties of RCC under flexural loading. Int J Mod Eng Res (IJMER). 2(6):4022–4025. www.ijmer.com
4. Slowik M (2019) The analysis of failure in concrete and reinforced concrete beams with different reinforcement ratio. Arch Appl Mech J 89:885–895
5. Prantik C, Santanu B (2017) Material characteristics and design parameters for limit state method of design as per IS: 456-2000 and IRC: 112-2011—a comparison
6. IS 456-2000: Code of practice for plain and reinforced concrete, New Delhi, India
7. Eurocode 2: Design of Concrete structures—part 1: general rules and rules for buildings; The European Standard EN1992-1-1
8. BS 8110: Part 2:1997: Structural use of concrete, part 2: code of practice for special circumstances. British Standard Institution, London
9. Indian Standard IS:1786 (2008) Specification for high strength deformed steel bars and wires for concrete reinforcement (fourth Revision)—First Reprint October 2008.Bureau of Indian Standards, Manak Bhavan, 9 Bahadur Shah Zafar Marg, New Delhi, India
10. Rasheed MHF, Agha AZS (2020) Minimum depth-span ratio of beams in order to control maximum permissible deflection. Open Civil Eng. J. 14:402–414
11. Ezeagu CA, Obasi KC (2016) Empirical elasticity models of concrete made with local aggregates. Int J Adv Res 4:1987–1997

12. Noguchi T, Tomosawa F, Nemati KM, Chiaia BM, Fantilli AP (2009) A practical equation for elastic modulus of concrete. ACI Struct J 106:690–696
13. Anbuvelan K, Subramanian K (2014) An empirical relationship between modulus of elasticity, modulus of rupture and compressive strength of M60 concrete containing metakaolin. Res J Appl Sci Eng Technol 8:1298
14. Baalbaki W, Benmokrane B, Chaallal O, Aitcin PC (1991) Influence of coarse aggregate on elastic properties of high-performance concrete. ACI Mater J 88:499–503

Understanding the Binder Chemistry, Microstructure, and Thermo-Mechanical Properties of Metakaolin Under Alkali, Acid Activation—A Comparative Study

G. Sharmila, N. Vanitha, Janani Karuppaiyan, and R. Jeyalakshmi

Abstract Geopolymers are appreciated for their good resistance to fire temperatures. This study investigates the results of the effect of elevated temperature exposure at 800 °C on the chemical and structural changes of metakaolin hardened in phosphoric acid (MP) and sodium hydroxide (MA) reactions at 800 °C. The cured specimen strength of MA is 15.08 MPa, whereas MP is 46.2 MPa, which meets suitable mechanical properties and a compact microstructure character. The thermal stability of the sample was investigated by TGA/DTA, followed by ex-situ analysis using XRD, XPS, and SEM with EDAX. The results revealed that the matrix of MP comprised amorphous phases of silico-aluminium phosphate, albeit with some non-reacted crystalline phases at ambient temperature, while in MA there was a notable crystalline phase of the zeolitic type embedded in gel. In the XPS line of MA^T, an upward shift of O1s BE from MA is due to the loss of Si–O–Na and oxygen in the sialate-siloxo structure. In MP^T, the partial inclusion of [PO_4] in [Si–O–T: T–Si/Al] results in desilication, finally resting in SiO_2 phases. Thus, the thermal treatment and annealing manifested a loss of three-dimensional networks in both samples. Still, the most heat-resistant phases were found to be nepheline in MA^T, aluminium phosphate, and the phospho cristobalite phase of SiO_2 in MP^T. Thus, the microstructural results are useful for predicting the strength retention mechanism of geopolymers.

Keywords Metakaolin · Sodium hydroxide · Phosphoric acid · Thermal resistance · Structural properties

G. Sharmila · J. Karuppaiyan · R. Jeyalakshmi (✉)
Department of Chemistry, College of Engineering and Technology, SRM Institute of Science and Technology, Kattankulathur, Tamil Nadu, India
e-mail: jeyalakr@srmist.edu.in

N. Vanitha
Department of Biochemistry and Chemistry, Jaya College of arts and Science, Thirunindravur, Tamilnadu, India

Y. K. Guruprasad et al. (eds.), *Recent Advances in Infrastructure Development and Management—Volume 1*, Lecture Notes in Civil Engineering 570,
https://doi.org/10.1007/978-981-96-1554-4_9

1 Introduction

Geopolymer is an amorphous inorganic polymer consisting of Si and Al in the tetrahedral network structure, and it is regarded as a green binding material alternative to the traditional binder, OPC. Geopolymers with a suitable feed stock containing silica and alumina, activated by alkali hydroxides, alkali silicates, and acids, exhibit vital improvements in their mechanical performance. These geopolymers are used in photocatalytic substrates, adsorbents for water treatment, encapsulation of radioactive wastes, and the ceramic industry, with the advantages of their easy fabric ability and better efficiency. Compared with traditional cement-based materials, geopolymer cement has more special characteristics, such as chemical resistance [1], high-temperature resistance [2], lower permeability and creep performance [3], and higher strength at early stages [4].

From the previous reports, the reaction with either NaOH or KOH can be explained as follows: Alkalination and tetravalent Al production in the side group sialate (Si–O–Al–$(OH)_3$–Na^+) comprise the first stage. [5, 6] After that, the base hydroxide (OH^-) forms a bond with the silicon atom and can expand its valence sphere to a penta-covalent state. This starts the process of alkaline dissolution. The intermediate silanol, Si–OH, and the basic siloxo, Si–O, that result from the cleavage of the siloxane oxygen in Si–O–Si due to the transfer of the electron from Si to O, can outline the subsequent course of the reaction. Subsequently, additional Si–OH groups are formed, and the ortho-sialate molecule is the key component of geopolymerisation which is isolated. Simultaneously, the basic Si–O reacts with the terminal bond Na^+ to form Si–ONa. The alkali NaOH is then released and reacts again, polycondensing into the Na-poly(sialate) Nepheline framework. This condensation arises between ortho-sialate molecules, the reactive groups Si–ONa and aluminium hydroxyl OH–Al, and the cyclo-tri-sialate structure developed [7, 8].

The primary challenges with alkali materials were shrinkage from loss of water and carbonation from excessive alkali. The exorbitant cost of activating solutions, particularly sodium silicate [9], further restricts their industrial expansion. Therefore, the development of novel activation methods that are affordable and have no adverse effects on the environment has been the main focus of the most recent research on this subject. After that, phosphoric acid was looked at as a medium to make geopolymer cement with the structural part of poly-phospho siloxo, or Si–O–P–O–Si–O geopolymers [10, 11]. Unlike alkali activation, here there are four steps involved: partial dissolution of metakaolin releases the aluminium as Al^{+3}, followed by condensation of PO_4 units and Al^{+3}, along with linkages of Si–O hydrated phases to Al. Further condensation steps led to silica gel, Al–O–P, and Si–O–Al networks [12]. Therefore, silico-alumino phosphate embedded with new mineral phases of aluminium phosphate, the berlinite type, is responsible for the enhancement of mechanical strength. Using silica fume as an alkali activator and mono-aluminium phosphate as an acid activator is also reported, other than conventional acid or alkali [13].

In distinction, it is clear that the replacement of alkaline solutions with phosphoric acid has led to a major change in the type of gels, and the obtained materials may have different physico-chemical and thermal properties. There have been extensive studies on metakaolin excited by alkali silicate geopolymers and their thermal behaviour [14], but none on NaOH-activatedness. Similarly, temperature-related studies on phosphate geopolymers are scarce. However, the refractory-type $AlPO_4$ reinforces the matrix potential at elevated temperatures, which is a primary advantage of the phosphate type [15, 16]. Hence, a comparison of the performance of alkali and phosphoric acid solutions separately as reactive ingredients that are prepared under optimised conditions could be very interesting. The aim of the present study is to elucidate the thermal behaviour and the related structural changes, phase transformations, and physical properties of the geopolymers with different basic structures. The molecular structure and pore connectivity in these gels are different due to their framework, which provides prerequisites for their use as a ceramic material. Therefore, fundamental studies on phase evolution and thermal chemical reactions of these chemically complex materials were investigated in this study to explore the corresponding mechanisms involved. The microstructural features of the hardened pastes before and after heat were explored using multiple techniques that include X-ray photoelectron spectroscopy and SEM–EDS.

2 Materials and Methods

Metakaolin is thus a study precursor obtained from EICL Private Limited, Kerala. Metakaolin is composed of 90% silica and alumina and traces of titanium dioxide and other metal oxides. Two types of activators were used in this study: alkaline sodium hydroxide (NaOH) of 10 M and orthophosphoric acid (H_3PO_4) (85%), procured from Southern India Scientific Corporation. The alkali geopolymer (MA) was formed from metakaolin and NaOH solutions (solid to liquid S/L—0.9) and acid (MP) using H_3PO_4 solution under S/L 1.0, respectively. Initially, metakaolin was mixed with this activating solution, and then slurry was mixed in a digi mortar mixer for 8 min, cast in a 50 mm × 50 mm cube, and kept on the vibrating table for 5 min to ensure the removal of air bubbles. Then, they were sealed with a polyethylene sheet for 24 h, followed by curing at 80 °C for 8 h in an oven. After the initial curing, they were stored at an ambient temperature for 28 days. This set of five samples was subjected to thermal treatment by heating in the furnace at 800 °C for 2 h at a heating rate of 50 °C per minute. Thermally treated samples are denoted as MA^T and MP^T, respectively.

The crushed paste samples in the form of powder were used for characterisation. The XRD was done using a PANalytical *X*-ray diffractometer [PANalytical, Netherlands] at a scanning rate of 2° per minute using CuKα radiation. Micro-Raman analysis was done in HORIBA, LabRam HR Evolution. Laser wavelengths: 633 nm; resolution: 0.4 cm^1 (@532 nm). A Gemini 300 SEM was used for SEM EDAX: maximum acceleration voltage (20 kV), magnification range (20–300000x) with a gold coating (4–7 nm thick), and aperture range (10–300 microns). A PHI5600S multi-technique system has been utilised for XPS, and an Al monochromatic *X*-ray was applied with 10 mA and 15 kV at 350 W of power. Carbon calibration enhanced the accuracy of the peaks. Using the NETZSCH STA 2500 Regulus, the TGA/DTA analysis was carried out in an environment of nitrogen at an elevated temperature of 10 °C per minute. The compressive strength of the sample was tested by following the standard test procedure (ASTM C109) using the compression testing machine (AIMIL) at a loading rate of 3 KN after the 28-day curing period. To check the water stability of the prepared geopolymer, oven-dried mature samples were soaked in water for 12 h and monitored for their physical integrity after drying in a hot air oven. The difference in mass is calculated in the samples under wet and dry cycles, in addition to the measurement of compressive strength.

3 Results and Discussion

3.1 Compressive Strength and Mass Analysis After Wet and Dry Cycles

The mechanical performance at room temperature was obtained, and the MA samples showed an average strength of 15.08 MPa and MP of 46.2 MPa at 28 days of curing age. After thermal exposure, the mechanical strength tends to be lowered and settles at 5 MPa in MA^T and 14.8 MPa in MP^T due to the evaporation of water as well as the chemical structural changes. The alkali cement lost its strength within three wet and dry cycles, whereas the phosphate cement retained 85% of its strength at the same treatment. Thus, water resistance is poor because sodium hydroxide excites metakaolin. It can be explained by the hydrolysis of Si–O–Na, which is a network-modifying part in the alkaline-activated samples. However, the hydrolysis of Si–O–P bonds into Si–OH and P–OH in MP slightly reduces the bulk strength. Water absorption for MA is around 9.13%, whereas for MP it is 4.08%, as it is comparatively high due to the lower densification of the matrix. The mass loss of strength is connected to the enlargement of pore size and is also responsible for the reduction of compactness in the samples.

3.2 XRD Analysis

The results of ex-situ *X*-ray diffraction analysis of both types of geopolymers presented before and after thermal treatment are shown in Figs. 1a, b. Alkali-activated geopolymer shows the lower value of the compressive strength test result (MA = 15.08 MPa), and it can be deduced that the low content of amorphous geopolymer gel resulted from the NaOH attack. While the phase analysis of this composition relating to zeolite phases observed at 12°,16°, 21°, 23°, 26°, 30°, 34°, 44°, 52°, and 68° degrees of 2θ values is not contributing to the strength, further noticeable

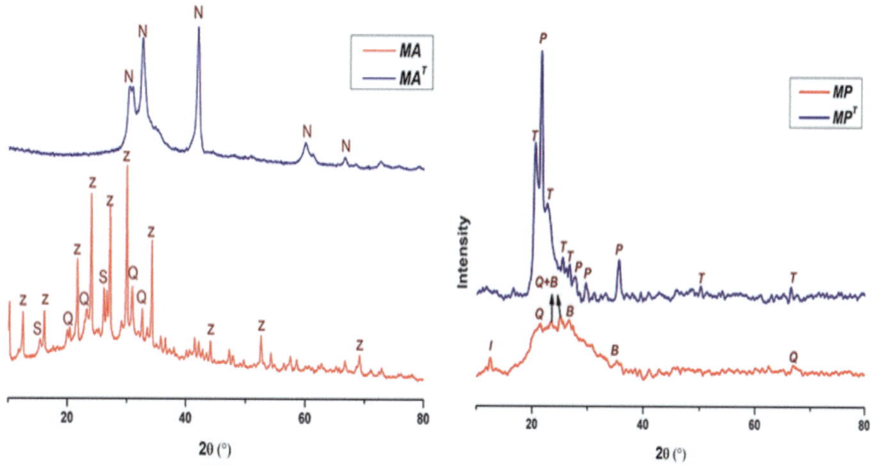

Fig. 1 **a** XRD of alkali geopolymer before (MA) after heat (MA^T), **b** XRD of phosphate geopolymer before (MP) after heat (MP^T) (Q- Quartz, Z- zeolite, S- sodalite, N- Nepheline, I- Illite, B- Berlinite, T- Phosphotridymite, P- Phosphocristobalite)

phase assemblages correspond to the sodalite phase found at 15° and 25° 2θ value. Here, sodium hydroxide is the major driving force for the desilication of metakaolin, followed by the rearrangement of aluminium silicate ions to zeolite. The role of these zeolite crystals in MA was tested after thermal treatment at 800 °C, as verified by XRD assemblages. The crystalline peaks observed in the cured samples disappeared, and broadening of the hump resulted. Further, new peaks related to nepheline phases ($2NaAlSiO_4$) were identified from the peaks located at 2θ values of 30°, 34°, 44°, 52°, and 68°. Thus, the use of NaOH as an alkaline activator is associated with a fast early stage at the rate of dissolution and precipitation, which may explain the presence of smaller crystals of zeolites as reported elsewhere [17].

The XRD patterns of the acid-elaborated samples before (MP) and after heat at 800 °C (MP^T) are shown in Fig. 1b. As cured samples of MP possessed a broad amorphous hump between 20° and 30° of 2θ degree, along with peaks corresponding to crystalline illite and quartz present in raw metakaolin, they were mostly inert. The reflections located at around 18°, 22°, and 27° correspond to silicic acid, amorphous silica, and geopolymeric networks, respectively. The polymorphs of $AlPO_4$, namely berlinite and the hydrated form of aluminium phosphate reflections, are identified at 2θ values of 20°, 25°, 27°, 29°, 35°, and 62°. After heat treatment, the samples were mostly crystallised, with a noticeable suppression of the amorphous peak of the geopolymer. At 800° C, $AlPO_4$ polymorphs such as berlinite (hexagonal), phosphotridymite (triclinic), and phosphocristobalite (orthorhombic) were deduced with the predominant phase of phosphotridymite. It was also noted that the disappearance of illite peaks at elevated temperatures. Hence, the thermo-chemical reactions of the metakaolin under acid and alkali treatment altered the mineral phases of the products formed during the activation [15]. The alkali system favoured the formation of more zeolitic products that are responsible for the compressive strength at room temperature and tend to transform into more of the nepheline phases also reduced the strength factor to half, whereas the crystalline phases of phosphates of aluminium and silica augment the strength of the phospo-geopolymer, and strength retention is nearly 68% after thermal load.

3.3 Micro-Raman Spectral Analysis

The Raman scattering assessment has been a widely used technique to track chemical reactions since it is quick, non-destructive, and precise when illustrating materials. The covalence of the silica tetrahedron in the raw material as well as in the hardened geopolymer showed a well-defined vibrational signature. The whole spectrum ranges from 50 to 1200 cm^{-1} analysed, apart from the stretch mode of OH bonds in the frequency region 3000–4000 cm^{-1} and 1700–1600 cm^{-1}, which are relevant to the hydrated phases of the geopolymer. The peak assignments generally followed in the raw material and geopolymer are as follows: (i) Range between 100 and 400, below 500 cm^{-1}, to T–O–T, Si–O–Si/Al bending. (ii) Between 500 and 580 cm^{-1} symmetry stretching of the Al–O_4 tetrahedral, Al–O bond. (iii) Various Q(nAl) environments cover the frequency range between 850 and 1200 cm^{-1}.

MK and both geopolymers have a well-known Si–O–Si and Al–O–Si molecular bond that can be assessed from the intensity of the band in the region 950–1100 cm^{-1}, and the shifting of this band might give information on the structural changes occurring during the thermo-chemical transformation [12]. In Fig. 2a, a sharp peak at 141 cm^{-1} in MK is related to the crystalline part of quartz, as O–Si–O vibration remains unaltered in cured and heated samples that are consistent with *X*-ray diffraction patterns. This band is present in NaOH-activated samples at 147 cm^{-1}, with an additional peak at 111 cm^{-1} in MA^T. The change is in Si–O stretching vibration. On the other hand, phosphate-elaborated MP showed the link of Si–O–Al at slightly higher wave numbers at 145 cm^{-1} and 141 cm^{-1} in MP^T. Upon heating, silica-phased crystalline phases are identified in XRD due to desilication, and hence, the Si–O–Si peak dominates the same wavenumber region, nearly at 141 cm^{-1}. Similarly, bands appearing in the range between 190 and 300 cm^{-1} are from Si–O stretching vibrations, seen in all the samples, in addition to vibration modes of T-O-T bonding (where T is Si or Al) in the range of 380–530 cm^{-1}. This peak characteristic is related to zeolite mineral phases. The bending vibration of Si–O–Si appeared at 634 cm^{-1} in MK and has shifted to 639 cm^{-1} in MA and MA^T. At the same time, silica gel also formed in acid activation (MP) by excess Si ions, and hence, the vibration of Si–OH conformed to the band of 605–609 cm^{-1}. Upon heating, these Si–OH groups became SiO_2; therefore, in MPT, this bond settles at 645 cm^{-1} wavenumbers, as it is attributed to SiO_2 [15, 18]. In alkali-activated samples before and after thermal exposure, a ring vibration of Si–O-Si/Al appeared at 735 and 751 cm^{-1}, and a vibration band arose due to Si in a tetrahedron at 835 cm^{-1}.

The bending vibration of Si–O–Si/Al at 738 and 741 cm^{-1} in the phospo-aluminate samples [19].

Various Q environments present in geopolymeric samples are highlighted in Figs. 2b and c, according to that, the Si–O–T frequency is almost in the range of 1004, 1014, and 1011 cm^{-1}, respectively, at MK, MA, and MP samples. $Q^0{}_{Si}$, $Q^2{}_{Si}$, $Q^3{}_{Si}$, and $Q^4{}_{Si}$ are all seen in the MA^T sample compared to $Q^0{}_{Si}$ and $Q^4{}_{Si}$, which are present in MA. It was also further notified that the $Q^0{}_{Si}$ frequency appeared as a strong, intense band at 843 cm^{-1} in both MP and MP^T. Al–O–P vibration at 1175 cm^{-1} in MP was lowered due to the thermal reaction and rested as SiO_2 in MP^T at 1111 cm^{-1}. An Al-O vibration band at 818 cm^{-1} appeared at MK and disappeared at MA and MP geopolymers. On the other hand, the octahedral Al, which has a bond with silico-alumino phosphate (MP), is described as Al/Si–O–P reflected at 270 cm^{-1}.

3.4 SEM–EDAX Analysis

The micro-morphological structure of the geopolymers is presented in Fig. 2. Energy-dispersive *X*-ray (EDX) analysis is given in the inserted table. Si, Al, Na, and P wt% are determined according to the reactions carried out with acid and alkali activation. A clearly distinguished phase appeared while MK reacted under acid or alkali hardener in terms of homogeneity, microcracks, and porosity. SEM microscopic picture of MA suggests the fully reacted homogeneous gel phase of sodium alumino silicate hydrated (N–A–S–H) gel phase with fewer micro cracks. Among MA and MA^T, one can observe pores with crystallites of different shapes in MA^T. The crack width was also widened, and pores were randomly distributed. On the other hand, homogeneity and compactness were strengthened in MP samples with acid excitation. The closer look revealed the crystallites of aluminium phosphate as reported in the literature and our own results. It can also be seen that a newly formed phase of silico-alumino phosphate gels with smooth morphology in cured MP. But in MP^T, there are more crystallite zones with a less amorphous phase, which is in accordance with XRD results. The main crystalline network consisted of berlinite phases, quartz, cristobalite, and phosphocristobalite, and explained their presence for their contribution to the compressive strength. EDX-layered images of O, Si, Al, and P or Na also evidenced the enhancement of P/Al in MP^T compared to MP (Fig. 3).

3.5 X-Ray Photoelectron Spectral Analysis

XPS surface investigation was carried out to study the nature of Si, Al, O, Na, and P on the surface of the matrix. We can find out that the increasing proportion of Si, Al, Na, P, and O represents their strong intensity peaks. Individual elemental BE lines were assessed from the survey graphs. In Fig. 4, the Si 2p line is located in

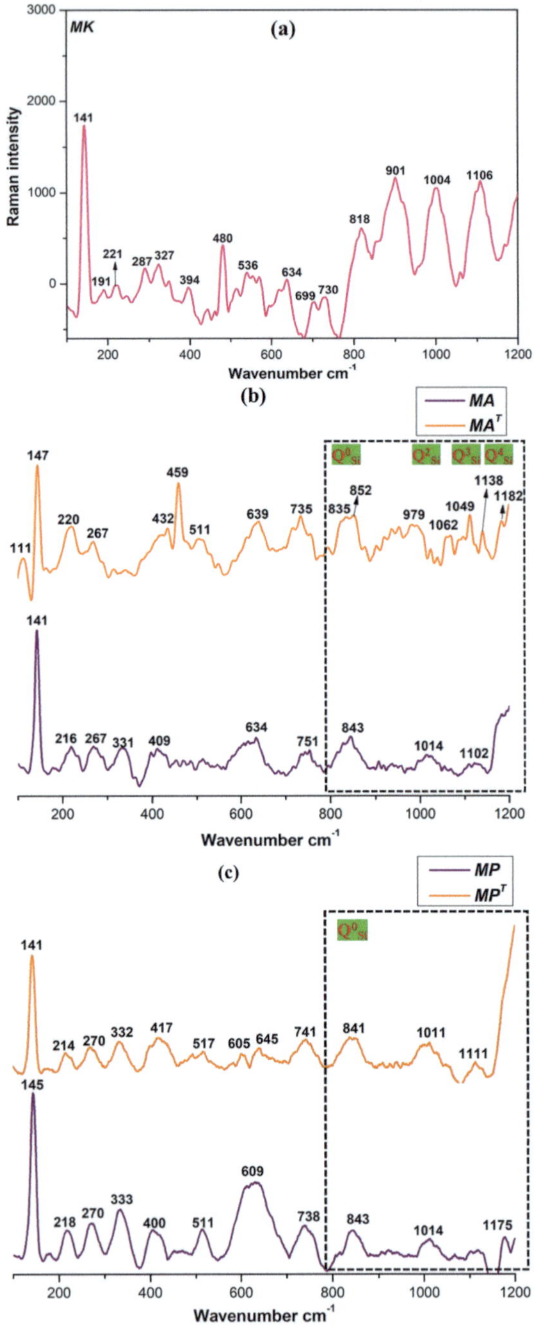

Fig. 2 Raman spectra of geopolymers **a** Raw MK, **b** MA, MA^T, **c** MP, MP^T

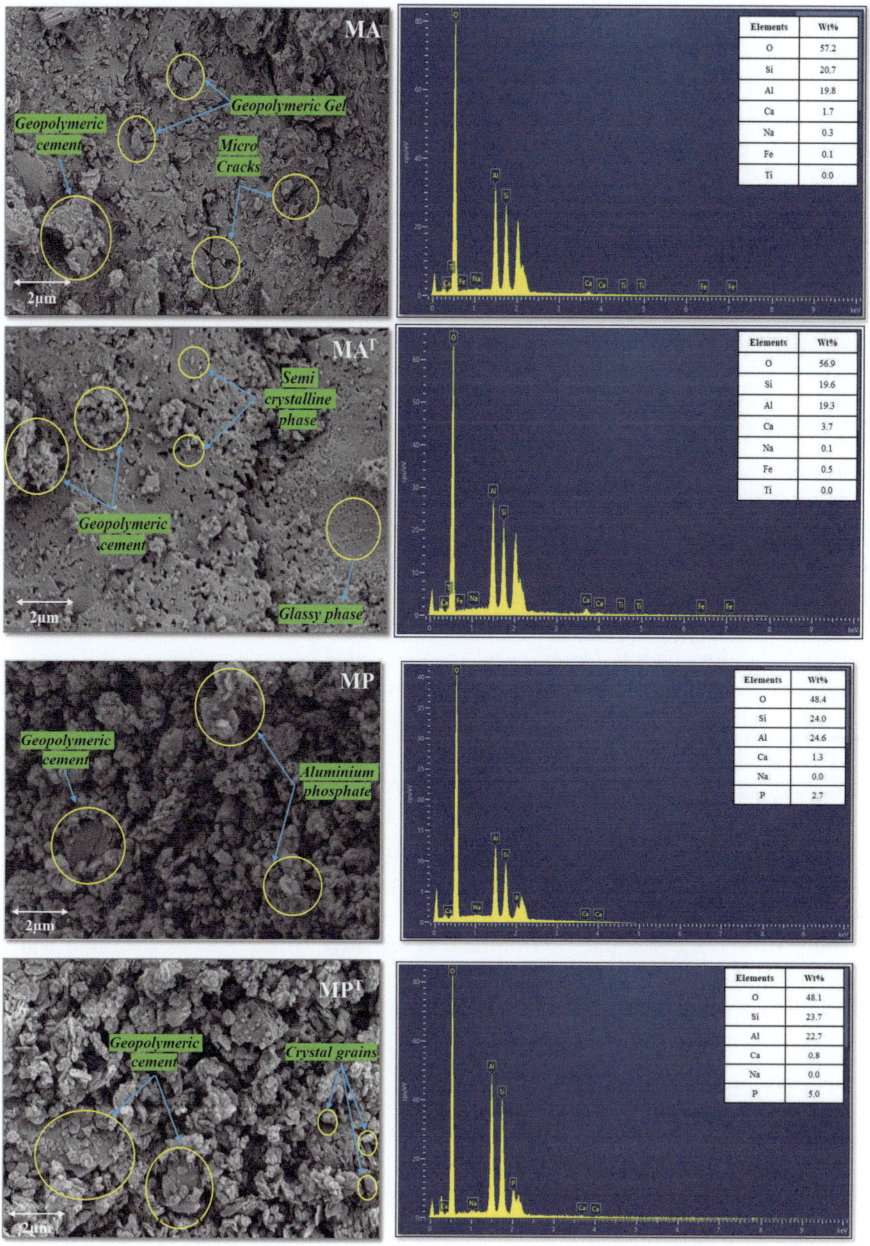

Fig. 3 Morphology of metakaolin alkali and acid geopolymer

alkaline-based samples before heating at 103.4 eV (geopolymer gel products), and after heating, it resolves into three peaks: 101.9 (alumino silicate products), 103.6 (geopolymer gel products), and 104.4 (nepheline). The Si 2p spectrum fits within 103 eV in MP and MP^T, which is attributed to Si 3/2 and Si 1/2 from SiO_2. Heating causes it to coincide with 103.5 eV, indicating that quartz is the source. Once the geopolymer gel hardened, the peaks of Al 2p were moved to 75.4 and 76.3 eV from their original location at 75.3 eV. Accordingly, there is just one contribution to the Al 2p line of phosphate geopolymer at around 75 eV, which is explained by both Al–O and Al–OH bonds. In MP and heated samples, the partial replacement of silica tetrahedron units with PO_4 is confirmed by the major peaks of the Al ion spectra, which show Al-O bonding at 74.21 $\pm$ 0.2 eV and $AlPO_4$ at 75.8 eV [20].

The O1s XPS spectra of both alkaline-activated geopolymers before and after heating samples were resolved into three peaks. In the unheated sample, BE 532.9 and 532.5 eV were generally assigned to the silanol group, as it is a Si–OH bond in basic groups. This peak shifted to 533.1 eV with high intensity, which corresponded to Si–O–Na, and the less intense peaks at 536.0 and 539.2 eV corresponded to Si–O either in zeolite or silica [21]. Basic electrostatic interactions provide an explanation for the variations in binding energies of Si–OH and Si–O–Na. Because the hydrogen atom has higher electron negativity than the silicon atom, there are, on average, fewer electrons at the oxygen atom in Si–OH than in Si–O–Si [22]. As a result, the electrons' binding energy increases as the electron–electron repulsion at the oxygen atom decreases. It also has a 533 eV energy level, which is typical of the Si–O bond in tetrahedral forms as bridging oxygen. The O1s peak seen in MP^T can be compared to the Si binding energy line of quartz [23].

The XPS spectra of Na 1 s show a significant decrease in peak intensity compared to the peak shift. Two peaks, 1069.1 and 1073.5 eV, split the unheated sample and represent the Si–O–Na bond. The heated sample is deconvoluted into three peaks at 1071.2, 1072.7, and 1073.7. The change of typically hydrated sites into unhydrated states of aluminium silicates during heating results in slightly higher binding energy. In other words, the sodium environment is sensitive to the loss of hydrated water [24]. The P binding energies, which are located in the MP geopolymer at 134.2 and 135.1 eV, represent the oxygen bridges of metaphosphate and $AlPO_4$. Metaphosphate is associated with the binding energy of 134.5 eV in MP^T, while orthophosphate correlates with an additional peak at 133.9 eV, which has a slightly lower energy shift. The peak at 136.2 eV corresponds to P1/2 of P-O-P, or pentavalent tetra-coordinated phosphorus (PO_4 tetrahedral) [21].

3.6 *Thermal Analysis*

The accompanying thermograph, which is shown in Fig. 5, indicates a constant mass loss at elevated temperatures. Thermal analysis (TGA) was conducted from ambient temperature to 1000 °C. Three different types of water can escape in the

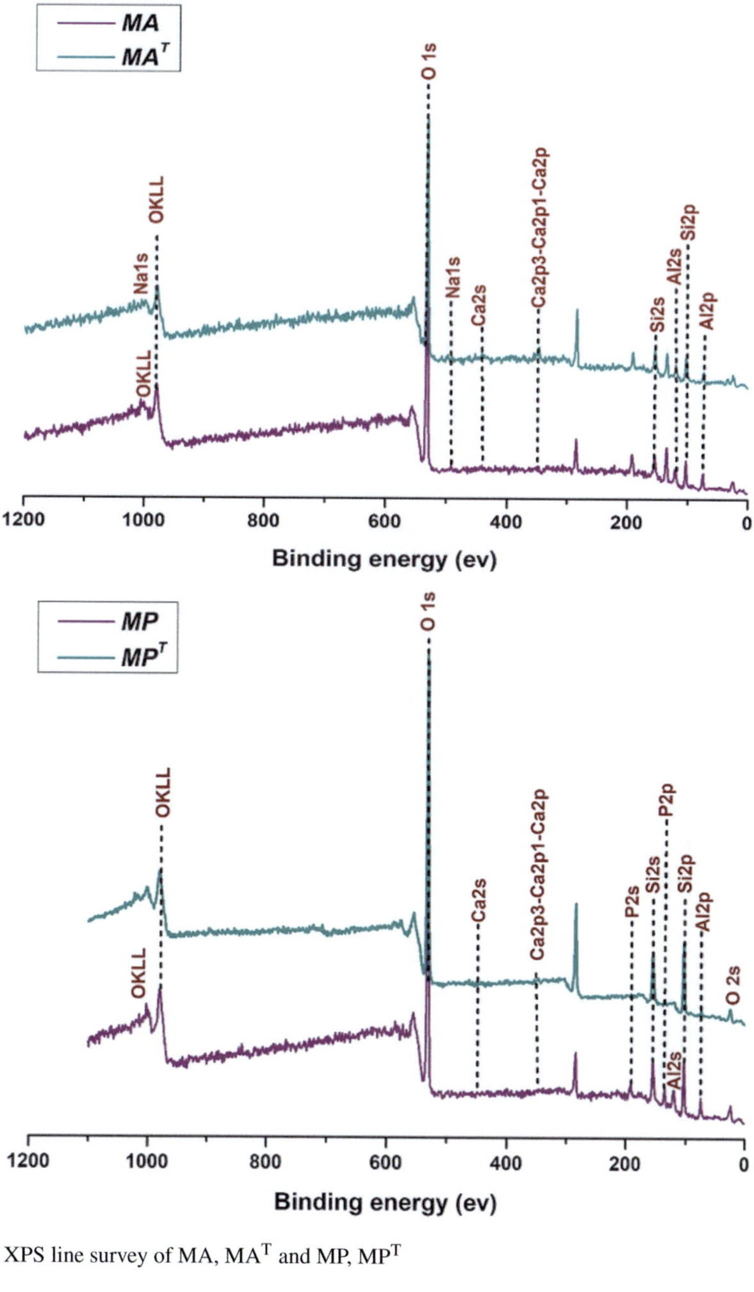

Fig. 4 XPS line survey of MA, MA^T and MP, MP^T

30–350 °C temperature range: hydroxyl groups, chemically bonded water, and physically bonded water. The processes that cause mass loss during heat treatment are most probably water evaporation and dehydroxylation. A wide range of endothermic processes is linked to weight losses in the low-temperature range of 100–230 °C. These phases include free water, which corresponds with hydrated aluminium phosphates in acid activation (a loss of 21%), and physical water in alkali activation (a loss of 31%). The dehydroxylation of Si–OH and Al–OH units was accomplished in the 200–500 °C temperature range. Due mostly to the properties of crystal water, the vast majority of the weight loss transpired before 240 °C. At temperatures above 300 °C, the hydroxyl groups that cause the loss of bound water evaporate. In other words, the formation of hydro/hydroxyl sodalite during the alkali mediation reaction, normally formed nearly at 400 °C from zeolitic phases at the immature stage, became dehydrated in the temperature range of 420–630 °C. This may even have led to restructuring of the gel and pore networks, as reported previously [25].

The weight loss trend in the acid-treated metakaolin weakened and was in line with the $AlPO_4.nH_2O$'s declining O–H content. Beginning at 600 °C and going up to 800 °C, all samples can be regarded as thermally stable. With a small mass loss of -1%, it stabilises around 600 °C and remains unchanged up to 1000 °C. If the temperature is raised above 900 °C, the carbonate by-products ($CaCO_3$ and Na_2CO_3) that are generated from an excess of unreacted NaOH and CO_2 will decarbonise and cause further weight loss in the MA sample [26]. The loss of free water involves approximately 10% shrinkage, whereas the loss of chemically bound water causes approximately 90% shrinkage. The residual mass is 60% for MA and 75% for MP. Thus, phosphate-based metakaolin geopolymers are more stable than alkali-based geopolymers.

The DTA thermogram made it clear that the structural changes in the geopolymer samples (MA, MP) resulted from the removal of barely discernible hydroxyl groups. The cause of dehydration occurs in relatively gradual processes, but it can be eliminated over a large temperature range. At 120 °C, the hygroscopic water forms present in geopolymers are eliminated. The peak's width indicates that the elimination of crystalline and hygroscopic water may overlap. Water that is chemically bonded starts to evaporate when the temperature rises above 300 °C. However, there is significant variation in the temperature and strength of the endothermic peaks in the MP sample. The dehydration reaction, which corresponds to the peak at 102 °C, intensifies, suggesting that phosphoric acid is eliminating water. The large number of hydrated phases that result from the gel formation, however, might be able to explain this. The creation of the $Al_2(H_2P_2O_7)$ phase corresponds to the next peak, which occurs around 135 °C. In the 400–600 °C temperature range, the endothermic wide effect implies the $Al(PO_3)_3$ phase emerges from the $Al(H_2PO_4)_3$.

Fig. 5 Thermal analysis of metakaolin activated with alkali and acid (TGA/DTA)

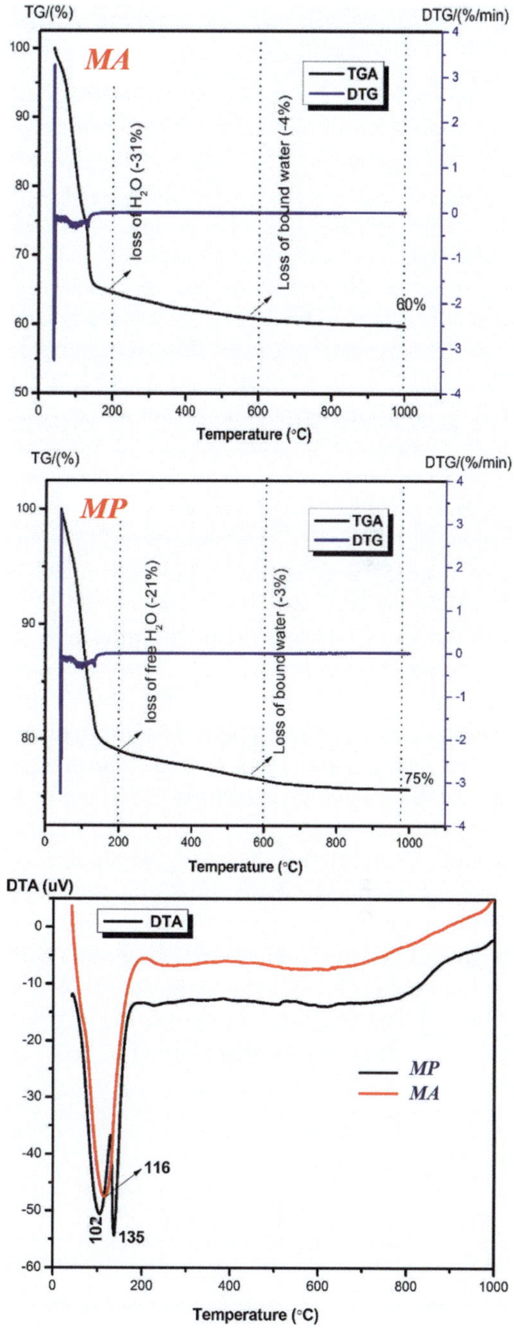

3.7 Implications of the Current Findings

Alkali mediation reaction is different from alkali silicate hardening in the sense that dissolution of aluminium is lessened under OH attack, and hence, interaction with silicate anions with hydrolysed aluminate species is incomplete. The secondary reaction of Si sheet in MK for the polycondensation reaction stage is partial. As a result, MK (without any hydrolysed silicate species) geopolymerisation yields more zeolite-type crystalline compounds with the disappearance of amorphous halo in the raw material. The structure of the metakaolin-gel phase in geopolymer is comprised of small clusters of precipitates, zeolitic phases, with pores dispersed in the structure of interstices observed from SEM morphological analysis. The content of soluble silicon is intimately linked to gel densification and porosity. Further notification of the crystalline *X*-ray reflections, including hydroxyl sodalite, a type of zeolite, particularly in lower-soluble silicon, and at the same time, the structural transformation from amorphous to crystalline or vice versa after thermal treatment was observed. The micro-Raman spectral frequencies appeared at 409 and 331 cm^{-1} in MA, which correspond to zeolite disappearing and being transformed into the nepheline structural features at 1650 cm^{-1}. The XPS shows the lower binding energy of the Si 2p line in the MA^T(103.6 eV) with a smaller, more intensely resolved zeolite peak at 104.4 eV, which represents the transformation of zeolite to nepheline. Similar evidence exists in the BE of O1s from the bonding to the nonbonding position. From the above points, it is clear that the stable nepheline phases are responsible for the binding action at elevated temperatures in alkali cement.

The synthesised silico-alumino phosphate-type gel effect of phosphate-based metakaolin geopolymer retained its crystalline phases, which reduced the deteriorating structural impact of water evaporation that passes through the material during heating. The enriched alumino phosphate and phosphocristobalite phases significantly counterattack the harmful effects of thermally induced volumetric changes. Further high-temperature stable crystalline form embedded in the silico-phosphate gel phases positively influences the strength performance of this type of geopolymer. The retention of bonds in Si–O–Si and Al–O–P was established from XPS test results. Based on the consolidated test data, the following mechanism is proposed for alkali and acid-excited metakaolin (Fig. 6).

4 Conclusions

This study examined the microstructural characterisation, thermal behaviour, and mineralogical composition of alkali and acid-activated metakaolin geopolymers independently and compared their thermal resistance behaviour based on the products formed. This research leads to a better understanding of the thermal behaviour of the geopolymer matrix in a fire situation. The main conclusions are as follows:

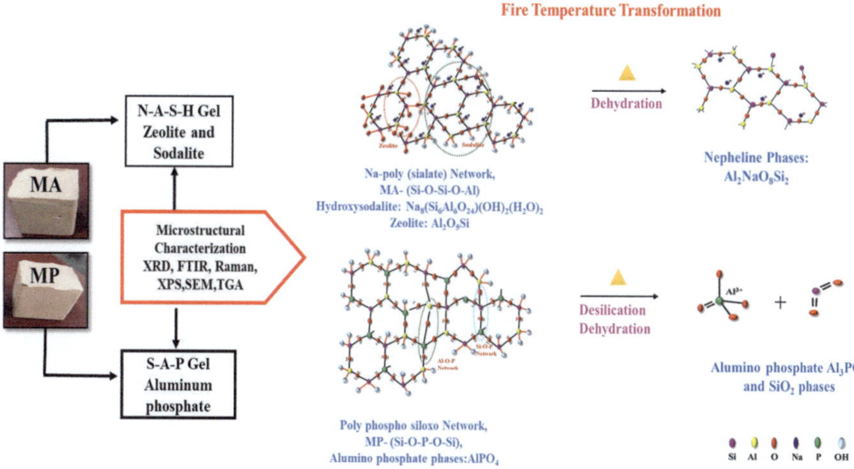

Fig. 6 Reaction mechanism of fire temperature transformation on MA, MP

- A correlation between mechanical strength and the primary and secondary products of geopolymerisation assisted by sodium hydroxide and phosphoric acid was individually established. The retention of strength at elevated temperatures was rationalised in terms of sodium alumino silicate gels with zeolitic crystalline phases in alkali geopolymers. The appearance of a new crystalline phase of alumina phosphates in acid-elaborated MP, known as berlinite and hydrated berlinite, significantly contributed to the compressive strength.
- Thermal stress destroyed the zeolitic network of geopolymers in alkali-activated metakaolin, thus drastically reducing the thermo-mechanical strength. On the other hand, the high level of aluminium phosphate phases embedded in the silico-aluminium phosphate gels reinforces the strength of MP^T.
- According to the XRD results of the NaOH-activated matrix, one can observe that zeolite of the tetrahedral ring with hydroxyl sodalite, a type of zeolite, is different from alkali silicate activated geopolymer. The intense peaks of nepheline retain even at 8000 °C with a structural collapse of zeolite, contributing to the strength and physical integrity of the samples. At the same time, the resistance to high temperatures in the aluminium phosphate network was triggered by a radical change in the network by the inclusion of a phosphate tetrahedron in the siloxo network. Desilication of silico-alumino phosphate resulted in stable crystalline phases like quartz, crystobolite, and phospo-crystabilite embedded in the in the matrix.
- Thermogram data revealed that the evaporation of free water and the dehydration of chemically bound water were mainly responsible for the mass loss in both geopolymers. Among these types, alkali geopolymer experienced heavier mass loss, as much as 40%, and the mass loss accounted for at stage I is much more than 80% of total mass loss. Acid geopolymers retain 78% of their residual mass,

and most of the mineral phases are intact after the removal of different amounts of free water at a temperature below 200 °C.

- XPS data implies the change of coordination of Al in the [SiO_4-T] unit with more nonbonding oxygen in the unheated samples of MA and the transformation of penta valence to tetra valence of P in the $AlPO_4$ phases in MP. Thermo-chemical reactions led to more bonding oxygen sites that were ascribed to the shifting of O1s at higher BE in both MA^T and MP^T.

4.1 Declaration

All the authors declare that there is no conflict of interest.

Contribution Statement.

Sharmila. G, Vanitha. N, and Janani Karuppaiyan: Investigation, Methodology, Data curation and original draft preparation.

Dr R. Jeyalakshmi: Conceptualisation, Editing and Supervision, Funding acquisition.

Acknowledgements This work was supported by the Department of Science and Technology, the Government of India (GOI), under the grant DST/TDT/WMT/2017 14/03/18, GOI, Kuttuva Silicates, Madurai. The authors greatly acknowledge the utilisation of facilities provided by DST-FIST, Department of Chemistry and SEM facility by NRC, SCIF SRM Institute of Science and Technology, Kattankulathur, India. Authors thank the authorities of SRMIST for funding to the research scholars.

References

1. Albitar M, Ali MM, Visintin P, Drechsler M (2017) Durability evaluation of geopolymer and conventional concretes. Constr Build Mater 136:374–385. https://doi.org/10.1016/j.conbuildmat.2017.01.056
2. Kong DL, Sanjayan JG (2010) Effect of elevated temperatures on geopolymer paste, mortar and concrete. Cem Concr Res 40(2):334–339. https://doi.org/10.1016/j.cemconres.2009.10.017
3. Singh B, Ishwarya G, Gupta M, Bhattacharyya SK (2015) Geopolymer concrete: a review of some recent developments. Constr Build Mater 85:78–90. https://doi.org/10.1016/j.conbuildmat.2015.03.036
4. Ravikumar D, Neithalath N (2013) Electrically induced chloride ion transport in alkali activated slag concretes and the influence of microstructure. Cem Concr Res 47:31–42. https://doi.org/10.1016/j.cemconres.2013.01.007
5. Provis JL, Van Deventer JSJ (2007) Geopolymerisation kinetics. 2. Reaction kinetic modelling. Chem Eng Sci 62(9):2318–2329. https://doi.org/10.1016/j.ces.2007.01.028
6. Rees CA, Provis JL, Lukey GC, Van Deventer JS (2008) The mechanism of geopolymer gel formation investigated through seeded nucleation. Colloids Surf A: PhysChemical Eng Asp 318(1–3):97–105. https://doi.org/10.1016/j.colsurfa.2007.12.019
7. Antonic T, Subotic B, Stubicar N (1997) Influence of gel properties on the crystallization of zeolites: part 1: influence of alkalinity during gel preparation on the kinetics of nucleation of zeolite A. Zeolites 18(4):291–300. https://doi.org/10.1016/S0144-2449(97)00004-3

8. Baerlocher C, McCusker LB, Olson DH (2007) Atlas of zeolite framework types. Elsevier
9. Roy S, Adhikari GR, Gupta RN (2007) Use of gold mill tailings in making bricks: a feasibility study. Waste Manage Res 25(5):475–482. https://doi.org/10.1177/0734242X07076944
10. Guo H, Zhang B, Deng L, Yuan P, Li M, Wang Q (2021) Preparation of high-performance silico-aluminophosphate geopolymers using fly ash and metakaolin as raw materials. Appl Clay Sci 204:106019. https://doi.org/10.1016/j.clay.2021.106019
11. Louati S, Baklouti S, Samet B (2016) Acid based geopolymerization kinetics: effect of clay particle size. Appl Clay Sci 132:571–578. https://doi.org/10.1016/j.clay.2016.08.007
12. Tchakouté HK, Rüscher CH (2017) Mechanical and microstructural properties of metakaolin-based geopolymer cements from sodium waterglass and phosphoric acid solution as hardeners: a comparative study. Appl Clay Sci 140:81–87. https://doi.org/10.1016/j.clay.2017.02.002
13. Zribi M, Samet B, Baklouti S (2020) Structural characterization of phosphate-based geopolymer. In: Advances in materials, mechanics and manufacturing: proceedings of the second international conference on advanced materials, mechanics and manufacturing (A3M'2018), Hammamet, Tunisia, pp 36–42. Springer International Publishing. https://doi.org/10.1016/j.clay.2016.08.007404
14. Mathivet V, Jouin J, Gharzouni A, Sobrados I, Celerier H, Rossignol S, Parlier M (2019) Acid-based geopolymers: understanding of the structural evolutions during consolidation and after thermal treatments. J Non-Cryst Solids 512:90–97. https://doi.org/10.1016/j.jnoncrysol.2019.02.025
15. Khabbouchi M, Hosni K, Mezni M, Zanelli C, Doggy M, Dondi M, Srasra E (2017) Interaction of metakaolin-phosphoric acid and their structural evolution at high temperature. Appl Clay Sci 146:510–516. https://doi.org/10.1016/j.clay.2017.07.006
16. Provis JL, Lukey GC, van Deventer JS (2005) Do geopolymers actually contain nanocrystalline zeolites? A reexamination of existing results. Chem Mater 17(12):3075–3085. https://doi.org/10.1021/cm050230i
17. Kaze CR, Jiofack SBK, Cengiz Ö, Alomayri TS, Adesina A, Rahier H (2022) Reactivity and mechanical performance of geopolymer binders from metakaolin/meta-halloysite blends. Constr Build Mater 336:127546. https://doi.org/10.1016/j.conbuildmat.2022.127546
18. Izci E (2014) Structural and dielectric properties of acid activated metakaolinite and kaolinite. In: 11th international GeoRaman conference, vol 1783, p 5097
19. Douiri H, Louati S, Baklouti S, Arous M, Fakhfakh Z (2016) Enhanced dielectric performance of metakaolin–H_3PO_4 geopolymers. Mater Lett 164:299–302. https://doi.org/10.1016/j.matlet.2015.10.172
20. Shimada H, Matsubayashi N, Imamura M, Sato T, Nishijima A (1996) Determination of external surface composition of zeolite particles by synchrotron radiation XPS. Catal Lett 39:125–128. https://doi.org/10.1007/BF00813741
21. Pu S, Zhu Z, Song W, Wang H, Huo W, Zhang J (2021) A novel acidic phosphoric-based geopolymer binder for lead solidification/stabilization. J Hazard Mater 415:125659. https://doi.org/10.1016/j.jhazmat.2021.125659
22. Elizondo-Villarreal N, Obregón-Guerra R, García-Méndez M, Sánchez-Espinoza AP, Alcorta-García MA, Torres-Barrera RO, Coello V, Castaño VM (2016) Nanomodification of a natural clinoptilolite zeolite. Rev Adv Mater Sci 47:74
23. Bare SR, Knop-Gericke A, Teschner D, Hävacker M, Blume R, Rocha T, Schlögl R, Chan AS, Blackwell N, Charochak ME, ter Veen R. Surface analysis of zeolites: an XPS, variable kinetic energy XPS, and low energy ion scattering study. Surf Sci 648:376–382. https://doi.org/10.1016/j.susc.2015.10.048
24. Black L, Garbev K, Stemmermann P, Hallam KR, Allen GC (2004) X-ray photoelectron study of oxygen bonding in crystalline C-S–H phases. Phys Chem Miner 31:337–346. https://doi.org/10.1007/s00269-004-0401-3

25. Dal Poggetto G, D'Angelo A, Blanco I, Piccolella S, Leonelli C, Catauro M (2021) FT-IR study, thermal analysis, and evaluation of the antibacterial activity of a MK-geopolymer mortar using glass waste as fine aggregate. Polymers 13(17):2970. https://doi.org/10.3390/polym13172970
26. Yu G, Jia Y (2021) Preparation of geopolymer composites based on alkali excitation. Arab J Geosci 14:1–8. https://doi.org/10.1007/s12517-021-06908-8

Study on Sisal Fiber Reinforced Concrete and Its Characteristics

Ashwinnkumar Ijari and Mohammed Umar Farooque Patel

Abstract The study investigates the impact of sisal fibers, ranging from 20 to 30 mm in length, in enhancing the properties of M20 and M30 concrete mixes at varying proportions (0.5, 1.0, 1.5, and 2%). The aspect ratio of sisal fibers was maintained at 30 mm. The addition of sisal fibers with proper aspect ratio demonstrates significant improvements in concrete's mechanical properties. Compressive strength and tensile strength peak at 1% sisal fiber content and decline thereafter. In both M25 and M30 grade concrete, normal sisal fibers enhance compressive strength by 4–14%, with soaked sisal fibers yielding a range of 25–34%. Parallel placed reinforcement further improves strength, ranging from 35.73 to 44.75%. Tensile strength follows a similar pattern, with the maximum achieved at 1.5% sisal fiber content. In both M25 and M30 grade concrete, normal sisal fibers enhance tensile strength by 5–17%, while soaked sisal fibers exhibit a range of 14–33%. Flexural strength peaks at 1.0% sisal fiber content, with normal sisal fibers improving strength by 4–38% in M25 and M30 grade and soaked sisal fibers yielding a range of 16–68% in M25 and M30.

Keywords Sisal fiber · Concrete · Fiber reinforced concrete · Compressive strength · Flexural strength · Tensile strength · And Na2CO3-treated sisal fiber

1 Introduction

Cement is a widely used material globally, with an annual consumption of around 2.86 billion tons. However, cement production is a significant source of carbon dioxide (CO_2) emissions, which are harmful to the environment. This is mainly due to the high carbon footprint associated with each ton of cement produced [1]. To

A. Ijari (✉) · M. U. F. Patel
Department of Civil Engineering, BLDEA V.P. Dr. P.G. Halakatti College of Engineering and Technology, Vijayapur, Karnataka, India
e-mail: ijariashwn@gmail.com

M. U. F. Patel
e-mail: cv.farooq@bldeacet.ac.in

Y. K. Guruprasad et al. (eds.), *Recent Advances in Infrastructure Development and Management—Volume 1*, Lecture Notes in Civil Engineering 570,
https://doi.org/10.1007/978-981-96-1554-4_10

address the environmental concerns, researchers have been exploring alternatives and innovations in the construction industry. One promising solution is fiber reinforced concrete (FRC), which is a composite material comprising cementitious composites and natural fibers. FRC is particularly valuable in low-cost housing projects [2]. Natural fibers like sisal, jute, coir, banana, and others are incorporated into cement concrete to enhance its properties. These fibers improve characteristics such as post-cracking behavior, strength, impact resistance, and tensile strength, making the concrete more durable and sustainable. Sisal, for instance, is a widely available natural fiber in Kenya and Brazil. It is used to reinforce concrete, creating sisal fiber reinforced concrete (SFRC) [3]. SFRC prevents crack propagation after yielding, enhancing the material's overall strength and durability. However, over time, the strength of fiber composites may decrease due to factors like alkaline attack and hydrogen movement. The extraction of natural fibers like sisal typically involves using hand extraction machines with serrated or non-serrated blades [4]. The size and quality of fibers extracted play a crucial role in the performance of the composite material. In experiments, fibers were cut into 30 mm lengths, and their aspect ratio and diameter varied. Despite these variations, slab samples with different fiber content and sizes showed similar total crack widths, indicating the potential for consistent performance. Additionally, natural fibers offer benefits such as impact resistance, sound absorption, and fire resistance, thanks to their low permeability [5]. Overall, the research and application of natural fiber-reinforced concrete aim to make construction materials more sustainable, eco-friendly, and suitable for various applications in the building industry.

2 Materials and Methodology

2.1 Methodology

1. This study is conducted to analyze the compressive strength, tensile strength, and flexural strength of concrete when the add-on of sisal fiber to concrete.
2. Collection of materials used in the experiment with 0, 0.5, 1.0, 1.5, and 2% of sisal fiber. Sisal fiber was cut into pieces as per aspect ratio.
3. Concrete mix is designed as per IS 10262-1982 and IS 456-2000 for nominal concrete. The grade of concrete we adopted was M30 and M25.
4. Cubes were casted with regular sisal fiber addition, sodium carbonate-soaked fiber, and also for parallel reinforcement of fiber.
5. Cylinders were cast according to normal and soaked NaCO3 solution fibers.
6. Prismatic beams were cast with random dispersion of fiber used and tested after 28 days.
7. Tests are being conducted on hardened concrete after 7, 14, and 28 days for cubes and cylinders.

8. The result obtained from the test is being analyzed and discussed with conventional concrete.

2.2 Materials Used for the Research

- **Cement**: The concrete utilized in the substantial blend configuration was Pozzonola Portland Concrete (PPC) of grade 43.
- **Fine Aggregate**: Fine aggregates M sand were used, having 2.8–3.0 specific gravity and also size passing from 4.75 mm.
- **Coarse Aggregate**: Coarse aggregate commonly known as crushed aggregate was used and has 2.4–3.0 specific gravity and a nominal size of 10–20 mm.
- **Water**: In accordance with IS 3025 (parts 22, 23), portable liquid water has been used to mix and cure concrete molds in the campus laboratory.
- **Superplasticizer**: In the current review, superplasticizer FOSROC CONPLAST SP 430 DIS was utilized to improve the usefulness of pressure strength of the blended cement and decrease water concrete proportion, denser and close finished concrete with diminished porosity and thus more sturdy cement is got.
- **Sisal Fiber**: The fiber of sisal came from ECO GREEN UNIT, which is based in Coimbatore, India. Sisal fiber exhibits exceptional characteristics, such as high strength and modulus, a straightforward structure, affordability, remarkable toughness, and recyclability, coupled with reduced susceptibility to support, wear, and tear. Additionally, it possesses a remarkable capacity for water absorption.

Tables 1, 2, and 3 represent the physical properties of fine aggregate, coarse aggregate, and sisal fiber, physical properties like specific gravity, fineness modulus, bulk density, water absorption, elongation of fiber, Young's modulus, etc.

Table 1 Physical properties of fine aggregate

Properties	Result
Fineness modulus	3.07
Specific gravity	2.44
Bulk density	1600 kg/m^3
Water absorption	4.5%

Table 2 Physical properties of coarse aggregate

Properties	Result
Specific gravity	2.72
Fineness modulus	6.3
Flakiness index	34.8%
Bulk density	1550 kg/m^3
Elongation index	44.9%
Water absorption	0.5%

Table 3 Physical properties of Sisal fiber

Properties	Result
Elongation	0.8–12 mm
Density	1.45 g/cm3
Young modulus	9–22 GPa

3 Results and Discussion

3.1 Compression Test on Cubes

The compressive strength of the cubes gives us the information of the potential strength of the concrete mix from which it is sampled. It helps in determining whether correct mix proportions of various mix proportions of various materials were used to get the desired strength. The cubes casted as per design and kept for curing and then tested for 28 days (Fig. 1; Table 4).

The 28-day compression test graph demonstrates a positive correlation between increasing fiber content and improved compressive strength for normal, soaked, and parallel reinforced conditions, with the optimal value at 1%. In comparison with conventional concrete, normal fibers show enhancements of 4.47%, 11.35%, 6.32%,

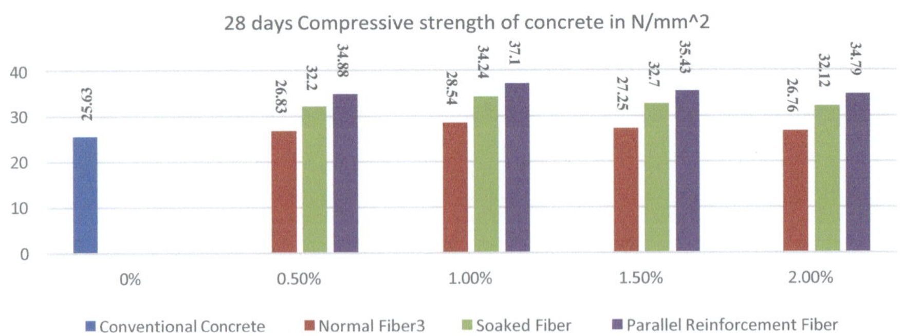

Fig. 1 Comparison of 28 days result of compressive strength of concrete for M25 grade

Table 4 Comparison of 28 days result of compressive strength of concrete for M25 grade

% of sisal fiber	28 days compressive strength in N/mm^2		
Conventional concrete			
0%	25.63		
Fiber reinforced concrete			
% of sisal fiber (%)	**Normal fiber**	**Soaked fiber**	**Parallel reinforcement fiber**
0.5	26.83	32.20	34.88
1.0	28.54	34.24	37.10
1.5	27.25	32.7	35.43
2.0	26.76	32.12	34.79

Table 5 Comparison of 28 days result of compressive strength of concrete for M30 grade

% of sisal fiber	28 days compressive strength in N/mm^2		
Conventional concrete			
0%	30.5		
Fiber reinforced concrete			
% of sisal fiber (%)	**Normal fiber**	**Soaked fiber**	**Parallel reinforcement fiber**
0.5	32.45	35.66	37.31
1.0	34.58	38.038	39.76
1.5	33.59	36.95	38.63
2.0	30.89	32.88	34.37

and 4.01%; soaked fibers exhibit improvements of 25.63%, 33.59%, 27.58%, and 25.51%; and parallel reinforced fibers demonstrate increases of 36.09%, 44.75%, 38.23%, and 35.73% at 0.5%, 1.0%, 1.5%, and 2% levels, respectively (Table 5).

The graph indicates a positive relationship between increasing fiber content and 28-day tensile strength for both normal and soaked fibers, with the optimal percentage being 1.5%, followed by a slight reduction. Compared to conventional concrete, normal fibers exhibit enhancements of 4.46, 11.81, 16.79, and 8.39% at 0.5%, 1.0, 1.5, and 2%, while soaked fibers show improvements of 12.07, 20.20, 25.45, and 16.27% at the same percentages (Figs. 2 and 3).

3.2 *Split Tensile Test*

The split tensile strength test of cylinder gives us the information of the potential strength of the concrete mix from which it is sampled. It helps in determining whether correct mix proportions of various mix proportions of various materials were used

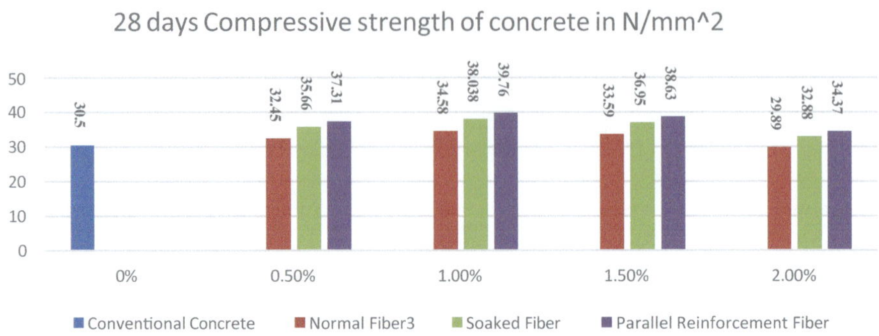

Fig. 2 Comparison of 28 days result of compressive strength of concrete for M30 grade

Fig. 3 Casting and testing of cubes

to get the desired strength. The cylinders are casted as per design and kept for curing and then tested for 28 days (Table 6).

The 28-day tensile strength test graph indicates a positive relationship between fiber percentage in concrete mix and tensile strength for both normal and soaked conditions, with optimal improvement at 1.5% fiber content followed by a slight

Table 6 Comparison of 28 days result of tensile strength of concrete for M25 grade

% of sisal fiber	28 days compressive strength in N/mm^2	
Conventional concrete		
0%	3.54	
Fiber reinforced concrete		
% of sisal fiber (%)	**Normal fiber**	**Soaked fiber**
0.5	3.89	4.18
1.0	4.12	4.43
1.5	4.36	4.69
2.0	3.74	4.02

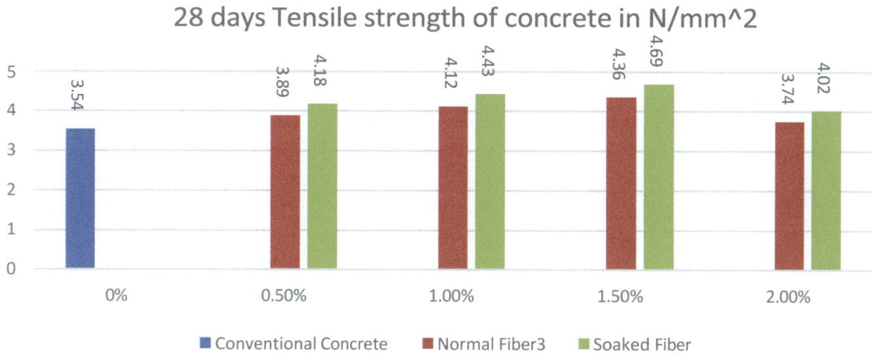

Fig. 4 Comparison of 28 days result of tensile strength of concrete (M25 grade)

Table 7 Comparison of 28 days result of tensile strength of concrete for M30 grade

28 days tensile strength in N/mm²		
Conventional concrete		
0%	3.81	
Fiber reinforced concrete		
% of sisal fiber (%)	**Normal fiber**	**Soaked fiber**
0.5	3.98	4.27
1.0	4.26	4.58
1.5	4.45	4.78
2.0	4.13	4.43

reduction. In comparison with conventional concrete, normal fibers show enhancements of 9.88%, 16.38%, 23.16%, and 5.64%, while soaked fibers exhibit improvements of 18.07%, 25.14%, 32.48%, and 13.55% at 0.5%, 1.0%, 1.5%, and 2% levels, respectively (Fig. 4; Table 7).

The graph indicates a positive relationship between increasing fiber content and 28-day tensile strength for both normal and soaked fibers, with the optimal percentage being 1.5%, followed by a slight reduction. Compared to conventional concrete, normal fibers exhibit enhancements of 4.46, 11.81, 16.79, and 8.39% at 0.5, 1.0, 1.5, and 2%, while soaked fibers show improvements of 12.07, 20.20, 25.45, and 16.27% at the same percentages (Figs. 5 and 6).

3.3 Flexural Strength

The flexural strength test of beams gives us the information of the potential strength of the concrete mix from which it is sampled. It helps in determining whether correct mix proportions of various mix proportions of various materials were used to get the

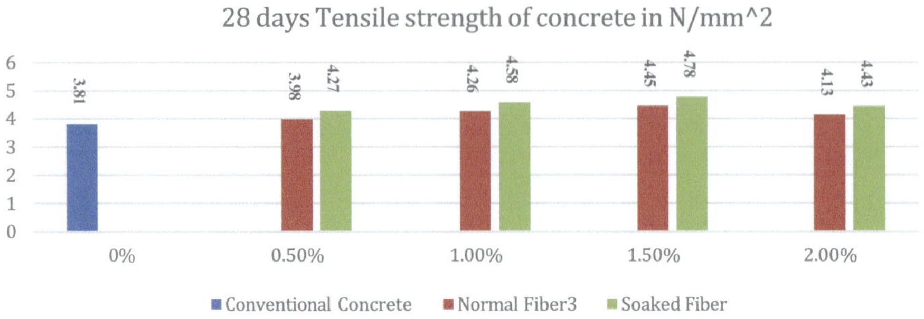

Fig. 5 Comparison of 28 days result of tensile strength of concrete (M30 grade)

Fig. 6 Casting and testing of cylinder

desired strength. The cylinders are casted as per design and kept for curing and then tested for 28 days (Table 8).

In figure, a clear positive correlation is observed between fiber content and 28-day flexural strength in both normal and soaked fibers, with the optimum at 1%. Compared to conventional concrete, normal fibers show improvements of 30.64, 45.43, 37.09, and 18.01% at 0.5, 1.0, 1.5, and 2%, while soaked fibers exhibit enhancements of 49.73, 67.47, 56.98, and 32.79% at the same percentages (Fig. 7; Table 9).

The graph illustrates a positive correlation between increased fiber content in concrete mix and improved 28-day flexural strength for both normal and soaked fibers, with the optimal value at 1%. Compared to conventional concrete, normal fibers show enhancements of 4.61, 37.13, 17.96, and 8.73% at 0.5, 1.0, 1.5, and 2% respectively, while soaked fibers exhibit improvements of 16.01, 48.54, 29.36, and 12.2% at the same percentages (Figs. 8 and 9).

Table 8 Flexural strength test of concrete with partial adding of sisal fiber with cement for normal M25 grade

Flexural strength in N/mm^2				
% of sisal fiber (%)	Unsoaked fibers		Soaked fibers	
	Load value (kg)	Flexural strength (N/mm^2)	Load value (kg)	Flexural strength (N/mm^2)
0	950	3.72	950	3.72
0.5	1240	4.86	1420	5.57
1.0	1380	5.41	1590	6.23
1.5	1300	5.101	1490	5.84
2.0	1120	4.39	1260	4.94

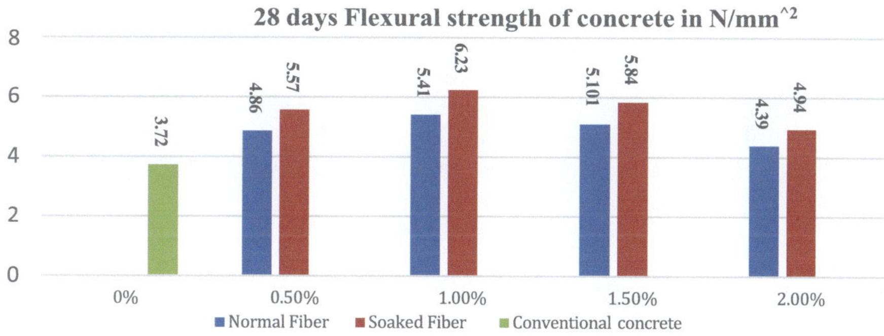

Fig. 7 Comparison of 28 days result of flexural strength of concrete (M25 grade)

Table 9 Flexural strength test of concrete with partial add-on of sisal fiber with cement for normal M30 grade

Flexural strength in N/mm^2				
% of sisal fiber (%)	Unsoaked fibers		Soaked fibers	
	Load value (kg)	Flexural strength (N/mm^2)	Load value (kg)	Flexural strength (N/mm^2)
0	1050	4.12	1050	4.12
0.5	1100	4.31	1220	4.78
1.0	1440	5.65	1560	6.12
1.5	1240	4.86	1360	5.33
2.0	1080	4.21	1130	4.35

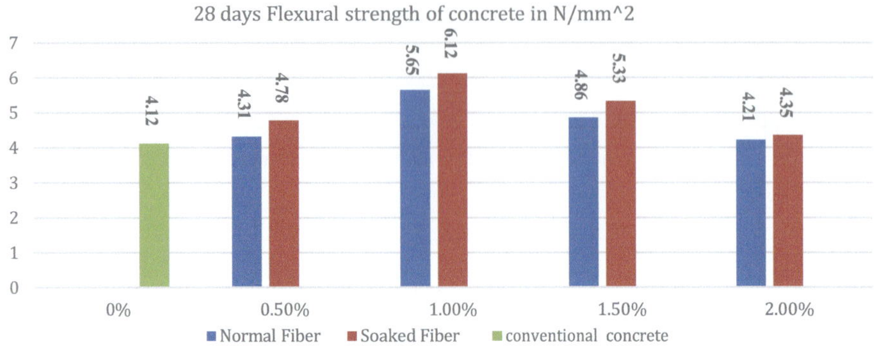

Fig. 8 Comparison of 28 days result of flexural strength of concrete (M30 grade)

Fig. 9 Casting and testing of beams

4 Conclusion

- As we can see, due to increase in the percentage of sisal fiber, the compressive stress, tensile strength, and flexural strength parameter can increase up to some extent, but after that it will decrease.
- The above result from compressive test shows that the 28-day characteristics strength of concrete is maximum at 1% sisal fiber content, and henceforth, it keeps on decreasing. In comparison with conventional concrete, the normal sisal fiber ranges from 4.01 to 11.35% and soaked sisal fiber ranges from 25.51 to 33.59% and for parallel placed reinforcement 35.73–44.75% in case of M25 grade. In comparison with conventional concrete, the normal sisal fiber ranges from 1.27 to 14% and soaked sisal fiber ranges from 8 to 25% and for parallel placed reinforcement 12–30% in case of M30 grade.
- The above result from tensile strength test shows that the 28-day characteristics strength of concrete is maximum at 1.5% sisal fiber content, and henceforth, it keeps on decreasing. In comparison with conventional concrete, the normal sisal fiber ranges from 5 to 17% and soaked sisal fiber ranges from 14 to 33% in case of M25 grade. In comparison with conventional concrete, the normal sisal fiber

ranges from 5 to 17% and soaked sisal fiber ranges from 12 to25% in case of M30 grade.

- The result from flexural strength test shows that the 28-day characteristics strength of concrete is maximum at 1.0% sisal fiber content, and henceforth, it keeps on decreasing. In comparison with conventional concrete, the normal sisal fiber ranges from 18 to 36% and soaked sisal fiber ranges from 32 to68% in case of M25 grade. In comparison with conventional concrete, the normal sisal fiber ranges from 4 to + 38% and for soaked sisal fiber it ranges from 16 to 48% in case of M30 grade.

References

1. Izquierdo IS, Izquierdo OS, Ramalho MA, Taliercio A (2017) Sisal fiber reinforced hollow concrete blocks for structural applications: testing and modeling. Constr Build Mater 151:98–112. https://doi.org/10.1016/j.conbuildmat.2017.06.072
2. Nayak JR, Bochen J, Gołaszewska M (2022) Experimental studies on the effect of natural and synthetic fibers on properties of fresh and hardened mortar. Constr Build Mater 347:128550. https://doi.org/10.1016/j.conbuildmat.2022.128550
3. Thomas BC, Jose YS (2022) Impact of sisal fiber reinforced concrete and its performance analysis: a review. Evol Intell 15(2):865–875. https://doi.org/10.1007/s12065-019-00230-9
4. Castañeda D, Silva G, Salirrosas J, Kim S, Bertolotti B, Nakamatsu J, Aguilar R (2020) Production of a lightweight masonry block using alkaline-activated natural pozzolana and natural fibers. Constr Build Mater 253:119143. https://doi.org/10.1016/j.conbuildmat.2020.119143
5. Pacheco-Torgal F, Jalali S (2020) Cementitious building materials reinforced with vegetable fibers: a review. Constr Build Mater.https://doi.org/10.1016/j.conbuildmat.2020.07.024

Enhancing Concrete with Partial Replacement of Granite Powder as Fine Aggregate

Shobhit Vishwanath Mareguddi, Nayeemulla Inamdar, and Arun Bagi

Abstract The study aimed to explore the feasibility of utilizing GP (granite powder), a waste material from granite cutting processes, as a partial replacement for fine aggregate in concrete. The main objective was to investigate its impact on concrete properties, particularly compressive strength. The research involved conducting a series of experiments with varying percentages of GP replacement (0, 10, 20, 30, 40, and 50%) to determine the optimal replacement ratio. Concrete mixes were prepared according to the specified proportions, and the beams were cast and cured for a designated period. The results revealed that the incorporation of GP into concrete showed significant promise. Particularly, when 20% of the fine aggregate was replaced with GP, the concrete exhibited notable enhancement in compressive strength without compromising structural integrity. This finding not only highlights the potential for utilizing granite waste but also contributes to sustainable construction practices by recycling and repurposing industrial by-products. The study's outcomes suggest that partial replacement of fine aggregate with GP offers an environmentally friendly solution, reducing waste generation from the granite cutting industry. It also promotes resource optimization and sustainable practices in the construction sector. The positive results at 20% replacement open up opportunities for further research and optimization to identify the most suitable replacement percentages for specific construction applications.

Keywords Granite powder · Granite cutting waste · Compression strength · Flexural strength · High strength concrete · M50 grade · Mechanical properties

---

S. V. Mareguddi (✉) · N. Inamdar · A. Bagi
Department of Civil Engineering, BLDEA V.P. Dr. P.G. Halakatti College of Engineering and Technology, Vijayapur, Karnataka, India
e-mail: shob-hitmareguddi4@gmail.com

Y. K. Guruprasad et al. (eds.), *Recent Advances in Infrastructure Development and Management—Volume 1*, Lecture Notes in Civil Engineering 570,
https://doi.org/10.1007/978-981-96-1554-4_11

1 Introduction

The industrial by-products and garbage are increasing on a daily basis, the entire globe is confronted with the difficulty of disposing of them. Both the increased quantity of garbage and the difficulty in disposing of it are compelling researchers to utilize it in a sustainable way. The concrete industry, from a different perspective, is having a greater negative influence on the environment because it uses a lot of sand also cement, two common conventional materials. The Govt. of India put restrictions on mining of river sand in numerous states because the ongoing harvesting river sand has turned into a serious environmental problem. Hence, these by-products are being repurposed in the concrete industry as an alternative to conventional materials as a resource of managing industrial waste and reducing the utilization of conventional materials.

However, the evaluation of GP integration within concrete as a substitute to fine-grained aggregate remains unexplored. Because of its excellent flowability, full compaction may be achieved without the need of any external vibration. It provides a number of advantages, including faster construction and lower labor costs, flexibility in design, and a homogenous concrete matrix free of honeycombing. Thanks to the aforementioned useful properties, concrete's involvement in the construction sector is also progressively rising with each passing day. Therefore, our research focused on determining the effect of GP on concrete production. The GP is varied from 0 to 100% by adding 20% by mass of fine-grained aggregate.

2 Experimental Investigation

This study focuses on the preparation of M50 grade concrete and investigating the effect of granite powder on strength parameters of concrete. This deals with collection of materials, testing of materials, casting of specimens, and their testing. Total 36 specimens of cubes, cylinders, and prismatic beams are casted for each test, i.e., compression, split tensile, and flexure test, out of which 18 are for 7 days strength test and remaining 18 are for 28 days strength test. The GP of 0, 10, 20, 30, 40, and 50% by the weight of fine aggregate are added during the time of mixing.

2.1 Materials

General—This section addresses the topic of the materials utilized in the existing work such as cement, aggregates, and also the procedure involved in casting of test specimens and testing of specimens. It also includes the various parameters utilized in the existing work.

Cement—In this study, 53-grade cement (OPC) was employed as the binding agent. The physical attributes and elemental compositions of the cement were examined, with elemental compositions determined through energy-dispersive X-ray (EDX) analysis. Fine-grained aggregate in the form of river sand with a largest dimension of 4.75 mm and coarse aggregate in the form of basalt stone with a largest dimension of 10 mm were employed. Prior to usage in this research, both aggregates were meticulously dried. The physical characteristics of the fine and coarse aggregates, along with GCW, are deliberated upon. The particle size distribution of the fine and coarse aggregates, along with GCW, is discussed.

Fine Aggregate—Fine-grained aggregate is the essential constituent in concrete that consists of natural sand or crushed stone. The hardened characteristics of concrete are heavily influenced by the quality and fine-grained aggregate density. If you choose fine-grained aggregate based on grading zone, particle form and surface texture, abrasion and skid resistance, and absorption, you may make the concrete or mortar mixture more durable, stronger, and less expensive.

Coarse Aggregate—It traverses through 63 mm I.S sieve and retained on 4.75 micron aggregates which are small pieces of broken stones in irregular shapes and sizes. In construction work, the cement is mixed with inert strong and durable hard materials.

Water—Clean and potable water devoid of organic contaminants and oil, readily accessible within the college premises, was employed for concrete mixing.

Granite cutting waste—GCW was obtained in its wet state from a nearby granite cutting industrial sector. It was dried before use in this investigation, as shown. The physical parameters of fine and coarse aggregates, as well as GCW, are presented (Tables 1, 2 and 3).

Table 1 Section properties of building models

Characteristics	Data	As per IS-12269-2013
IST (Initial setting time) (min)	48	30
FST (Final setting time) (min)	390	600
Compressive strength (28 days) (MPa)	54	≥ 53
Std. consistency (%)	32	30–35
Specific gravity	3.15	3.10–3.16

Table 2 Characteristics of fine-grained aggregate

Characteristics	Data	As per IS-383-2016
Zone	II	–
Type	M Sand	–
Specific gravity	2.64	2.3–2.9
Fineness modulus	2.68	2.6–2.9

Table 3 Characteristics of coarse aggregate

Characteristics	Data	As per IS-2386-1936 part III
Fineness modulus (F.M)	6.16	6.0–6.9
Nominal size	20 mm	–
Specific gravity	2.72	2.5–3.0

Table 4 Mix proportion of M50 grade concrete for 1 m^3

% of GF	Cement	FA (kg)	CA (kg)	Fly ash (kg)	Water (kg)	WC	SP (kg)
0	417.6	356	644	46.4	197.2	0.36	3.46
10	417.6	356	644	46.4	197.2	0.36	3.46
20	417.6	356	644	46.4	197.2	0.36	3.46
30	417.6	356	644	46.4	197.2	0.36	3.46
40	417.6	356	644	46.4	197.2	0.36	3.46
50	417.6	356	644	46.4	197.2	0.36	3.46

2.2 *Mix Proportion*

The concrete blend was formulated following the calculated mix design with the correct ratios, and the necessary amount of granite powder was incorporated until a uniform consistency was reached (Table 4).

The cube dimensions are 150 mm × 150 mm × 150 mm. The cylinder has a diameter of 150 mm and a height of 300 mm. The beam is characterized by dimensions of 500 mm × 100 mm × 100 mm.

3 Results and Discussions

3.1 *Compression Test*

The assessment of concrete strength through compressive strength testing of concrete cubes is a broadly recognized and established procedure. The following procedure outlines the steps to conduct the compression-strength trial for concrete cubes that have been enhanced by moderately replacing the fine-grained aggregate with GP (Table 5; Fig. 1).

Table 5 Compression strength test on cube at 28 days

S. No.	Trail 1	Trail 2	Trail 3	Average (N/mm^2)
GP0	54.5	57.6	56.7	55.86
GP10	56.2	58.5	57.9	57.24
GP20	58.8	60	59.72	59.45
GP30	57.9	59.1	58.7	58.36
GP40	56.8	58.1	57.7	57.5
GP50	55.7	56.9	56.6	56.26

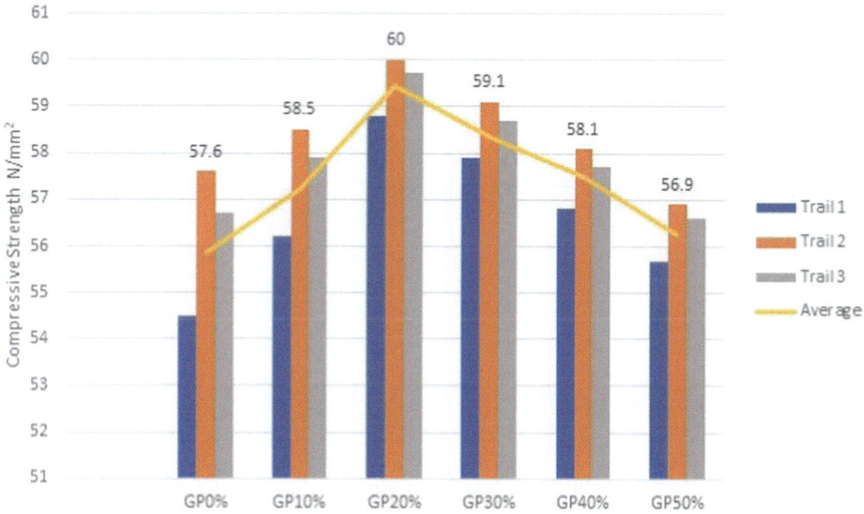

Fig. 1 Compression strength test on cube at 28 days graph

3.2 *Flexure Strength Test*

Testing flexural strength is a vital aspect of evaluating the physical characteristics of concrete. Flexural strength testing can provide appreciated insights into the behavior of the concrete mixture and the properties of the fractional replacement of GP as a fine- grained aggregate. To examine the flexural strength of a beam for improving concrete by partly changing fine-grained aggregate with GP, various replacement percentages must be utilized in the trials (Table 6; Fig. 2).

Testing the Flexural Strength

a. After the curing period, remove the beams from the curing tank and allow them to reach room temperature.
b. Take the dimensions (length, width, and depth) of each beam using a caliper or ruler.

Table 6 Flexural strength test on beam at 28 days

S. No.	Trail 1	Trail 2	Trail 3	Average (N/mm^2)
GP0	6.4	6.6	6.8	6.56
GP10	6.7	6.9	6.8	6.84
GP20	7.2	7.5	7.4	7.32
GP30	7.4	7.5	7.6	7.48
GP40	7	7.4	7.2	7.18
GP50	6.8	7.1	6.9	6.96

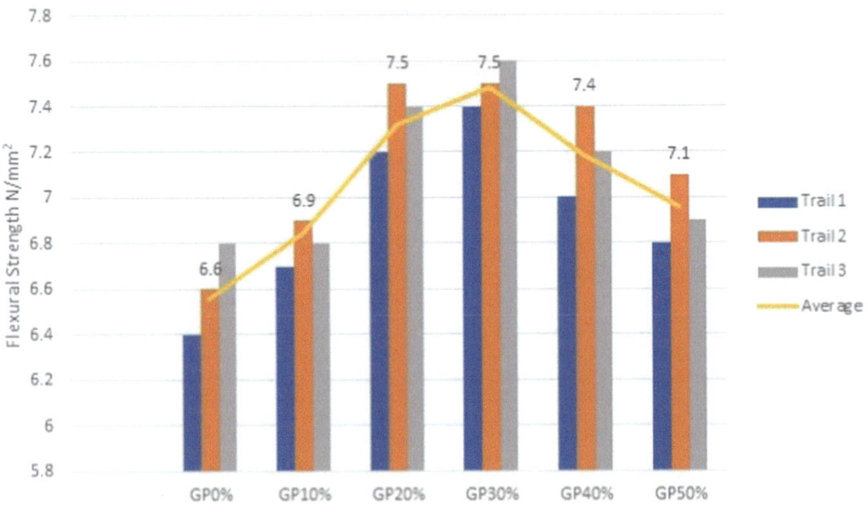

Fig. 2 Flexural strength test on beam at 28 days graph

c. Set up the Universal Testing Machine for three-point bending test.
d. Place the beam on the testing machine with the bottom side facing upwards and ensure the load is applied at the middle of the beam.
e. Gradually apply the load until the beam fails, and record the extreme load (P) at failure.

3.3 *Split Tensile Test*

Split tensile strength, also known as indirect tensile strength, is a measure of a material's resistance to cracking under tensile stress. It is determined by applying a compressive load to the cylindrical or cubical specimen along its length and measuring the tensile strength of the specimen.

Table 7 Split tensile test on cylinder at 28 days

SL. No	Trail 1	Trail 2	Trail 3	Average (N/mm^2)
GP0	4.3	4.4	4.6	4.46
GP10	4.4	4.6	4.7	4.58
GP20	5.4	5.7	5.6	5.56
GP30	5.2	5.5	5.4	5.34
GP40	5	5.4	5.2	5.2
GP50	5	5.4	5.1	5.18

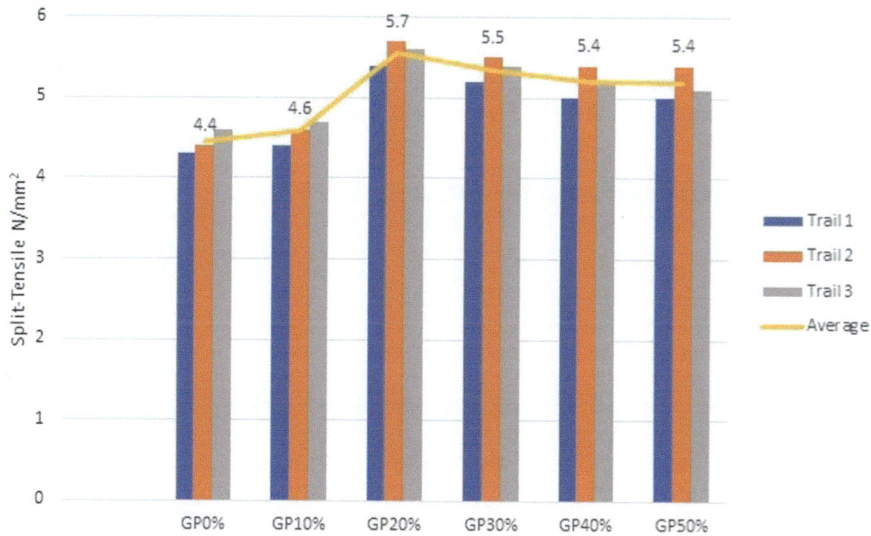

Fig. 3 Split tensile test on cylinder at 28 days graph

Conducting split tensile strength tests on concrete cylinders containing differing proportions of granite powder as an additional for fine-grained aggregate can produce significant visions into the effect of granite powder incorporation on the concrete's tensile strength. The subsequent outline delineates the process for performing split tensile tests across various levels of granite powder substitution (Table 7; Fig. 3).

4 Conclusion

1. The experiment aimed to enhance concrete by replacing a portion of fine-grained aggregate with granite powder.
2. When 20% of the fine-grained aggregate was replaced with granite powder, concrete showed 59.45 N/mm^2 of compressive strength.

3. Concrete exhibited satisfactory flexural strength up to 7.48 N/mm^2 substituting 30% of the fine-grained aggregate with granite powder.
4. The concrete demonstrated acceptable split tensile strength of 5.56 N/mm^2 when 20% of the fine-grained aggregate was replaced with granite powder.
5. The positive outcome at 20% replacement suggests the potential to utilize a significant amount of granite waste.
6. Structural integrity of the concrete is not compromised with this level of replacement.

References

1. Bhikshma V, Reddy N (2019) Self-compacting concrete with partial replacement of granite cutting waste as fine aggregate. Int J Innov Technol Explor Eng 8(12):1378–1381
2. Gupta S, Saxena M (2020) Utilization of granite cutting waste in the production of self-compacting concrete. J Clean Prod 267:122110
3. Khajuria H, Chandel S, Singh AK (2018) Properties of self-compacting concrete with partial replacement of fine aggregate by granite cutting waste. Int J Eng Technol 7(4.25):41–43
4. Panda B, Dey UK (2019) Sustainable use of granite waste in self-compacting concrete production: an overview. J Clean Prod 215:127–137
5. Singh S, Singh RP (2018) Self-compacting concrete with partial replacement of fine aggregate by granite cutting waste. Int J Civ Eng Technol 9(8):120–128
6. Taha R, Fatah AA, Ajeel AH (2020) Utilization of granite cutting waste as fine aggregate in self-compacting concrete. Int J Emerg Technol Adv Eng 10(11):128–135
7. IS 10262 (2019) Concrete mix proportioning—guidelines
8. IS 456 (2000) Plain and reinforced concrete—code of practice
9. IS 383 (1970) Specification for coarse and fine-grained aggregates from natural sources for concrete
10. IS 1489-1 (1991) Specification for Portland pozzolana cement. Is 269 for OPC cement

Retrofitting of RCC Structure Using Fluid Viscous Dampers

R. Gayathri, T. S. Malleshiah, Vatsala, M. Harshitha, and R. Darshan

Abstract Commercial buildings have the potential to improve their energy efficiency and long-term usefulness through retrofitting, reconstruction, and renovations. Retrofitting typically entails making adjustments to already-built commercial structures that could boost their durability and strength. In addition to addressing the issue of building replacement, the expansion of rehabilitation, and the urban characteristics of the city as a whole, reconstruction with additional floors can develop into a more affordable option to demolition. The owner's requirements, urban planning issues, and population density all have an impact on the decision to build more stories. In this project, we will use ETABS 2019 to study an existing G + 8 building both statically and dynamically, try to add more stories, and assess the building's sustainability. The same building was retrofitted utilizing the FVD technology to help the structure sustain the weight of the extra floors because the incremental floors caused failure of some concrete frames. The results led to the examination of a number of variables, including base shear, story drift, and maximum displacement.

Keywords Retrofitting · Incremental of stories · Fluid viscous dampers (FVD)

1 Introduction

The method that is most frequently utilized to improve older structures is retrofitting. It is defined as the modification of existing structures to improve their stability and structural integrity. Thus, a building is less likely to sustain permanent damage.

R. Gayathri · T. S. Malleshiah · M. Harshitha (✉) · R. Darshan (✉)
Department of Civil Engineering, The Oxford College of Engineering, Bengaluru, Karnataka, India
e-mail: harshitham885@gmail.com

R. Darshan
e-mail: darshanr7100@gmail.com

Vatsala
Department of Civil Engineering, Bangalore Institute of Technology, Bengaluru, Karnataka, India

Y. K. Guruprasad et al. (eds.), *Recent Advances in Infrastructure Development and Management—Volume 1*, Lecture Notes in Civil Engineering 570,
https://doi.org/10.1007/978-981-96-1554-4_12

However, in order to make structures code compliance, there is a requirement for upgrading due to the updating and altering of seismic codes. There are occasions when traditional retrofitting techniques like installing a shear wall, wing walls, bracing, and jacketing of beams and columns become ineffective and impossible. This is when other revolutionary approaches, such as the energy dissipation system, enters the scene. Energy dissipation system are claimed to be the active safety system for attaining the functional purposes of modern codes.

2 Fluid Viscous Dampers

This damper uses a viscous fluid inside a cylinder to dissipate energy. Due to their wide range of sizes, simplicity of installation, adaptability, and cooperation with other members, this damper is widely employed in design and retrofitting. They are frequently used to reduce structural sway in buildings. By serving as a passive energy dissipation mechanism, FVD improves seismic performance while also increasing the effective stiffness of both new and existing structures.

3 Literature Review

Gayathri et al. [1]: The G + 9 RCC construction, which is close to Tumakuru, was employed by the authors of this study. They have examined the building and modified it with two types of earthquake protection, namely shear walls and larger columns, in preparation for the impending earthquake that is predicted. They used ETABS to examine the structure both statically and dynamically. Instead of enlarging the column, they found that the shear wall was a superior option.

Mahakalkar et al. [2]: In order to understand the responses of multi-storied RC structures, this research uses a G + 15 story RC structure with FVD and friction dampers. The multi-story RC building's dynamic response has been studied using the program ETABS 2017. The time-history and response-spectrum method is used to estimate the dynamic response of the structure. They got to the conclusion that using square columns over rectangular columns was better in terms of seismic response and that FVD is a more effective seismic behavior regulator than friction dampers.

Hangal et al. [3]: This study compared a symmetrical plan with and without a shear wall, x bracing, and FVD by situating them at each corner. Response-spectrum analysis was carried out utilizing ETABS as the analysis method. The outcome shown that a shear wall decreased displacement more effectively than X-bracing and FVD. However, all three systems lessened the structural harm brought on by lateral forces. They also got to the conclusion that FVD is a good alternative since they perform well when displacement and stiffness parameters are taken into account.

Yadav et al. [4]: The plan having lateral irregularities along with fluid viscous dampers (FVD500) underwent a seismic evaluation. The time-history method of analysis was used to assess the RCC framed construction. They came to the conclusion that the natural period of all plans is secure and within Indian standards.

Priyanka et al. [5]: This study examines the seismic retrofitting of an RC structure with FVD250 in various zones and found that reduced displacement, acceleration, and base shear improved the seismic performance. Additionally, Zone V was designed to be the least protected area. Additionally, it is suggested to build the same building using FVD500 and to investigate the study with different irregularities as a future scope.

4 Objective and Methodology

4.1 *Objective*

For an existing building, to gradually add stories.

To determine whether the structure can support the added weight.

To increase strength and make it resistant to lateral loads by retrofitting the failing concrete members with FVD (Tables 1 and 2).

Table 1 Building parameters

Member	Size (mm)
Model 1	
Column	600 × 600 (1–3 floors) 500 × 500 (4–6 floors) 400 × 400 (7–9 floors)
Beam	300 × 300 (1–3 floors) 300 × 250 (4–6 floors) 300 × 250 (7–9 floors)
Slab	150
Wall thickness	200 mm
Model 2	
Column	350 × 350 (10–11 floors)
Beam	300 × 250 (10–11 floors)
Model 4	
Column	300 × 300 (12th floor)
Beam	250 × 250 (12th floor)

Table 2 General input data

Grade of concrete	M40, M30
Grade of steel	Fe 550, Fe 500, Fe 415
Live load	3 kN/m^2
Floor finish	1.5 kN/m^2
FVD	250 kN capacity

4.2 Methodology

Model 1: Existing structure without retrofitting (G + 8).
Model 2: Structure with two-increased stories without retrofitting (G + 10).
Model 3: The same model 2 is retrofitted using fluid viscous dampers.
Model 4: The model 3 will be increased with one more story (G + 11).
Model 5: The above model (model 4) with G + 11 is retrofitted using FVD.

4.3 Procedure Followed

Step 1**:** Model and analyze the asymmetrical G + 8 building using the data that is currently available. (Model 1).
Step 2: Add two more stories to the same model and then examine it. (Model 2).
Step 3: Verify the safety of a building.
Step 4: Use FVD to retrofit the failed model (Model 3).
Step 5: Add an additional level to model 3, and evaluate it once more (Model 4). Check the outcome.
Step 6: Repeat the process of retrofitting the model 4 with the FVD to increase its durability (Model 5).
Step 7: Decide how many additional floors can be taken into consideration, and then, compare the results of the models.

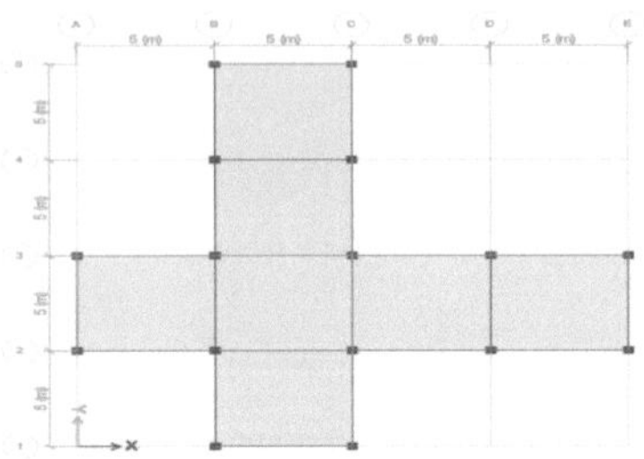

v.

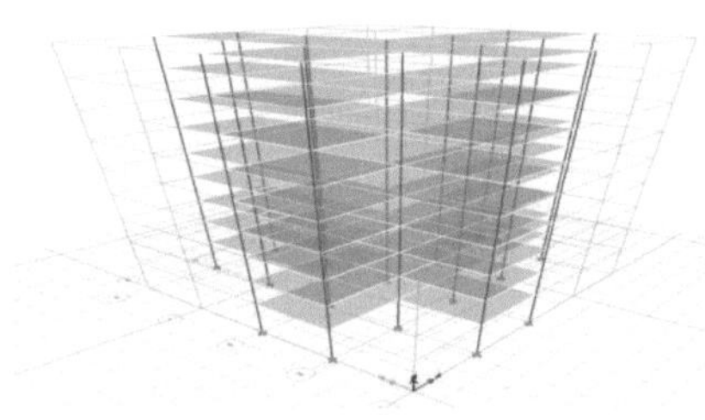

Plan and 3D View of model 1

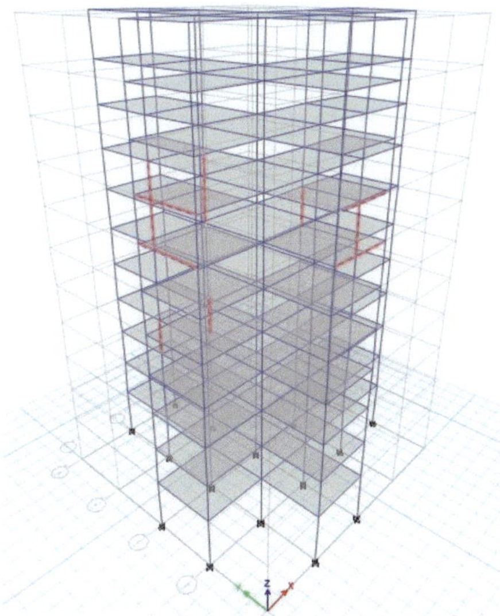

3D View of Model 2

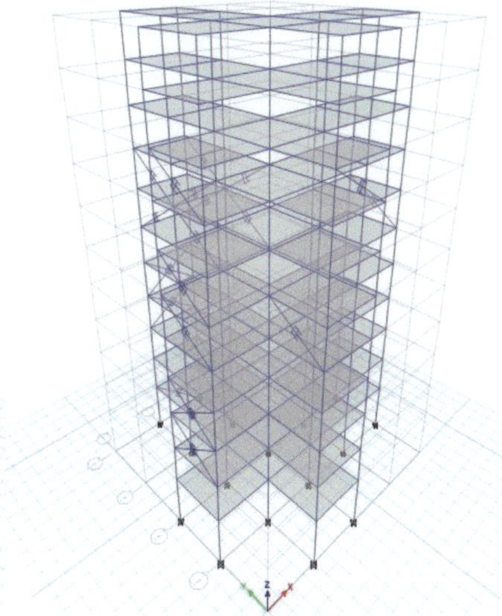

3D View of Model 3

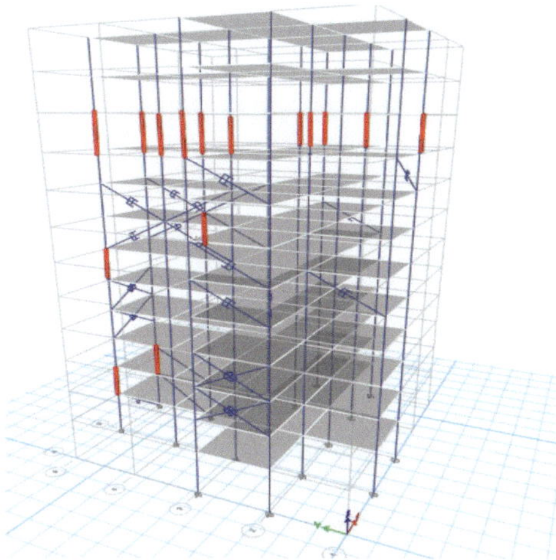

3D View of Model 4

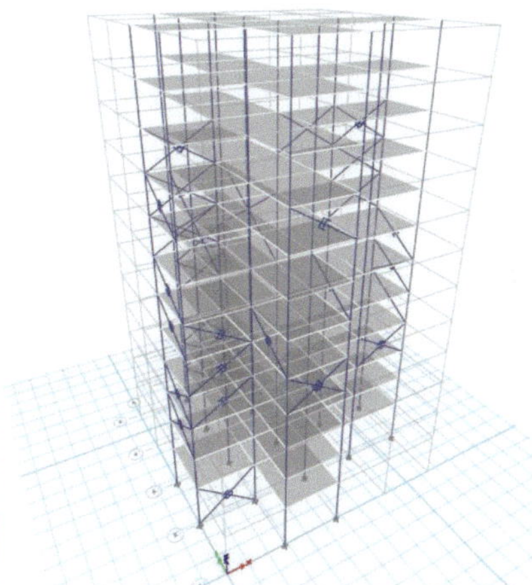

3D View OF Model 5

5 Analysis and Results

Maximum Story Displacement: Story displacement is the term used to describe the movement that occurred at each story level. Multi-story building's upper levels will exhibit the most story displacement. As the height increases, the story displacement will have the largest value.

Maximum storey displacement

Type of building	Equivalent-static method		Response-spectrum method	
	EQX	EQY	RSX	RSY
Model 1	27.7154	27.7154	23.5339	23.5339
Model 2	36.4752	36.4752	29.4501	29.4501
Model 3	33.0504	27.5597	27.0693	18.4532
Model 4	30.9049	24.5083	30.5680	22.9708
Model 5	13.9621	19.7999	17.2500	15.2907

Comparison of Max storey displacement

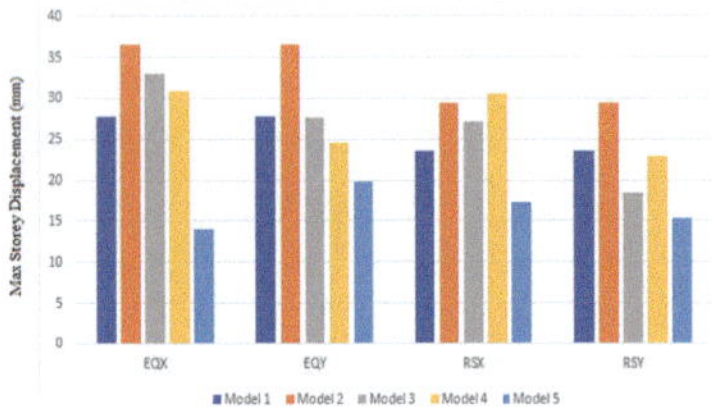

Story Drift: Story drift was the term used to describe the lateral movement of a floor with respect to the floor below. By dividing story drift by story height, one may determine the story drift ratio. There is not a unit for story drift.

Storey drift

Type of building	Equivalent-static method		Response-spectrum method	
	EQX	EQY	RSX	RSY
Model 1	0.001226	0.001226	0.001098	0.001098
Model 2	0.001462	0.001462	0.001248	0.001248
Model 3	0.00151	0.001699	0.001178	0.00138
Model 4	0.001384	0.00146	0.001189	0.001638
Model 5	0.001104	0.001824	0.001604	0.00138

Comparison of Max storey drift

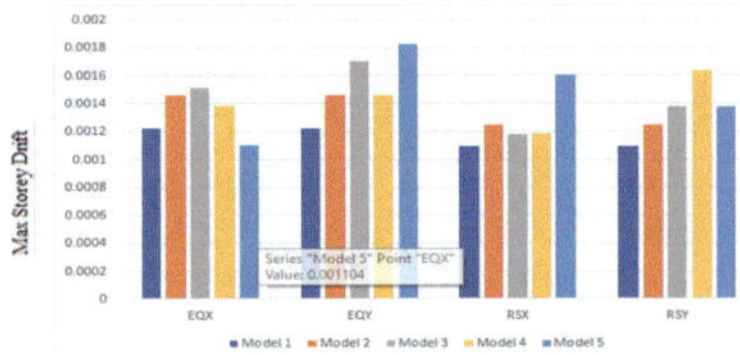

Base Shear: The largest expected lateral force from seismic activity into the base of the building is measured as base shear.

Base shear

Type of building	Equivalent-static analysis		Response-spectrum analysis	
	EQX	EQY	RSX	RSY
Model 1	383.755	383.755	382.616	382.616
Model 2	361.584	361.584	361.584	361.584
Model 3	140.411	130.602	188.277	80.749
Model 4	128.203	110.861	177.631	77.577
Model 5	306.188	68.978	245.226	57.178

Comparison of base shear

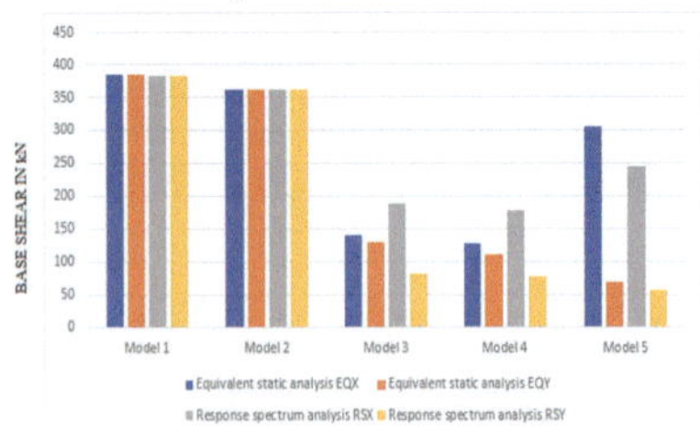

6 Results, Discussion, and Conclusions

Figure 1 compares the displacement values for all the five models stated above using both the equivalent-static approach and the response-spectrum method. The bar graph demonstrates that displacement has increased in model 2 in comparison with model 1 due to model 2's building being taller than model 1. As a result of the addition of FVD, the model 3 has less displacement than the model 2, which was altered from the model 2. Additionally, the displacement in model 4 is higher in the static analysis than it is in the dynamic analysis.

Figure 2 compares the story drift of 5 models using equivalent-static analysis and response-spectrum analysis. The bar graph shows that the building (model 3) has a little higher story drift than model 2. When comparing models 4 and 5, there is an increase in the x-direction and a decrease in the y-direction for story drift in the dynamic analysis. The graphs for story stiffness of model 3 and 5 are also shown in Figs. 3 and 4. Figure 5 illustrates a comparison of base shear for the five models in equivalent-static analysis and response-spectrum analysis.

This chapter can be concluded as follows:

- The strength of the existing G + 8 building has risen as a result of the use of FVD as a retrofitting technology.
- The maximum story displacement of the retrofitted structure is lower than the displacement of the structure without retrofit.
- When compared to the structure that has not been retrofitted, the retrofitted structure's story drift has somewhat risen.
- Everywhere the FVD was used, its overall stiffness was enhanced in the retrofitted model.
- Models with the retrofit have a lower base shear value as compared to models without it.
- FVD installation is a simple process, but it still needs labor that is qualified and trained.

References

1. Gayathri R (2023) Comparative study of seismic retrofitting methods for durability gain of an existing old building. IJONS 14. ISSN 0976-0997
2. Mahakalkar P, Pawade P, Dabhekar K (2023) Analysis of seismic effect of dampers as energy dissipation system in multi-storied RCC building. In: AIP conference proceedings, vol 2747, no 1. AIP Publishing
3. Hangal DR, Sri Raghu ME (2022) Comparative analysis of multi-storey building without and with a shear wall, x- bracing and FVD. IRJET 9. e-ISSN: 2395-0056, p-ISSN: 2395-0072
4. Yadav BK, Sahani S (2022) Comparative study on seismic behaviour of different shapes of RCC- structure with help of viscous damper
5. Priyanka N, Thivya J, Vijayaraghavan J (2019) Seismic study of multi-storey structure with FVD using ETABS. Dimension 1000:12–18. IRJET 6. e-ISSN: 2395-0056, p-ISSN: 2395-0072

6. Subasini Y, Sivakumar CG (2020) Analytical study on seismic-response of steel and RC structures with fluid-viscous damper. IOP Publishing
7. Dasari SG, Srinivasa Rao K (2019) Seismic evaluation of RCC framed buildings with and without passive energy dissipators. JES 10. ISSN: 0377-9254
8. Sah MK, Mallik RK, Motra GB (2021) Seismic performance of RC frame building with FVD. In: IOE graduate conference, vol 10. ISSN: 2350-8914
9. Al Agha W, Umamaheswari N (2020) Comparative study on seismic-performance of reinforced concrete building without and with fluid-viscous dampers. IJERM. ISSN: 2349-2058
10. Kotecha SP, Patil RS, Gandhe GR (2022) Earthquake response study of multi-storied RC building with FVD on sloping ground. IJSREM 6. ISSN: 2582-3930
11. Naveen J, Nagasai A, Selwynbabu J et al (2018) Comparative study of unsymmetrical buildings without and with dampers by Etabs. IJRAT. e-ISSN: 2321-9637
12. Rajput PS, Rahul S (2022) Analysis of a tall structure considering two different type of dampers using analysis tool. IJSRCE 6. ISSN: 2456-6667
13. Indian Standards (2016) Criteria for earthquake resistant design of structures, sixth revision. IS 1893 (Part 1), New Delhi
14. Indian Standards (1897) Code of practice for design loads (other than earthquake) for buildings and structures, second revision. IS 875 (Part 2), New Delhi
15. Fluid viscous dampers: general guidelines for engineers. Taylor Devices Inc.

Investigating the Effect of Glass Fibers on the Strength Development of Concrete

Mohammed Hussain Choudhari, Nayeemulla Inamdar, and Arun Bagi

Abstract This research report investigates the impact of glass fiber on concrete strength development. Concrete of the M50 grade is prepared in accordance with IS codes. Total 30 specimens of cubes, cylinders, and beams are casted, out of which 15 are for 7 days strength testing and remaining 15 are for 28 days strength testing. Concrete specimens with varying glass fiber content (0.5, 1, 1.5, and 2%) by weight of cement were tested for compressive, tensile, and flexural strength at different curing ages. Results showed significant strength improvements with glass fiber inclusion, enhancing compressive, tensile, and flexural properties. However, an optimum fiber content was identified. This study provides valuable insights for designing and applying fiber-reinforced concrete in construction projects, enhancing the durability and resilience of structures. In conclusion, the investigation carries substantial possibilities for improving concrete structure's performance. By examining mechanical properties of GFRC, the research strives to offer valuable insights capable of improving our knowledge for composite materials in construction engineering.

Keywords Glass fiber · AR-GF · Glass fiber-reinforced concrete · FRC · High strength concrete · M50 grade · Mechanical properties · Silica fume · Fly ash

1 Introduction

The characteristics of concrete have enhanced over the time to meet the ever-increasing demands of modern construction practices. Concrete is a widely used material for the construction due to availability, low cost, and durability. It is a brittle material compared to compressive strength, and the flexural and tensile strengths are relatively low. In plain concrete, the structural cracks develop due to drying shrinkage and cause failure to the structure. The materials with improved mechanical properties

M. H. Choudhari (✉) · N. Inamdar · A. Bagi
Department of Civil Engineering, BLDEA V. P. Dr. P. G. Halakatti College of Engineering and Technology, Vijayapura, Karnataka, India
e-mail: mdhussain0041@gmail.com

Y. K. Guruprasad et al. (eds.), *Recent Advances in Infrastructure Development and Management—Volume 1*, Lecture Notes in Civil Engineering 570,
https://doi.org/10.1007/978-981-96-1554-4_13

increase the toughness and tensile strength of the concrete [1]. Concrete is a widely used building material but is subject to many limitations in engineering applications because of poor ductility, high brittleness, low tensile strength, low tensile strain, and poor impact toughness [2].

One such approach is the incorporation of reinforcing fibers into concrete, resulting in the generation of composite materials known as fiber-reinforced concrete (FRC). Most of the carried out research on mechanical behavior of FRC were made with steel, carbon, and natural fibers. However, few attempts were made with glass fibers [3]. In modern times, there is a lot of interest in the glass fibers (GF), among various types of reinforcing fibers. Glass fiber-reinforced concrete (GFRC) is the material where glass fibers in the form of short and discontinuous are dispersed throughout the concrete matrix. By incorporating these fibers into the concrete, its tensile strength and flexibility are enhanced, resulting in heightened resistance against cracking and breaking. It has been shown that GFRC specimens exhibit a significant improvement in bending strength [4]. The objective of this research is to know how glass fibers influence the progress of concrete's strength. The investigation will involve the preparation of concrete samples of varying proportions of glass fibers. The concrete specimens will be tested at various curing ages. The investigation carries substantial possibilities for improving concrete structure's performance. By examining mechanical properties of GFRC, the research strives to offer valuable insights capable of improving our knowledge and applications for composite materials in construction engineering.

2 Experimental Investigation

Concrete of the M50 grade is produced in accordance with IS codes. For each test, i.e., compression, split tensile, and flexure test, total 30 specimens of cubes, cylinders, and beams are casted, out of which 15 are for 7 days strength testing and remaining 15 are for 28 days strength testing. The GF of 0.5, 1, 1.5, and 2% by weight of cement are added during the time of mixing of concrete.

2.1 Materials

Cement plays a big role in providing strength, durability, and cohesion to concrete structures. OPC 53 grade cement is used. The main purpose of using fine aggregate is to occupy the empty spaces between the larger coarse aggregates, effectively filling the voids, providing stability, cohesion, and workability to the mixture. Manufactured sand (M sand) is used. 20 mm size of coarse aggregate, conveniently accessible within the college campus, was employed. The aggregate underwent several examinations, such as specific gravity, and other parameters, adhering to the specifications stated in IS: 2386-1963 and IS: 383-2016. White color, chopped stranded type AR-GF are obtained from Deogiri Sanitary, Vijayapura. Glass fibers of length 12 mm are used. Silica fume is obtained from the Fiber Region, Chennai. It is made up of extremely tiny particles, which have an average size of about 0.1 μm. Fly ash is obtained from NTPC Kudgi (Tables 1, 2, 3, 4 and 5).

Table 1 Properties of cement

Characteristics	Results	According to IS 12269-2013
Initial setting time (min)	42	30
Final setting time (min)	340	600
Specific gravity	3.15	3.10–3.16
Standard consistency (%)	30	30–35
Compressive strength (MPa)	53.8	≥ 53

Table 2 Properties of fine aggregate

Characteristics	Results	According to IS 383-2016
Specific gravity	2.62	2.3–2.9
Fitness modulus	2.61	2.6–2.9
Zone	II	–

Table 3 Properties of coarse aggregate

Characteristics	Results	According to IS 2386 part III-1936
Specific gravity	2.70	2.5–3
Water absorption	0.5%	0.5–1%
Nominal size	20 mm	–
Fineness modulus	6.12	6.0–6.9

Table 4 Properties of glass fiber

Properties	Results
Fiber length	12 mm
Colour	White
Type	Synthetic Fiber, chopped strands
Usage	For concrete mould and industrial
Tensile strength	1700 MPa
Diameter	14 μm
Material	Alkali resistance glass
Aspect ratio	875
Specific gravity	2.68 g/cc
Modulus of elasticity	72 Gpa

Table 5 Properties of silica fume

Properties	Results	As per ASTM C1240
Appearances	Powder	-
Sp. gravity	2.22	2.2–2.3
Color	White	–

Table 6 Mix proportion of M50 grade concrete for 1m^3

% of GF	Cement	FA (kg)	CA (kg)	Fly ash (kg)	Silica fume (kg)	Water (kg)	W/C	SP (kg)
0	358	643	1200	67	22	152	0.373	4.07
0.5	358	643	1200	67	22	152	0.373	4.07
1	358	643	1200	67	22	152	0.373	4.07
1.5	358	643	1200	67	22	152	0.373	4.07
2	358	643	1200	67	22	152	0.373	4.07

2.2 *Mix Proportion*

The mixture of concrete was prepared as per calculated mix design in the correct proportions, and required quantity of glass fibers were added until a uniform consistency is achieved (Table 6).

Cube size: 150 mm × 150 mm × 150 mm;
Cylinder size: 150 mm dia, 300 mm height;
Beam size: 500 mm × 100 mm × 100 mm.

3 Results and Discussions

3.1 *Workability Test*

The workability of concrete is a crucial property that influences its ease of handling, placing, and compaction during construction. Workability is often measured using the slump test. A higher slump value generally indicates better workability. The workability of the concrete mix with different % of glass fibers is conducted, and results are tabulated in Table 7. For the concrete mix with 0% GF, the slump value is 96 mm. This serves as a baseline or reference point for the workability of conventional concrete without any glass fibers. A slump of 96 mm suggests good workability. A slight decrease in slump from the baseline may indicate a mild reduction in workability. However, a slump of 93 mm still suggests that the concrete mix remains relatively workable. The slump further decreases with the addition of 1% glass fibers. This

Table 7 Slump values

% of GF	Slump value (mm)
0	96
0.5	93
1	91
1.5	88
2	85

reduction may be attributed to the influence of the fibers on the flowability of the concrete mix. Despite the decrease, a slump of 91 mm indicates that the mix remains workable, though potentially requiring slight adjustments during placement. The slump continues to decrease as the glass fiber content reaches 2%. This suggests that the workability is affected, and the concrete mix may require more effort during placement and compaction to achieve the desired characteristics.

3.2 Compression Test

The compressive strength of concrete is a critical parameter that reflects its ability to withstand axial loads and is typically assessed at the age of 28 days. Table 8 shows that as the percentage of fiber increases, the compression strength of cube is also increased up to 58.10 N/mm^2 at 1% GF, and later on further increase in fiber content showed decrease in strength.

The addition of 0.5% glass fibers results in an increase in compressive strength compared to the 0% GF. This improvement suggests a positive influence of the glass fibers on the concrete's ability to resist compressive forces. The compressive strength continues to increase up to 58.10 N/mm^2 with the addition of 1% glass fibers. This suggests that the fibers are contributing to enhanced strength characteristics in the concrete mix. While a slight decrease in strength is observed compared to the 1% GF mix, the compressive strength at 1.5% glass fibers remains relatively high. This indicates that there might be an optimal range for glass fiber content, and exceeding it could lead to diminishing returns in terms of compressive strength.

Table 8 Compression strength test results of cubes

Grade of concrete	% of GF	Avg. strength at 7 days (N/mm^2)	Avg. strength at 28 days (N/mm^2)
M50	0	32.33	54.77
M50	0.5	34.42	56.60
M50	1	35.50	58.10
M50	1.5	34.80	57.55
M50	2	34.10	56.80

Fig. 1 Crack pattern of cube at 0% GF

Fig. 2 Crack pattern of cube at 1% GF

There is a further decrease in compressive strength with the addition of 2% glass fibers. This decrease suggests that beyond a certain point, the addition of glass fibers may have diminishing returns on compressive strength or may even lead to some adverse effects. As illustrated in Figs. 1 and 2, the cube specimen exhibits more cracks at 0% GF than it does at 1% addition of fibers. Comparing the specimens to the 1% GF, they fail earlier. Early crack occurrence leads to low strength (Fig. 3).

3.3 Split Tensile Strength Test

In the investigation of the mechanical properties of M50 grade concrete, the focus extends to the split tensile strength at the critical age of 28 days, considering the

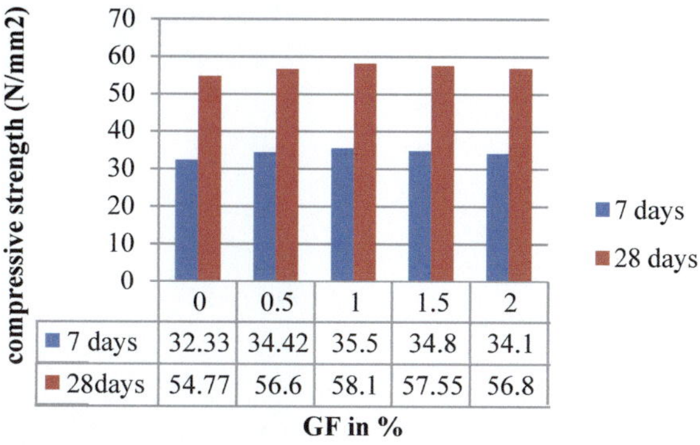

Fig. 3 Bar graph of strength of cubes

incorporation of varying percentages of glass fibers (GF) into the mix, as provided in Table 9. The normal concrete mix, representing 0% glass fibers, exhibited a split tensile strength of 4.8 N/mm^2. Upon introducing 0.5% GF, there was a notable enhancement, resulting in a split tensile strength of 5.2 N/mm^2. Further increments in glass fiber content led to a progressive increase in split tensile strength, reaching 5.9 N/mm^2 for 1% GF. However, a slight reduction was observed at 1.5% GF, with a recorded split tensile strength of 5.7 N/mm^2. Beyond this point, at 2% GF, there was a decrease in split tensile strength to 5.0 N/mm^2. From Fig. 6, it can be said that at 28 days, 23% of cylinder strength is increased when glass fiber are added up to 1%. And when 2% GF are added to the mix, the strength increased only 4.2%. Figure 4 shows the crack pattern of the cylinder with 0.5% GF, and the spread of cracks in this specimen is less compared to normal concrete mix specimen, and more compared to the 1% GF specimen. As the % of fibers further increases above 1% GF to 2% GF, more cracks were observed as seen in Fig. 5. From this, it can be explained that the fibers arrest the cracks at micro level and increase load absorbing capacity of the material.

Table 9 Split tensile strength test results of cylinders

Grade of concrete	% of GF	Avg. strength at 7 days (N/mm^2)	Avg. strength at 28 days (N/mm^2)
M50	0	3.5	4.8
M50	0.5	3.52	5.2
M50	1	3.85	5.9
M50	1.5	3.73	5.7
M50	2	3.34	5.0

Fig. 4 Crack pattern of cylinder at 0.5%

Fig. 5 Crack pattern of cylinder at 2%

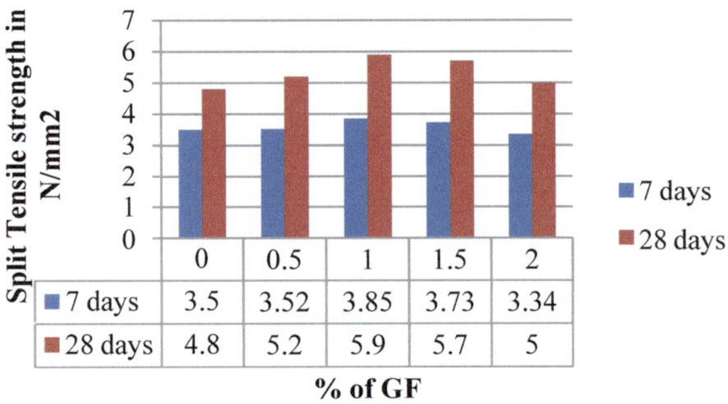

Fig. 6 Bar graph of strength of cylinders

3.4 Flexure Strength Test

The evaluation of flexural strength in M50 grade concrete at 28 days, incorporating varying percentages of glass fibers (GF) into the mix, unveils intriguing insights into the mechanical behavior of the material. The normal concrete, representing 0% GF, displayed a flexural strength 7.11 N/mm^2 of prismatic beam as shown in Table 10. With the introduction of 0.5% GF, there was a notable increase, resulting in a flexural strength of 7.84 N/mm^2, suggesting a positive impact on the material's ability to resist bending forces. The trend continued to demonstrate improvement as the GF content was raised to 1%, reaching a remarkable flexural strength of 9.31 N/mm^2. At this stage, the cracks are considerably controlled resulting in high flexure strength. However, a slight reduction was noted at 1.5% GF, registering a flexural strength of 7.65 N/mm^2, possibly indicating an optimal range for glass fiber content. Beyond this range, with 2% GF, there was a decrease in flexural strength to 6.89 N/mm^2, and at this stage, the beam could not bear the load much longer compared to the 1% GF.

Around 31% of the strength is increased, when 1% of GF are added to the mix as per Fig. 9. This is because the fibers arrest the crack propagation leading to increase in the load carrying capacity. Figure 8 shows the crack pattern of the prismatic beam specimen at 0.5% GF. It was observed that at 0% GF the specimen failed very early and the cracks are more; sudden failure of the beam is observed with loud noise. At 0.5% GF as shown in Fig. 8, the specimen failed slightly late as compared to the 0% GF and complete failure of beam is observed. At 1% GF, the cracks on the beams are very less as compared to the other mix.

Table 10 Flexure strength results

Grade of concrete	% of GF	Avg. strength at 7 days (N/mm^2)	Avg. strength at 28 days (N/mm^2)
M50	0	4.41	7.11
M50	0.5	5.34	7.84
M50	1	6.81	9.31
M50	1.5	5.10	7.65
M50	2	4.39	6.89

Fig. 7 Crack pattern of beam at 1%

Fig. 8 Complete failure of beam at 0.5%

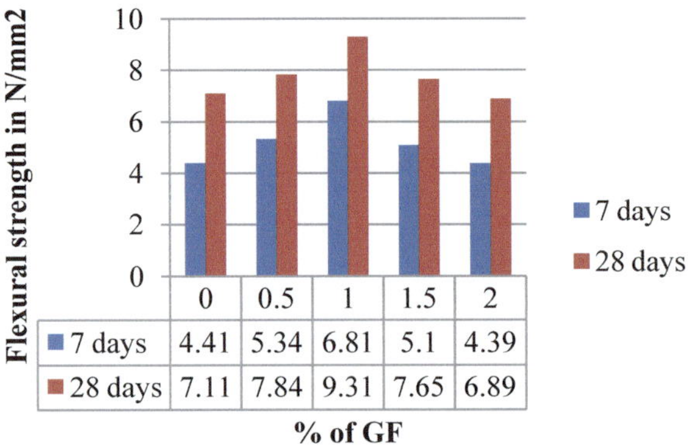

Fig. 9 Bar graph of strength of beam

4 Conclusion

1. It was observed that addition of GF can slightly reduce the workability of the concrete mix. However, by making proper mix design adjustments and appropriate admixtures, the workability can be maintained within acceptable limits.
2. The presence of GF enhances the resistance to cracking, compared to the concrete specimens without GF.
3. The addition of glass fibers to concrete has shown significant improvements in the mechanical characteristics of the material. Max compressive strength was achieved at 1% of GF and obtained 6.1% strength increase.
4. Split tensile strength increased up to 5.9 N/mm^2 at 1% GF, and flexural strength increased up to 9.31 N/mm^2 at 1% GF.
5. The investigation revealed that the optimal dosage of glass fiber in concrete lies in the range of 1% by weight of cement. At lower fiber contents, there is a noticeable improvement in strength, while higher fiber contents results in diminishing returns.
6. It was observed that the % of strength increased in flexural strength and split tensile strength due to addition of fibers is much higher than compressive strength.
7. The use of glass fiber in concrete depends on several factors, including availability and cost, desired performance, and specific project requirements. While there may be slight increase in material costs because of inclusion of fibers, the enhanced durability and reduced maintenance costs associated can lead to long-term cost savings.

Further research and testing are encouraged to explore the optimization of mix design, fiber orientation, and long-term performance of glass fiber-reinforced concrete in various applications.

References

1. Sanjeev J, Sai Nitesh KJN (2020) Mater Today Proc. https://doi.org/10.1016/j.matpr.2020.03.208
2. Yuan Z, Jia Y (2021) Constr Build Mater 266:121. https://doi.org/10.1016/j.conbuildmat.2020.121048
3. Hilles MM, Ziara MM (2019) Eng Sci Technol Int J 22:920–928. https://doi.org/10.1016/j.jestch.2019.01.003.
4. Madhkhan M, Katirai R (2019) Constr Build Mater 225:146–158. https://doi.org/10.1016/j.conbuildmat.2019.07.128
5. Güneyisi E et al (2019) Constr Build Mater 211:349–362. https://doi.org/10.1016/j.conbuildmat.2019.03.087
6. Anandaraj S et al (2018) Constr Build Mater. https://doi.org/10.1016/j.conbuildmat.2018.06.090
7. Kelestemur O etal (2014) Constr Build Mater 60:17–24. https://doi.org/10.1016/j.conbuildmat.2014.02.061

8. Ganta JK et al (2020) Structures 28:956–972. https://doi.org/10.1016/j.istruc.2020.09.042
9. Ahmad S et al (2017) Procedia Eng 173:807–813
10. IS 383 (2016) Specification for coarse and fine aggregates from natural sources for concrete. Bureau of IS, New Delhi, India
11. IS 456 (2000) Plain and reinforced concrete. Bureau of Indian Standards, New Delhi, India
12. IS 10262 (2019) Concrete mix proportioning—guidelines. Bureau of Indian Standards, New Delhi, India
13. IS 516 (1959) Methods of tests for strength of concrete. Bureau of Indian Standards, New Delhi, India
14. IS 2386 (Part 3) (1963) Methods of test for aggregates for concrete. Bureau of IS, New Delhi, India
15. IS 15388. Silica fume specification. BIS, New Delhi, India
16. IS 12269 (2013) Ordinary Portland cement, 53 grade-specification. Bureau of Indian standards, New Delhi, India

Incorporation of Agro Industrial Wastes in Sustainable Slag-Based Alkali-Activated Pavement Concretes

Akhila Sheshadri, Shriram Marathe, and Łukasz Sadowski

Abstract The utilization of alkali-activated binders has emerged as a promising avenue for sustainable construction practices, offering an environmentally friendly alternative to conventional cement-based systems. This study investigates the use of two major agro-wastes, namely rice husk ash (i.e., RHA) and sugarcane bagasse ash (i.e., SBA), as ingredients in producing slag-based alkali-activated concrete, with an emphasis on workability and mechanical performances. The study extends its scope to encompass a comprehensive mix designed to meet the requirements of Pavement Quality Concrete. The investigation involves the systematic replacement of the total binder, ground granulated blast furnace slag (i.e., GGBS), with varying dosages of SBA and RHA, ranging from 0 to 20% at increments of 5%. The findings reveal that the incorporation of SBA up to 10% and RHA up to 15% satisfies not only the mechanical design specifications for Pavement Quality Concrete but also demonstrates the efficacy of agro-industrial waste utilization. This research significantly contributes to sustainable construction practices by effectively reusing agro-industrial waste products through alkali activation, through which it supports the global initiatives to reduce environmental impact and promote eco-friendly construction methods. This study advocates for a paradigm shift toward sustainable and environmentally responsible pavement construction possibilities, fostering a greener and more resilient future for the concrete pavement construction through the integration of agro-industrial wastes in successfully producing superior pavement quality sustainable concretes.

Keywords Alkali activation · Sustainability · Sugarcane bagasse ash · Rice husk ash · Pavement quality concrete

A. Sheshadri · S. Marathe (✉)
Department of Civil Engineering, Nitte (Deemed to be University), NMAM Institute of Technology (NMAMIT), Karkala, Udupi, Karnataka, India
e-mail: ram.nmamit@gmail.com

S. Marathe · Ł. Sadowski
Department of Materials Engineering and Construction Processes, Wrocław University of Science and Technology, Wrocław, Poland

Y. K. Guruprasad et al. (eds.), *Recent Advances in Infrastructure Development and Management—Volume 1*, Lecture Notes in Civil Engineering 570,
https://doi.org/10.1007/978-981-96-1554-4_14

1 Introduction

On a global scale, agricultural and food wastes account for 44% of the total waste generated, while industrial and other wastes contribute to the remaining 56%. For several decades, agro-industrial waste materials rich in silica, including ground granulated blast furnace slag (GGBS), rice husk ash (RHA), silica fume (SF), sugarcane bagasse ash (SBA), and fly ash, have been extensively used in concrete production [1]. The agro and industrial sectors produce large amounts of spin-off waste, which are often discarded in landfills. The practice of burning agricultural waste has been a long-standing environmental issue due to its adverse impact on air quality and the associated health risks. Through a controlled combustion process known as calcination, conducted at temperatures between 700 and 900 °C, the unstructured silica in agro-industrial waste is converted into extremely reactive silica, while the residues are preserved. However, heating silica above 900 °C results in the formation of non-reactive crystalline silica [2, 22]. Studies have shown that these by-products offer numerous advantages, including improved durability and strength properties, reduced construction costs through decreased cement usage, environmental benefits by lowering carbon dioxide emissions, and facilitating the easier disposal of waste materials that would otherwise affect the eco-system [1, 23].

Conversely, "Ordinary Portland Cement," i.e., OPC which is a binding material, has been utilized in the construction of concrete and other applications for approximately 20 decades. Over time, the popularity of OPC has grown, resulting in increased production and subsequently causing environmental issues [4]. The environmental issues arise from the significant contribution of cement production toward global CO_2 emissions, accounting for approximately 8–10% of the total. Consequently, any future increase in cement production will result in a corresponding rise in atmospheric CO_2 levels. Given the environmental challenges posed by cement production, it is imperative to reduce the secretion of CO_2 into the atmosphere. This can be accomplished by efficiently employing various industrial and agro by-products. Incorporating these wastes helps minimize the disposal of industrial and agricultural waste, while also fostering the production of sustainable goods in the global market [10, 18]. In line with the conventional OPC-based cement system, alkali-activated cements or binders (i.e., AAB) are now being employed, exhibiting similar properties to their traditional counterparts [20, 19]. Through alkali activation, various raw materials (especially, industrial waste materials) would be renovated into helpful cementitious complexes with minimal environmental impact [3]. Alkali-activating chemical materials (such as NaOH, KOH, Na_2SiO_3, and K_2SiO_3) create an alkaline environment that facilitates the reaction and formation of low CO_2 cementing binders in AAB. Alkalis are readily accessible, and researchers have discovered abundant sources of SiO_2 and Al_2O_3 in diverse agro and industrial waste materials [24]. Extensive research has been conducted and continues to explore the use of agro-industrial wastes in concrete production. Integrating these wastes into the construction industry offers a substantial opportunity to condense or repurpose the waste generated by both industrial and agro segments [15, 14]. This method provides a sustainable solution to the challenges coupled with the disposal of agro-industrial wastes in landfills [16].

The construction industry typically prioritizes locally available raw materials over those located farther away to minimize transportation costs. Hence, it is highly advisable to source agro and industrial wastes from close-by locations for use in construction. The collective use of these by-products fosters a symbiotic relationship between the agro and industrial sectors [10]. This approach represents a sustainable and viable option, particularly in the context of AAB, which involve a combination of industrial and agro-waste materials to produce high strength concretes. While extensive studies exist on the occupy of industrial and agro wastes as pozzolanas in OPC-based concretes, the utilization of these products solely in AABs remains an area with limited exploration. With this gap in mind, the research objectives are formulated to provide scientific insights into the engineering properties of "Pavement Quality Alkali-Activated Concrete" (PAAC) developed using GGBS (derived from slag, an industrial waste) as main precursor material which is an industrial by-product, and separately, RHA and SBA as agro by-products. This article presents and discusses the the fresh and hardened properties of the novel and sustainable AAC designed for highway pavement applications.

2 Materials and Methodology

A procuration of the GGBS for this investigation was made from Jindal Steel Works (JSW), situated in Toranagallu, Karnataka. The substance in question functions as the primary precursor binder matter in the making process of AAC. The specific gravity of 2.89 was determined through testing. The source of the SBA and RHA, which had respective specific gravities of 2.28 and 2.19, was Jamnagar, Gujarat. The specific surface areas of GGBS, RHA, and SBA were 0.36 m^2/g, 139 m^2/g, and 1.79 m^2/g, respectively. Coarse and fine aggregates were secured from crushed granite and river bed, respectively, both of which were sourced locally in Nitte village. In accordance with the specifications of IS 383 [6] and with a specific gravity of 2.60, fineness modulus of 3.98, and water absorption of 3.21%, the fine aggregate (FA) belonged to zone II. The specific gravity of coarse aggregates (CA) was 2.68, their water absorption was 0.82%, and they possessed a fineness mod. of 6.13. Coarse aggregates utilized in this study consisted of a utmost size of 20 mm, complying with IRC:44–2017 [5]. Additionally, all aggregates, both coarse and fine, were maintained in a saturated surface dry (i.e., SSD) state throughout the study. Every necessary examination of the materials was carried out in adherence to the pertinent Indian Standards. Local suppliers in Mangaluru city provided water glass (Liquid Sodium Silicate, Na_2SiO_3, or "LSS") and solid sodium hydroxide (NaOH) flakes with an approximate 97% purity. LSS possessed a specific gravity of 1.57 and an activator modulus (Ms) of 2.23, which denotes the ratio of SiO_2 to Na_2O. In order to attain the intended Ms value, the alkaline activating chemical was geared up by dissolving

NaOH flakes in LSS. To adjust the water-to-binder ratio, laboratory tap water was subsequently introduced [15].

The mixture proportion design for standard cement concrete used in highways was designed using the methods recommended for building conventional OPC-based Pavement Quality Concrete (PQC) with an intended compressive strength of 55 MPa and a workability range of 25–75 mm slump, adopting the requirements of IRC: 44–2017. This process is in harmony with the typical specifications for concrete formulation as per the Indian standard IS: 10262. Subsequent modifications were made to the initial mix ratios to formulate the necessary slag-inclusive Pavement Quality Alkali-Activated Concrete (PAAC) mix, drawing upon existing scholarly literature works [12, 13, 17]. The study maintained a constant Na_2O dosage of 4.0% of total binder, Ms value of 1.25 for the alkaline activator chemical, and the final "water-to-binder" (i.e., w/b) proportion was set at 0.36. The initial formulation specifics for the slag-enriched AAC mix are presented in Table 1, which satisfies both the strength and workability requirements as per the mix design.

Subsequent research aimed to levy the impact of select agro-wastes (RHA and SBA) on the properties of PAAC mixtures. The standard binder (GGBS) quantity was decreased, substituting it with incremental additions of agro-waste in 5% intervals, varying from 0 to 20%. Tests for workability, as measured by the slump test, density, water absorption, and compressive strength assessments were performed at the end of 28th day to determine the optimal agro-waste content for each type independently. PAAC mixes containing RHA were denoted by "R," and those with SBA were designated as "S," with the subsequent number indicating the percentage replacement of GGBS by the corresponding agro-waste. The specifics of this nomenclature alongside preliminary results are presented in Table 2. The chosen agro-waste dosages were then used for additional studies, which included measuring split tensile and flexural strengths. These measured properties were subsequently compared with the baseline PAAC mix containing 100% GGBS as a binder.

According to the mix proportion design that was established, the materials were carefully weighed and batched before being completely combined to achieve a uniform combination. The specimens employed in the aforementioned study include cubes of face 10 cm, cylinders with the dimensions of 10 cm (dia.) × 20 cm (ht.), and beams measuring 10 cm (ht.) × 10 cm (breadth) × 50 cm (length). The removal of the cast samples was conducted twenty-four hours following the completion of the casting process [21. The specimens made of hardened cast PAAC were subsequently exposed to a period of moist curing for the earliest three days, chased by air curing under laboratory conditions awaiting the time of testing. Once the specimens had completed the curing process, the cubes and beams underwent testing to

Table 1 Preliminary mix proportion for slag-based alkali-activated PQC

Notation	GGBS	Ms	w/b	LSS	NaOH	CA	FA	Additional water
Unit	kg	Ratio	Ratio	kg	kg	kg	kg	kg
PAAC	438	1.25	0.36	75.14	11.18	1137	662	118.2

Table 2 Preliminary test results of investigations with RHA and SBA in PQC mixes

Mix ID	GGBS (%)	SBA (%)	RHA (%)	Slump (mm)	Compressive strength (MPa)	Density (kg/m^3)	Water absorption (%)
PAAC	100	0	0	50	82	2480	4.32
S-5	95	5	–	43	71	2421	5.28
S-10	90	10	–	31	62	2408	6.68
S-15	85	15	–	18	56	2390	7.79
S-20	80	20	–	9	52	2371	9.21
R-5	95	–	5	47	74	2469	5.12
R-10	90	–	10	41	68	2412	6.27
R-15	85	–	15	32	64	2432	7.08
R-20	80	–	20	21	59	2401	8.52

determine their compressive strength and 4-point flexural strength, respectively, in accordance with the specifications outlined in IS 516 [8]. Additionally, the cylinders were subjected to testing to assess their split tensile strength, following the guidelines provided in IS 5816 [9]. The test results were recorded during a period of 28 days of curing, with each data point representing the average of three trials.

3 Results and Discussions

3.1 Results of Preliminary Investigations on Agro Waste Materials

Table 2 meticulously outlines the initial test outcomes for the AAC mixes, showcasing various proportions of agro-wastes used to replace GGBS. Rigorous quality control measures were upheld throughout the experimental procedures, resulting in minimal standard deviations in the test samples' results. Specifically, variations in compressive strength outcomes were modest, ranging from 1.22 to 5.85%, comfortably within the permissible deviation threshold of 15% as stipulated by IS: 456–2000 [7]. Furthermore, variations in water absorption tests were below 3%, and those for slump tests ranged from 0 to 3.1%.

The reference mix PAAC demonstrated suitable workability with a 52 mm slump However, introducing SBA reduced workability significantly, with the S-20 mix recording a slump of only 9 mm. The S-10 mix, with a slump of 31 mm, falls within the design criteria of 25–75 mm, indicating acceptable workability for practical applications. Further, the target compressive strength for this study was set at 63 MPa (as the intended strength was 55 MPa). While PAAC exceeded this with 82 MPa, the S-10 mix achieved a near-target strength of 62 MPa. This suggests that a 10% replacement of GGBS with SBA is close to meeting the structural strength requirements and is the most promising candidate among the tested mixes for further investigation.

This can be linked to the mechanism described in the literature, where the increased constancy, set-time, and reduced soundness and flowability are influenced by the hygroscopic nature and bigger surface area of SBA. Additionally, the compressive strength was found to improve with age but decrease with higher SBA content at early ages [11].

The density of the concrete dwindled from 2480 kg/m^3 in the PAAC mix to 2371 kg/m^3 in the S-20 mix, indicating a lighter concrete with increased SBA content. Concurrently, water absorption increased from 4.32 to 9.21%, suggesting a correlation between SBA addition and porosity. Particularly, the S-10 mix, at 6.68%, shows an increased rate compared to PAAC but may still be manageable within certain construction parameters. Overall, considering all preliminary aspects, the S-10 mix emerges as a compelling option for more detailed examination. It not only meets the target strength but also complies with the required slump range, suggesting a balanced compromise between sustainability through partial GGBS replacement and maintaining necessary concrete properties.

Further, the results from the mixes with RHA replacement indicate that upon gradual inclusion of RHA, the slump decreased, with the R-15 mix displaying a slump of 32 mm which is still within the design parameter of 25–75 mm, suggesting adequate workability for pavement construction targets. The R-15 mix shows a compressive potency of 64 MPa, composing it a viable candidate for further investigation due to its proximity to the desired strength level. Further, a decrease in density from 2480 kg/m^3 in PAAC to 2432 kg/m^3 in R-15 was observed, hinting at the potential to achieve lighter concrete structures. Additionally, water absorption increased to 7.08% in the R-15 mix but remained within manageable limits. However, when additional RHA was added, we noticed a reduction in strength beyond acceptable limits and a decrease in slump values. This can be linked to the mechanism discussed earlier [1], where the configuration of alkali-silica reaction gels caused spreading out and cracking, leading to the observed decrease in strength and workability beyond the addition of RHA more than a desirable limits. Hence, through the consideration of the target strength and acceptable slump range, the R-15 mix emerges as the most suitable composition for further detailed studies.

3.2 Results of Mechanical Properties of Selected Pavement Quality Mixes

Strength evaluations at 7 and 28 days were performed on the reference Pavement Quality Alkali-Activated Concrete blend (i.e., PAAC mix), along with two chosen mixes: the S-10 mix comprising 10% SBA and 90% GGBS, and the R-15 mix containing 15% RHA and 85% GGBS from the preliminary investigations. The

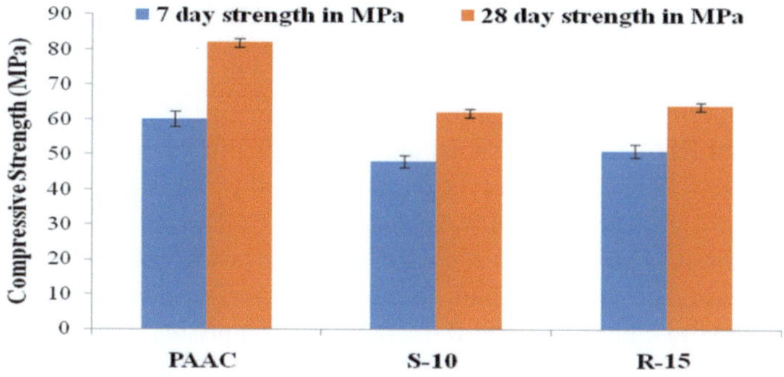

Fig. 1 Compressive strength of selected PQC mixes

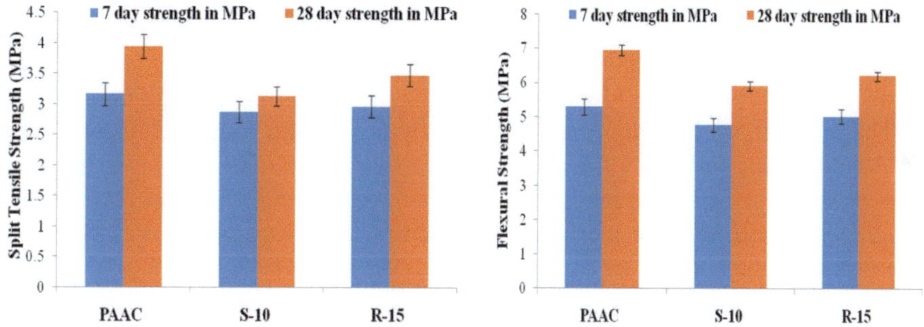

Fig. 2 Splitting tensile and flexural behavior of selected PQC mixes

outcomes of the flexural, split tensile, and compressive potency assessments are illustrated in Figs. 1 and 2.

The experimental results reveal important insights into the early-age and matured mechanical properties of concrete mixes under consideration. The S-10 mix demonstrated notable improvements in mechanical properties over time. The compressive strength showed a 29.17% increase from 7 to 28 days, while the split tensile might exhibited a 9.06% increase over the same period. Additionally, the flexural strength displayed a 23.85% increase from 7 to 28 days. However, at 28 days, there was a 24.39% decrease in compressive strength, a 20.66% decrease in split tensile strength, and a 14.82% decrease in flexural potency judged to the reference PAAC mix. In contrast, the R-15 mix displayed consistent and promising performance in mechanical properties. It exhibited a 25.49% increase in compressive power from seven days to twenty-eight days, along with a 17.57% boost in split tensile power and a 23.71% raise in flexural strength over the same period. At 28 days, there was a 21.95% decline in compressive strength, an 11.79% decrease in split tensile strength, and a 10.65% decrease in flexural power compared to the allusion PAAC mix.

4 Conclusions and Future Scopes

The present experimental investigation thoroughly examined the potential utilization of the SBA and RHA as partial substitutes for GGBS in the alkali-activated PQC systems. The key discoveries from the experimentations are as follows:

The S-10 mix, incorporating 10% SBA and 90% GGBS, demonstrated a notable improvement in mechanical properties over time. However, at 28 days, there was a decrease in flexural compressive, and split tensile vigors compared to the reference PAAC mix. On the other hand, the R-15 mixture, comprising 15% RHA and 85% GGBS, exhibited consistent and encouraging mechanical properties. The material demonstrated notable enhancements in mechanical (engineering) strengths between the seven-day and twenty-eight-day periods, indicating its potential for further investigation and utilization in pavement construction for highway applications.

The utilization of agricultural by-products, specifically SBA and RHA, not only aids in the preservation of the environment through the recycling of agro-based wastes but also aligns with the worldwide initiative for sustainable construction practices.

In light of the promising results, future studies should focus on investigating the durability properties of these concrete mixes to assess their long-term durability performance and also the microstructural investigations to enhance the understanding. Furthermore, the incorporation of fibers and other composites could be explored to augment the mechanical and functional behavior of the developed concretes, paving the way for advanced applications in high-quality pavement construction projects. Overall, the study signifies the potential of utilizing SBA and RHA in GGBS-based alkali-activated cementitious pavement concrete systems, underlining their viability as sustainable alternatives to traditional concretes and pointing toward a promising avenue for future research and development in pavement construction materials.

Acknowledgements The experimental work was conducted in the Civil Engineering Department laboratory at NMAMIT, Nitte. This research was financially supported by the NITTE Education Trust through the NURF research grant (Sanction number: *N/RG/NUFR1/NMAMIT/2022/09* dated 28.12.2022). The authors extend their honest gratitude to the Head of the Department (Civil), the Dean of Research and Development, and the Principal of NMAMIT-Nitte. Appreciation is also extended to the Directorate of Nitte (DU), and Nitte Education Trust (R) for their timely financial and logistical support, which was essential for the successful completion of this research work. Also, the authors love to thank the undergraduate students (B.E. Civil) for their help in conduction of the part of laboratory experimental investigations.

References

1. Blesson S, Rao AU (2023) Agro-industrial-based wastes as supplementary cementitious or alkali-activated binder material: a comprehensive review. Innov Infrastruct Solut 8:4. https://doi.org/10.1007/s41062-023-01096-8
2. Clark MW et al (2017) High-efficiency cogeneration boiler bagasse-ash geochemistry and mineralogical change effects on the potential reuse in synthetic zeolites, geopolymers, cements, mortars, and concretes. Heliyon 3:4. https://doi.org/10.1016/j.heliyon.2017.e00294

3. Falah M et al (2020) Effects of activator properties and curing conditions on alkali-activation of low-alumina mine tailings. Waste Biomass Valorization 11(9):5027–5039. https://doi.org/10.1007/s12649-019-00781-z
4. Gupta S, Kashani A (2021) Utilization of biochar from unwashed peanut shell in cementitious building materials—effect on early age properties and environmental benefits. Fuel Process Technol 218. https://doi.org/10.1016/j.fuproc.2021.106841.
5. IRC:44 (2017) Guidelines for cement concrete mix design for pavements
6. IS:383 (2016) Coarse and fine aggregate for concrete—specification
7. IS:456 (2000) Plain and reinforced concrete—code of practice
8. IS 516-1959 (1959) Indian standard methods of tests- for strength of concrete. Bureau of Indian Standards, New Delhi
9. IS 5816-1999 (1999) Splitting tensile strength of concrete—method of test (first revision). Bureau of Indian Standards, New Delhi
10. Jittin V et al (2021) Synergistic use of typical agricultural and industrial by-products for ternary cement: a pathway for locally available resource utilisation. J Clean Prod 279. https://doi.org/10.1016/j.jclepro.2020.123448
11. Kathirvel P et al (2020) Effect of partial replacement of ground granulated blast furnace slag with sugarcane bagasse ash as source material in the production of geopolymer concrete. Medziagotyra 26(4):477–481. https://doi.org/10.5755/j01.ms.26.4.23602
12. Marathe S et al (2021) Durability and microstructure studies on slag-fly ash-glass powder based alkali activated pavement quality concrete mixes. Constr Build Mater 287(123047):1–19. https://doi.org/10.1016/j.conbuildmat.2021.123047
13. Marathe S et al (2020) Performance of slag-fly ash based alkali activated concrete for paver applications utilizing powdered waste glass as a binding ingredient. Int J Pavement Res Technol. https://doi.org/10.1007/s42947-020-0173-2
14. Marathe S, Nieświec M, Gronostajska B (2024b). Alkali-Activated Permeable concretes with agro-industrial wastes for a sustainable built environment. Materials, 18(1):87. https://doi.org/10.3390/ma18010087
15. Marathe S, Sheshadri A, Sadowski Ł (2024a). Agro-industrial waste utilization in air-cured alkali-activated pavement composites: Properties, micro-structural insights and life cycle impacts. Cleaner Materials 14:100281. https://doi.org/10.1016/j.clema.2024.100281
16. Memon MJ et al (2021) Production of eco-friendly concrete incorporating rice husk ash and polypropylene fibres. Environ Sci Pollut Res 28(29):39168–39184. https://doi.org/10.1007/s11356-021-13418-3
17. Mithun BM, Narasimhan MC (2015) Performance of alkali activated slag concrete mixes incorporating copper slag as fine aggregate. J Clean Prod 112:837–844. https://doi.org/10.1016/j.jclepro.2015.06.026
18. Poudyal L, Adhikari K (2021) Environmental sustainability in cement industry: an integrated approach for green and economical cement production. Resour Environ Sustain 4. https://doi.org/10.1016/j.resenv.2021.100024
19. Sheshadri A, Marathe S, Sadowski Ł (2024b) Development of sustainable high strength slag based alkali activated pavement quality concrete using agro-industrial wastes: properties and life cycle analysis. Int J Pavement Eng 25(1). https://doi.org/10.1080/10298436.2024.2410953
20. Sheshadri A, Marathe S, Bettadapura Manjunath M, Jayasimhan A, Sadowski Ł (2024a) Effective utilization of foundry waste as aggregates in developing eco-friendly alkali-activated and conventional concretes for sustainable pavement infrastructure. Pract Period Struct Des Constr 29(3). https://doi.org/10.1061/PPSCFX.SCENG-1501
21. Sheshadri A, Marathe S, Lobo S, Sadowski Ł (2025). Production of eco-friendly alkali-activated concrete for sustainable highway infrastructure by effective utilization of recycled construction and foundry wastes. Comptes Rendus Chimie 28(G1):185-198. https://doi.org/10.5802/crchim.380
22. Thomas BS et al (2021) Biomass ashes from agricultural wastes as supplementary cementitious materials or aggregate replacement in cement/geopolymer concrete: a comprehensive review. J Build Eng 40. https://doi.org/10.1016/j.jobe.2021.102332

23. Thomas BS (2018) Green concrete partially comprised of rice husk ash as a supplementary cementitious material—a comprehensive review. Renew Sustain Energy Rev 82:3913–3923. https://doi.org/10.1016/j.rser.2017.10.081
24. Usha S et al (2014) Geopolymer binder from industrial wastes: a review. Int J Civ Eng Technol 5(12):219–225

Innovative Stabilization of Blended Soils for Sustainable Road Foundations Using Chemically Activated Binder

B. G. Megha Mayuri and Shriram Marathe

Abstract In the context of road subgrade stabilization, this research explores into the innovative application of geopolymer cement (GC) as chemical activator for enhancing the engineering properties of a blended soil composition prevalent in coastal Karnataka region, namely Lateritic Lithomarge, comprising lateritic soil and lithomargic (shedi) soil in various levels. The investigation draws inspiration from prior studies on lateritic and shedi soils in the region, recognizing their unique challenges and susceptibilities. Utilizing blended soil, the study explores the effectiveness of GC, synthesizing slag and fly ash as integral binder precursors. Geopolymerization, facilitated by an aqueous solution of sodium silicate-hydroxide mix, serves as the chemical activating agent. This stabilization process involves cementation and chemical alterations within the soil matrix induced by GC. Enhanced compaction properties, unconfined compressive strength (UCS), and California Bearing Ratio (CBR) values are meticulously examined and elucidated. By the trials, the optimal GC dosage was found to be 15–20% for the selected blended soil. Furthermore, the study extends its insights into pavement composition design (both flexible and rigid pavement), aligning with Indian standards, presenting a comparative analysis of the performance disparities between stabilized and unstabilized pavement subgrades. By focusing on the GC as a sustainable solution, the study contributes to the evolving landscape of sustainable pavement design in the realm of road infrastructure. The incorporation of blast furnace slag and fly ash in GC underscores its sustainable attributes, aligning with global efforts toward eco-friendly construction practices and resource optimization.

Keywords Alkali activation · Sustainability · CBR · Lateritic-Shedi · Pavement design

B. G. Megha Mayuri
Department of Civil Engineering, Mangalore Institute of Technology and Engineering (MITE), Moodabidre, Karnataka, India

S. Marathe (✉)
Department of Civil Engineering, Nitte (Deemed to Be University), NMAM Institute of Technology (NMAMIT), Karkala, Karnataka, India
e-mail: ram.nmamit@gmail.com

Y. K. Guruprasad et al. (eds.), *Recent Advances in Infrastructure Development and Management—Volume 1*, Lecture Notes in Civil Engineering 570,
https://doi.org/10.1007/978-981-96-1554-4_15

1 Introduction

Conventional bituminous roads are deteriorating more and more as a result of the volume of heavy traffic and the harsh weather. The pavements' design life will be shortened as a result of these new challenges. As a result, efforts are being made to find improved construction technologies that can address current issues and guarantee the pavement's good performance. When it comes to the general stability of the pavement, the soil subgrade course is universally acknowledged as being the most crucial layer [10]. Subgrade stabilization is the prevailing technique employed to alter the properties of vulnerable soil subgrades and enhance the operational efficiency of civil engineering platforms [3]. The majority of the stabilizing agents utilized are chemical substances, industrial residues, cementing agents, and so forth; the utilization of waste readily available in the vicinity for stabilization purposes could alleviate the "solid dissipate disposal" issues faced by the local government. Planning officials in engineering sector consider industrial waste management to be a significant issue due to the growing volume of demolition detritus, the dearth of suitable waste disposal sites, the escalation of transportation and waste expenses, and most significantly, the mounting apprehension regarding effluent and ecological degradation [8, 16, 22]. Considerable research attention has been devoted to the production of a type of "Alkali-Activated Cement", called "Geopolymer Cement", i.e., GC from industrial by-products due to the potential to use these substances to produce an inexpensive and environmentally sustainable cement-like stabilization agent. A chemical process known as the activation of waste materials in alkaline media, including fly ash and slag, enables the transformation of partially or completely amorphous and/or metamorphic vitreous structures into an exceptionally compact solidified stabilizing material. The present research focuses on a novel method of soil modification that is friendly to the surrounding ecosystem.

In the current study, Fly-ash and Slag industrial waste materials are employed to produce geopolymetric cementing agents for the stabilization of Lithomargic Soil (commonly called as Shedi soil, here it is referred as SH) and Lateritic Soil (i.e., LS) that are found in the coastal Karnataka region. In abundance, along the western coast of southern India is Shedi soil. This soil is troublesome because it is especially vulnerable to water and loses a considerable percentage of its strength when it is wet [11]. This makes the soil difficult to work with. With a greater silt content, these soils are typically classified as "sandy silt" or "silty sand" kinds. These high levels of silt contribute to a wide range of problems, including foundation failures, slope failures, embankment collapses, uneven settlements, and many more. Soil stabilization of various types is therefore necessary to fulfill the engineering needs. Soil stabilization is an age-old technique for improving poor soils and making them match specific engineering project criteria. Sand, cement, pond ash, and several other additions are frequently utilized in the process of achieving stabilization [19].

In tropical and subtropical regions that are characterized by copious amounts of rainfall and substantial erosion, lateritic soil is frequently found. Many regions of the continents, including Australia, South and Southeast Asia, Africa, and South

America, have a lot of lateritic soil. The geographic region between the Tropics of Cancer and Capricorn contains the great bulk of land regions that are home to laterites. In particular, lateritic soil is found in a number of nations, including the nation of Thailand, Vietnam, Laos, the Philippines, Cambodia, India, Indonesia, the republic of Malaysia, and Brazil. This phenomenon is usually found in areas with a strong monsoon conditions, where scorching temps and a lot of thunderstorms encourage the massive chemical transformation of the foundational rocks, which causes silica and various other mineral deposits to dissolve out. Consequently, a dense layer consisting of hydroxide compounds, iron, and oxides of aluminum is left over. This thick blanket of soil known as "laterite soil" is easily recognized by its reddish-brown color and tendency to consolidate into hard lumps [21]. In developing nations, these LS is widely used for important building projects. These soils are the predominant foundational structure of the soil in the coastal part of Karnataka region [13]. In most of the cases in the field, LS is available as a blend with SH are observed at highway building sites in these areas. When building pavement, it is necessary to have a thorough awareness of the properties and performances of these types of soils [4]. There are limited studies on the stabilization of blended soils of this area, and hence, an attempt was made to prepare and stabilize these blended soils in the current scope of the work.

Limited research has been done on the application of GC for stabilizing the soil under examination. Therefore, an attempt was made to use ground granulated blast furnace slag (GGBS) and fly ash (FA) as cementing admixures in the present study and chemically activate the binders to create GC. The ground profile is shown in Fig. 1, with pure lateritic, blended, and pure shedi soil strata located in the top, middle, and bottom sectors, respectively. Numerous studies have demonstrated the use of Portland Cement (OPC) as a basic additive to stabilize locally available lateritic and lithomargic soils and effectively carry out the application to the appropriate levels [9, 13, 17, 18]. OPC primarily causes particular soil particles to bind together, which results in the "chemical stabilization" of the original soil [14]. This process is also similar to in the case of GC [2, 16, 19]. Hence, the main objective of this study is to investigate the effect of geopolymeric cement in the stabilized shedi-lateritic soil blends through geotechnical investigations.

2 Materials and Methods

The pure LS and SH for this investigation were procured locally. The Udupi Power Corporation Limited (UPCL) Plant provided FA, while Jindal Steel Bellary provided GGBS. The necessary chemicals for geopolymerization, namely NaOH and LSS (Na_2SiO_3), were purchased from a local supplier. These source materials' specific gravities 1.40, 2.32, and 2.89 for the aqueous activating chemical solution, FA, and GGBS, respectively, were measured. In addition, the applicable BIS code of practice was followed when conducting the experiments to evaluate the stabilizer's effect on geotechnical properties.

Fig. 1 Cutting showing a typical profile of lateritic and shedi soil of south Canara

For the current investigation, three types of blended soils were prepared in the laboratory by using the procured, pure LS and SH soils through manual mixing. Accordingly, 75% LS + 25% SH, 50% LS + 50% SH, and 25% LS + 75% SH were the respective designations. Energy dispersive X-ray (EDX) results of the soil samples revealed the presence of oxygen (about 52%) and silicon (approximately 24%) elements in these mixtures. The core geotechnical parameters of the blended LS and SH are offered in Table 1. Figure 2 represents the particle size distribution of all the blended soils.

First, the produced LS-SH blends were subjected to preliminary testing to evaluate the soil properties. The optimal dosage of "activator modulus" ("Ms value", i.e., "SiO_2/Na_2O ratio") and the "Na_2O-to-binder" ratio were selected with the help of the literatures [1, 20]. The Ms value remained constant at 1.25, and the ratio of "Na_2O-to-binder" was consistently maintained at 4.0% for all types of blended soils throughout the study [12]. Furthermore, an examination of the compaction parameters (specifically, maximum standard proctor density) was conducted subsequent to

Table 1 Fundamental geotechnical properties of soil blends

Soil features	Lateritic-shedi soil blends			Test reference (BIS code)
	75% LS + 25% SH	50% LS + 50% SH	25% LS + 75% SH	
Sp. Gravity of solids (G)	2.55	2.51	2.45	IS_2720-3:1980
Shrinkage limit (%)	21.3	22.2	22.2	IS_2720-5: 1985; IS_2720-6: 1972
Plastic limit (%)	26.4	27.7	28.3	
Liquid limit (%)	49.3	50.9	52.4	
Standard proctor test				
(i) OMC	16.70%	17.60%	19.80%	IS_2720-7: 1980
(ii) MDD	1.78 g/cc	1.63 g/cc	1.59 g/cc	
Modified proctor test				
(i) OMC	14.7%	15.7%	16.3%	IS_2720-8: 1983
(ii) MDD	1.88 g/cc	1.81 g/cc	1.74 g/cc	
Soil classification (IS)	SM	MH	MH	IS1498: 1970
UCS (kPa)	151.0	90.0	67.0	IS2720-10: 1991
Permeability (k) cm/s	1.79×10^{-7}	1.44×10^{-7}	0.94×10^{-8}	IS2720-17:1987
CBR (soaked) (%)	8.00	6.5	3.8	IS2720–16 1987

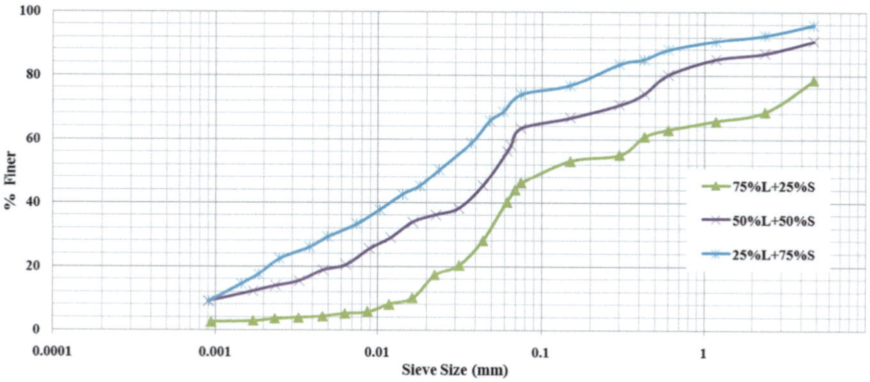

Fig. 2 Grain size distribution of lateritic-shedi blended soils

the adjustment of binder dosages in increments of 5% at regular intervals, ranging from 0 to 25%. In the context of nomenclature M, the numerical suffix indicates the proportion of binder incorporated in relation to the soil's weight. To illustrate, the designation M-10 signifies the inclusion of 10% GC in the soil mixture. In each mixture, the proportion of GGBS in the overall binder content was consistently maintained at 75%, while the proportion of FA was kept at 25%. As an illustration, in the composition denoted as M-10, the proportions of GGBS and FA were 7.5% and 2.5% correspondingly. The "CBR value" and "UCS" values were determined for selected combinations, and the results were reported as the average of three consecutive sample test results. The design recommendations for pavement were implemented in accordance with the specifications for low-volume roads, as outlined in the IRC-SP:62 standards for low-volume concrete pavements [7] and IRC:SP-72 for low-volume flexible pavements [6].

3 Results and Discussions

Standard proctor compaction test findings (mean of 3 test results) for the GC-treated blended soil mixes are shown in Fig. 3. The results show that when the GC measurement increases, the maximum dry density (MDD) of the mix increases up to a certain point, and then, additional GC increment leads to a decrease in the MDD. The Optimum Moisture Content (OMC) value of the soil that was produced after GC stabilization, another compaction metric, did not exhibit any clear pattern either. Due to the fact that the increase in the binder dosage caused the soil particles to have a tendency to bind themselves from the separated single grained structure, there was an uneven requirement for the amount of water needed to achieve the highest level of firmness that was observed during the compaction testing. According to the findings of the standard proctor test, the optimal MDD was achieved at a dosage of 15% GC, whereas the optimal dosage for the other two blended soils was 20% GC. By comparing the blends with and without stabilizer dosages, the MDD was increased by 23% for the 75% LS + 25% SH mixture. Likewise for "50% LS + 50% SH" and "25% LS + 75% SH" blend, there was a respective growth of 29% and 35% of preliminary MDD value. As a result, the relevant dosages that were selected were taken into consideration for the corresponding blended soils in order to conduct further CBR and UCS studies.

Furthermore, only the best three combinations were subjected to CBR and UCS testing. Optimal mixes, for example, 15% binder (i.e., M-15 mix) for 75% LS + 25% SH blend and 20% binder (i.e., M-20 mix) for "50% LS + 50% SH" and "25% LS + 75% SH" blend are distinguished from reference mixes with 0% GC (with 0% binder) by the higher percentages of the target materials. The outcomes of the UCS and CBR testing are presented in Table 2. According to the results of the tests, the CBR value in 75% LS + 25% SH blended soil could be increased to 213% and the UCS value to about 117% by incorporating GC; in 50% LS + 50% SH blended soil, the CBR and UCS value could be increased up to 188% by

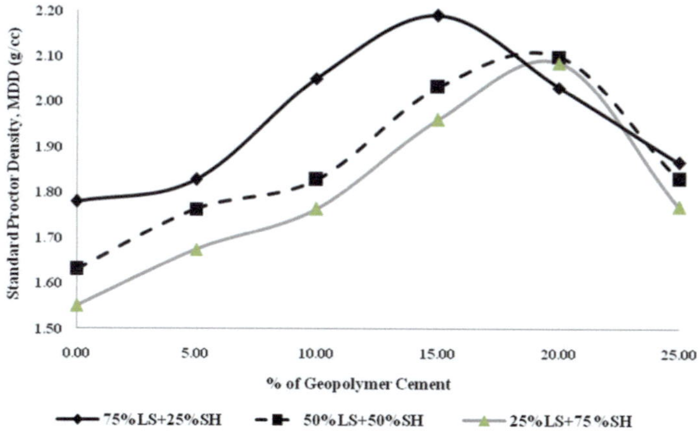

Fig. 3 Variation of standard proctor density with GC content for blended soils

incorporating 20% GC; and in the soil blend designated as 25% LS + 50% SH, the CBR and UCS values could be increased by 289% and 210% by incorporating 20% of GC. Previous studies on specific types of soil have reported an increase in the material's "post-peak strength" in response to a decrease in the stiffness merit, and they additionally discovered a reduction in the strength accomplishment upon an increase in the dosage of GC, with the latter finding being attributed to a reduction in the "elastic energy capacities" of the resulting modified soil [23]. Previous research has suggested that the reaction of the CaO content of the binder with "silicates and aluminates" obtainable from geopolymeric activating solvents may be responsible for the increased strength observed with GC stabilization. This results in the formation of "calcium aluminosilicate hydrates" [5, 15, 24].

In the context of rigid pavement design applications, the determination of the modulus of subgrade reaction (k) can be achieved by utilizing the CBR value (soaked) of the subgrade soil under consideration as per the guidelines of IRC-SP-62 recommendations [7]. Given the observed increase in the average California Bearing Ratio of blended soil subgrades amended with GC, it is reasonable to explore the potential for developing a sustainable and cost-effective design thickness for concrete pavements for low-volume roads. The results of k-value obtained by the linear interpolation and the corresponding % improvement have been shown in Table 3 for the ready reference.

Table 2 CBR and UCS results of selected GC modified blended soil

Mix ID	% GC	75% LS + 25% SH		50% LS + 50% SH		25% LS + 75% SH	
		CBR (%)	UCS (kPa)	CBR (%)	UCS (kPa)	CBR (%)	UCS (kPa)
M-0	0	9.00	151	6.50	90.0	3.80	67
M-15	15	28.20	327.8	–	–	–	–
M-20	20	–	–	18.70	260	14.70%	208

Table 3 Effect on modulus of subgrade reaction (*k*-value) upon stabilization [7]

% GC	75% LS + 25% SH		50% LS + 50% SH		25% LS + 75% SH	
	k-value (MPa/m)	% increase	*k*-value (MPa/m)	% increase	*k*-value (MPa/m)	% increase
0	46.41	–	40.7	–	34.54	–
15	90.24	94.44	–	–	–	–
20	–	–	68.5	68.43	59.42	72.03

In accordance with the guidelines outlined in IRC:SP-72, a proposed pavement composition for both unmodified and GC-modified blended soil subgrades in the case of flexible pavement is presented in Table 4. This composition is intended for a design traffic volume of 1.0 million standard axles, falling within the traffic category of T-8 as defined by IRC-SP 72 [6]. The OGPC wearing course of 20.0-mm thickness is regarded to be standard for all the works of compositions. Also, WBM (grade-3) with 75.0-mm depth is also expected to be employed in "low-volume" pavements. The utilization of the GC modifier facilitates comprehension of the composition of the pavement design, as depicted in Table 4.

The overall thickness of the bituminous pavement has the potential to be reduced by around 30–38%. Nevertheless, while doing research on the prevailing conditions in the field, it becomes feasible to formulate the composition design and ultimately get the precise configuration of both flexible and rigid pavements. The findings yielded positive outcomes, indicating potential for significant reductions in the consumption of natural materials during pavement construction. This, in turn, could result in cost savings and contribute to the overall goal of achieving sustainable practices within the pavement construction industry. The result of this knowledge also provides a sustainable alternative to the dumping issues associated with the removal of blast furnace slag and fly ash, which would otherwise end up in a perilous mass as landfills.

Table 4 Typical design composition of flexible pavement as per IRC:SP-72 [6]

% GC	Blended soil subgrade	Total thickness (mm)	Design composition (mm)				
			SG	GSB	GB	WBM	OGPC
0	Pure 75% LS + 75% SH	675	300	120	150	75	20
15	Stabilized 25% LS + 75% SH	440	100	100	145	75	20
0	Pure 50% LS + 50% SH	745	300	200	150	75	20
20	Stabilized 50% LS + 50% SH	465	100	120	150	75	20
0	Pure 75% LS + 25% SH	795	300	200	200	75	20
20	Stabilized 75%LS + 25%SH	560	200	120	145	75	20

Note GB: Granular Base course; WBM: Water Bound Macadam course; OGPC: Open Graded Premix Carpet; GSB: Granular Sub-base; SG: Soil Subgrade

4 Conclusions of the Investigation

It is commonly known in the construction industry that alkali-activated binders and geopolymers are more environmentally friendly and less dependent on virgin resources than traditional cement-based binders since they use industrial by-products as major predecessors. Furthermore, using them will result in fewer carbon emissions than using traditional cement, which will support a more ecologically conscious and sustainable building industry. This study employed geopolymer cement (GC) to stabilize blended lateritic-shedi soil subgrades commonly seen in the coastal districts of Karnataka state. The production of GC involved the utilization of fly ash (FA) and ground granulated blast furnace slag (GGBS) as binding materials, which were chemically activated through the addition of sodium silicate solution and NaOH flakes. The geotechnical characteristics of these marginal blended soils were examined and enhanced using an innovative stabilizing approach. The following are the key findings of the study:

- Study indicates that there is a positive correlation between the dose of GC and the MDD value of the soil blends. However, this relationship is only observed up to a certain threshold, beyond which higher levels of GC content lead to a decrease in the MDD value.
- The MDD of all three blended soils shows a notable increase when subjected to GC stabilization, in contrast to the mixes that were not stabilized. The CBR and UCS tests exhibited significant enhancements with the incorporation of GC with over 20% increment in test values. The enhanced soil strength can be attributed to the development of calcium aluminosilicate hydrates.
- The findings indicate that the use of GC has a notable impact on reducing the pavement thickness designed considering rural pavement infrastructure, due to the improvements in subgrade characteristics of stabilized soils, thereby presenting a viable and environmentally friendly alternative to conventional stabilizing techniques. This act of diminishing pavement thickness renders GC as cost-effective and environmentally sustainable, trustworthy stabilizer.

Overall, the incorporation of a slag-fly ash-based GC stabilizer as a modifier in the stabilization process of blended lithomargic-lateritic soil subgrades exemplifies an environmentally conscious method in the field of pavement construction. The significant decrease in pavement thickness, along with the possible cost savings in natural materials, enhances the sustainability and cost-effectiveness in pavement constructions. Moreover, the utilization of GC presents an ecologically sustainable option for tackling the disposal difficulties linked to industrial waste substances associated with the production of such stabilizers.

References

1. Amulya S et al (2020) Stabilisation of lithomargic clay using alkali activated fly ash and ground granulated blast furnace slag. Int J Pavement Eng 21(9):1114–1121. https://doi.org/10.1080/10298436.2018.1521520
2. Anburuvel A et al (2023) Characteristic evaluation of geopolymer based lateritic soil stabilization enriched with eggshell ash and rice husk ash for road construction: an experimental investigation. Constr Build Mater 387:131659. https://doi.org/10.1016/j.conbuildmat.2023.131659
3. Benrebouh I et al (2025) Effect of sulfate resistance cement treatment on saline soil (Sebkha) subgrades. Indian Geotech J 1–19. https://doi.org/10.1007/s40098-024-01156-5
4. Bhagyashree et al (2019) Mineralogical evaluation of lateritic soil of selected zones in Udupi district, Karnataka, India. Int J Civ Eng Technol 10(2):836–853
5. Firdous R, Stephan D (2019) Effect of silica modulus on the geopolymerization activity of natural pozzolans. Constr Build Mater 219:31–43. https://doi.org/10.1016/j.conbuildmat.2019.05.161
6. IRC:SP-72 (2015) Guidelines for the design of flexible pavements for low volume rural roads, 1st Rev
7. IRC:SP:62 (2014) Guidelines for design and construction of cement concrete pavements for low volume roads
8. Malegole SB, Pranab P (2021) Sustainable use of recycled concrete aggregates and waste rubber shreds in drainage layer of landfills. In: Japanese geotechnical society special publication. Mangaluru, pp 221–225. https://doi.org/10.3208/jgssp.v09.cpeg066
9. Marathe S et al (2015) Stabilization of lithomargic soil using cement and randomly distributed waste shredded rubber tyre chips. Int J Eng Trends Technol 23(6):284–288. https://doi.org/10.14445/22315381/ijett-v23p253
10. Marathe S, Shankar AUR (2023) Investigations on bio-enzyme stabilized pavement subgrades of lateritic, lithomargic and blended soils. Int J Pavement Res Technol 16(1):15–25. https://doi.org/10.1007/s42947-021-00107-0
11. Marathe S et al (2024) Stabilized lithomargic soil subgrades for low volume road design using industrial wastes. Int J Pavement Res Technol 17(6):1543–1554. https://doi.org/10.1007/s42947-023-00317-8
12. Marathe S et al. (2025) Properties of slag-based geopolymer-stabilized Indian lithomargic soil using sugarcane bagasse ash for sustainable pavement design. Studia Geotechnica et Mechanica Sciendo. https://doi.org/10.2478/sgem-2025-0003
13. Naik PA et al (2023) Properties of WFS incorporated cement stabilized lateritic soil subgrades for rural pavement applications. Int J Geosynth Gr Eng 9(4):1–17. https://doi.org/10.1007/s40891-023-00460-z
14. Patel A (2019) Soil stabilization. In: Geotechnical investigations and improvement of ground conditions. Woodhead Publishing Series in Civil and Structural Engineering, pp 19–27. https://doi.org/10.1016/B978-0-12-817048-9.00003-2
15. Phoo-Ngernkham T et al (2015) Effects of sodium hydroxide and sodium silicate solutions on compressive and shear bond strengths of FA-GBFS geopolymer. Constr Build Mater 91:1–8. https://doi.org/10.1016/j.conbuildmat.2015.05.001
16. Ramezani SJ et al (2023) Utilization of glass powder and silica fume in sugarcane bagasse ash-based geopolymer for soil stabilization. J Mater Civ Eng 35(4):1–20. https://doi.org/10.1061/(asce)mt.1943-5533.0004704
17. Sekhar DC et al (2017) Influence of granulated blast furnace slag and cement on the strength properties of lithomargic clay. Indian Geotech J 47(3):384–392. https://doi.org/10.1007/s40098-017-0228-8
18. Sekhar DC (2017) Studies on lithomargic clay stabilized using granulated blast furnace slag and cement. National Institute of Technology Karnataka, Surathkal, India
19. Shankar AUR et al (2012) Experimental investigation on lithomargic clay stabilized with sand and coir. Indian Highw 40(2):21–31

20. Shivaramaiah A et al (2020) Utilization of lateritic soil stabilized with alkali solution and ground granulated blast furnace slag as a base course in flexible pavement construction. Int J Pavement Res Technol 13(5):478–488. https://doi.org/10.1007/s42947-020-0251-5
21. Tardy Y (1997) Petrology of laterites and tropical soils. Petrol Laterit Trop Soils
22. Vunnam V et al. Construction and demolition waste utilisation for recycled products in Bengaluru: challenges and prospects. https://smartnet.niua.org/content/8dc4e2a3-9ddc-42e8-82fd-c9f1787a8dba
23. Yadav JS, Tiwari SK (2017) A study on the potential utilization of crumb rubber in cement treated soft clay. J Build Eng 9:177–191. https://doi.org/10.1016/j.jobe.2017.01.001
24. Yi Y et al (2015) Alkali-activated ground-granulated blast furnace slag for stabilization of marine soft clay. J Mater Civ Eng 27(4):1–7. https://doi.org/10.1061/(asce)mt.1943-5533.0001100

Incorporation of Emulsion in Semi-dense Bituminous Concrete for Sustainable Cold Mix-Road Construction

N. Akshaya Krishna, Shriram Marathe, and I. R. Mithanthaya

Abstract This research explores the sustainable application of bituminous emulsions in cold mix asphalt (CMA) technology, contributing to innovation in sustainable road development. CMA technology, known for its eco-friendly attributes, eliminates the need for heating during construction, making it suitable for regions with varied climates. Bituminous emulsions offer a sustainable alternative with lower energy requirements, reduced greenhouse gas emissions, and improved workability. Experimental investigations involve the preparation of cold mix semi-dense bituminous concrete, i.e., CMSDBC with the use of bituminous emulsion instead of conventional hot asphalt bituminous binder. The outcomes highlight favorable mechanical properties, including stability, density, and air voids. The incorporation of ground granulated blast furnace slag (GGBFS) as a filler enhances the stability of the cold mix. The research objectives include studying the mechanical properties of bituminous emulsion, assessing the strength behavior of bitumen concrete produced with bituminous emulsion, exploring the application of emulsion in production of CMSDBC with additives, and evaluating the stability of the same in highway construction. Experimental results reveal that the SDBC mix with 8.50% optimum bituminous emulsion content achieves a maximum Marshall stability value of 8.06 kN. The addition of GGBFS as a filler enhances stability, reaching a maximum Marshall stability value of 13.76 kN, by satisfying all the other necessary requirements as per IRC guidelines. These results prove the potential of bituminous emulsion in CMA technology as an eco-friendly and sustainable solution for road construction, aligning with the objectives of innovation and environmental responsibility in infrastructure development.

N. Akshaya Krishna
Department of Civil Engineering, Mangalore Institute of Technology and Engineering (MITE), Moodabidre, Karnataka, India

S. Marathe (✉) · I. R. Mithanthaya
Department of Civil Engineering, Nitte (Deemed to Be University), NMAM Institute of Technology (NMAMIT), Karkala, Karnataka, India
e-mail: ram.nmamit@gmail.com

Y. K. Guruprasad et al. (eds.), *Recent Advances in Infrastructure Development and Management—Volume 1*, Lecture Notes in Civil Engineering 570,
https://doi.org/10.1007/978-981-96-1554-4_16

Keywords Cold mix asphalt · Sustainability · Marshall stability · SDBC · Bituminous emulsion

1 Introduction

As a result of its improved performance and longevity, hot mix asphalt (HMA) has become the most popular and commonly used form of bituminous pavement material. It is one of the older paving materials that has been discovered, and it has withstood the test of time. Nevertheless, despite its widespread use, this bitumen (HMA) has a number of shortcomings, including deterioration of the environment, greater use of energy, a boost in carbon dioxide emissions, decreased yield for mix fabrication, decreased laying job in wet and chilly conditions, a limited duration of construction in a year, oxidative stiffening of binder, safety and health perils to work force, etc. Despite its popularity, HMA has a number of disadvantages [4, 9]. Pavement experts are currently working on developing paving materials that have a small environmental impact in response to the increasing demand for ecologically conscious solutions in contemporary society. One of these solutions is called *cold mix asphalt* (CMA), which has the prospective to surmount several of the constraints that are associated with HMA. CMA is generated first and foremost by mixing aggregates with bituminous (or asphalt) emulsion in the absence of heat, which results in a significant reduction in both the production of harmful volatile substances and the usage of fuels. In contrast to HMA, the bonding that forms amid the aggregate and the remaining bitumen takes place at room temperatures. CMA, on the other hand, has several drawbacks, such as the fact that it is challenging to compact, that the consolidated mixtures have a high air-voids content, that it has a low initial strength, and that it takes a longer period of time to get fully cured specimens for improved performance [11].

As a result of heating the binder and aggregates to temperatures ranging from 150 to 160 °C, conventional methods of road building that use hot mix techniques result in the release of hydrocarbons and *suspended particulate matter* (SPM). The Honorable Supreme Court of India was required to seize harsh measures to curtail pollution all the way back in the year 2012. These efforts included issuing directions to shut down hot mix facilities in and around Delhi (capital of India), which were identified as important sources of environmental deterioration. An alternative to this is the utilization of CMA technology, which call for a low amount of energy input and do not add to the pollution of the surrounding environment. The use of emulsions in the creation of bituminous surfaces for rural to medium-volume roads has shown to be an effective strategy, functioning as a labor-intensive method by doing away with the requirement for heating. Cold mix methods based on bitumen emulsion are widely used in road construction around the world. These technologies offer a number of benefits, including the removal of heating requirements for the binder and aggregate during the creation of the jumble, the covering of moist aggregate, energy conservation, and protection of the environment [9].

On the other hand, in regions with heavy precipitation, it is unfavorable to construct bituminous roads using conventional paving-grade bitumen, as the inconsistent rainfall that occurs throughout the year disrupts the formulation and application of the mixtures. Likewise, the laying of hot mix asphalt is hindered by the extremely cold pavement temperatures that prevail in places that are surrounded by snow. By eliminating the requirement to heat the binder and aggregate, the use of bitumen emulsion helps to reduce the depletion of the environment while also saving on energy costs. Bitumen emulsions have a wide variety of applications in the construction of roads around the world. These applications help to reduce the amount of heat generated during the preparation of mixes, coat wet aggregate, and protect the environment while simultaneously saving energy. In contrast, the usage of bitumen emulsion in nations, such as France, Switzerland, and other European countries, is significantly higher than it is in India (2.7% versus up to 43%). The discrepancy may be linked to a lack of competence in this technique, a paucity of superior bitumen emulsion, and an absence of appropriate machinery [3, 10].

Due to the liquid state of bituminous emulsion at room warmth, there is no requirement to apply reheat or undergo drying processes for the mineral aggregate. Cold mix is a beneficial material in regions characterized by significant distances between the construction site and the plant, as well as low to moderate temperatures (below 40 °C). Moreover, the adaptability of cold mix enables it to be blended on-site at the construction location, over and above at a centralized facility, and later conveyed to the designated construction site [12]. The current study involves the preparation of a cold bituminous mix utilizing emulsion. In the context of the current research subject matter, an effort was made to investigate the mechanical performance of the *cold mixed semi-dense bituminous composite* (CMSDBC) mix that was developed utilizing environmentally friendly cold mix technology by the use of bituminous emulsion. There have been a few studies done in recent years that have indicated that the incorporation of *ground granulated blast furnace slag* (GGBFS) as a stuffing into pavement grade mixes is advantageous to improve the mechanical performance of HMA mixtures [7], which is due to its ability to cement and maintain its stiffness regardless of very humid environments [5]. Therefore, the current study aimed to develop SDBC mixes with GGBFS mineral filler and analyze their strength properties. Optimal Emulsion Content (OEC) determination, Marshall performance [2] is among the parameters examined. The IRC standards were followed throughout the mixing, developing, and testing processes on ingredient materials and resulting bituminous emulsion concrete mixtures [6].

2 Materials and Techniques

2.1 Material Properties and Mix Design Details

Bituminous emulsion of SS-2 type was collected from local suppliers: Hindustan Colas Private Limited, Mangalore. The several key parameters of bitumen emulsion, comparing the test results against specified requirements as provided by the supplier are as follows. The residue on the 600-micron IS sieve, representing the particle size distribution, had a requirement of 0.05%, and the actual result was 0.006%, indicating excellent compliance. Viscosity, assessed by the Saybolt Furol viscometer, displayed a requirement range of 30–150 s, with the measured result at 41 s, falling within the acceptable limits. Coagulation of emulsion at low temperature was observed to be nil, meeting the specified criterion. Particle charge was evaluated as positive, aligning with the expected behavior of emulsion. Miscibility with water, a crucial aspect, exhibited no coagulation, indicating good compatibility with water. The residue by evaporation, set at a maximum of 60%, was measured at 61.26%, slightly exceeding the specified limit. However, this minor deviation might still be acceptable based on project considerations. Finally, the specific gravity requirement ranged from 0.9 to 1.02, and the measured value was 1.01, indicating compliance with the given specifications.

The aggregate properties were assessed to ensure compliance with IRC standards. The combined index for aggregate shape was determined to be 28.3%, just below the specified limit of 35%, indicating acceptable particle shape. Aggregate toughness, evaluated through the aggregate impact value, yielded a result of 11%, slightly under the 27% IRC limit. Similarly, the Los Angeles Abrasion test for aggregate hardness produced a value of 22.64%, meeting the IRC criterion of less than 35%. The average bulk specific gravity of aggregates was 2.66, falling within the acceptable range of 2.5–3.2. Aggregate absorption, determined by water absorption, was found to be 1.2%, below the IRC threshold of 2%, signifying good resistance to water penetration. The stripping value, an indicator of aggregate resistance to stripping of bitumen, was measured at 3.2%, surpassing the IRC requirement of less than 5.0%. Hence, the aggregate properties align closely with IRC limitations as indicated by IRC:SP:100-2014 [6], indicating favorable characteristics for the usage in the cold mix preparation. Further, the mineral filler used in the study was GGBFS which was obtained from Jindal Steel Bellary. The specific gravity of the GGBFS was found to be 2.88. The combined gradation of the mineral aggregates is illustrated in Fig. 1.

2.2 Methodology in Cold Mix Technology for SDBC

CMA design technology involves the optimization of bitumen emulsion for incorporation into the aggregate mixture. To achieve the specified aggregate gradation for SDBC, a combination of various aggregate sizes was utilized. The filler, i.e.,

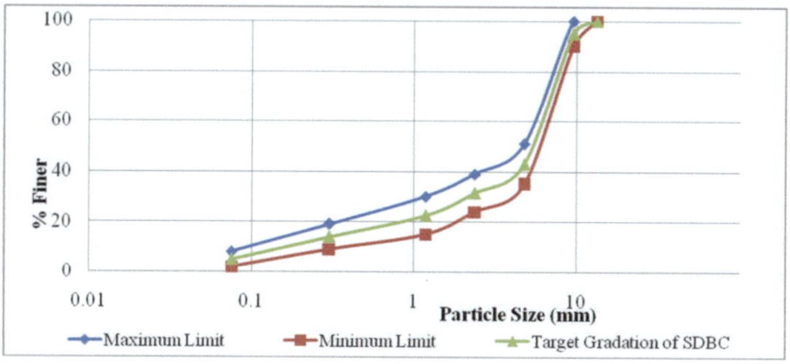

Fig. 1 Gradation of mineral aggregates for CMSDBC mix

the portion of materials in the gradation pertaining to the sieve size of less than 75 microns (IS sieve), was GGBFS, which was from 0 to 5% by the total mix weight at an interval of 1.0%. The graded aggregates were ensured to be in a saturated surface-dry condition by the use of laboratory water. The prepared aggregate-filler blend was then mixed with a calculated amount of bituminous emulsion. Hand mixing was done thoroughly for about 2 min to ensure the proper coating of aggregates with cold bitumen. Moreover, the CMA process employs standard Marshall Equipment to assess mix properties. Marshall test specimens, prepared in triplicate, featured varying binder content (bitumen emulsion). The cold mix was then inserted into the Marshall sampler, along with filter paper on the base cover and compacted using 50 blows from the standard Marshall hammer on both the sides. Following a period of 24 h, the Marshall specimens were removed out from the mold and exposed to a 72-h curing process in an air oven set at a temperature of 40 °C. Subsequent tests were performed on the Marshall specimens, including assessments of bulk density, stability, and flow value. The depicted sequence in Fig. 2 outlines the dry-mixing of graded mineral aggregates, incorporation of bituminous emulsion, preparation of CMSDBC mix through binder addition, the Marshall compaction sequence, and the compacted Marshall specimens. Diverse properties were evaluated, and additional design parameters, such as voids content and voids filled by emulsion binder, were premeditated.

The main purpose of the investigations was to examine the effects of different levels of bituminous binder (i.e., emulsion) in the CMSDBC mix. The range of binder percentages examined was from 7 to 10%, with increments of 0.5%. The focus has transitioned from optimum bitumen content (OBC) to optimum emulsion content (OEC), ascertained using density void analysis and stability flow tests. In this context, up to 5% of GGBFS was incorporated for investigation as mineral filler, and the Marshall Properties were investigated at the obtained OEC. All tests adhered to the standard methods outlined/adopted in the literature [1, 4]. For the selected type of mix, as per IRC [6] the Marshall stability value is required to be greater than 500 kg to ensure the structural integrity of the CMA. Additionally, the Marshall flow value, which measures the flow of the mix under a specified load, should be less than 8.0 mm, indicating a desirable level of deformation. The percentage of voids in the

Fig. 2 Dry-mixing of graded aggregates, bituminous emulsion, preparation of SDBC mix, Marshall compaction, compacted specimens

mixture binder is recommended to be inside the array of 6–10%, ensuring an optimal balance between stability and flexibility. Furthermore, the retained indirect tensile strength, assessed after conditioning, should be at least 75%, indicating the mix's ability to withstand tensile stresses over time [8]. These specifications, as per IRC guidelines, collectively establish the quality and performance standards expected from cold mix asphalt formulations in terms of stability, voids, flow, and strength.

3 Results and Discussions

3.1 *Marshall Performance and Optimum Bitumen (OEC) of CMSDBC Mix*

The results depicted in Fig. 3a highlight the influence of varying Bituminous Emulsion Content (BEC) on Marshall stability (MS). Notably, the maximum Marshall stability is achieved at 8.5% Bituminous Emulsion Content, reaching a peak value of 8.06 kN. Beyond this optimal point, a diminishing trend in stability is observed, emphasizing the critical role of selecting an appropriate emulsion content to maximize the performance of CMSDBC formulations. In Fig. 3b, the Marshall flow (MF) values corresponding to different BEC are presented. The results illustrate a gradual increase in MF as the emulsion content rises. Notably, at 8.5% BEC, the Marshall flow reaches 7.92 mm, indicating an optimal range for achieving desirable flow characteristics in CMSDBC.

In Fig. 4a, the bulk density values corresponding to different BEC percentages are presented. The results reveal a slight increase in bulk density as the BEC increases, reaching 2.169 g/cc at 8.5% BEC. The optimal BEC for achieving a balance between density and workability needs to be carefully determined, considering the specific

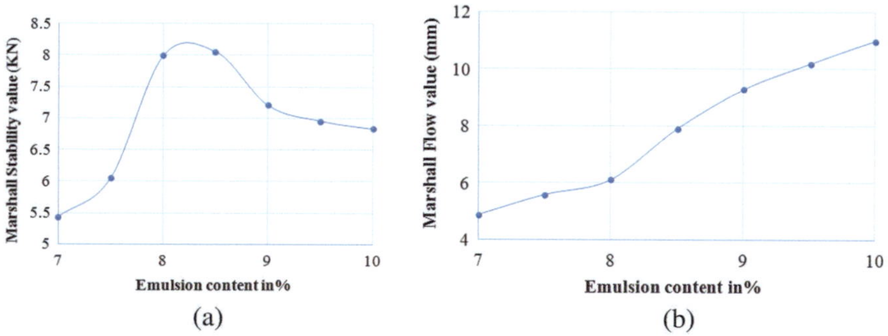

Fig. 3 Variations in Marshall stability, flow value with respect to emulsion content

requirements of the construction project. The values of voids filled with bituminous binder (VFB) for various percentages of BEC are shown in Fig. 4b. In accordance with the results, VFB consistently rises as BEC increases, achieving 77.53% at a trial dose of 8.5% BEC. This implies that a higher emulsion percentage helps fill up gaps in the mixture more effectively, which enhances the stability of the CMSDBC.

Further, the proportion of air voids (% AV) associated with various percentages of BEC is illustrated in Fig. 5a, whereas the variations in the voids in mineral aggregates (VMA) is shown in Fig. 5b. As the BEC increases, the results indicate that the % AV decreases, reaching at 6% at 8.5% BEC. The decrease in air cavities indicates that an increased emulsion concentration aids in the improvement of CMSDBC compactness and densification. Further, as the BEC increases, the results indicate that VMA increases gradually. This upward trend suggests that higher emulsion content promotes a greater volume of voids within the mineral aggregates, indicating improved workability and coating efficiency of the emulsion in the CMSDBC. Overall, the analysis of Marshall Properties for studied CMSDBC indicates that the OEC is achieved at 8.50% of BEC, showcasing peak values for stability, flow, and void characteristics. This optimal composition aligns with desired mechanical and

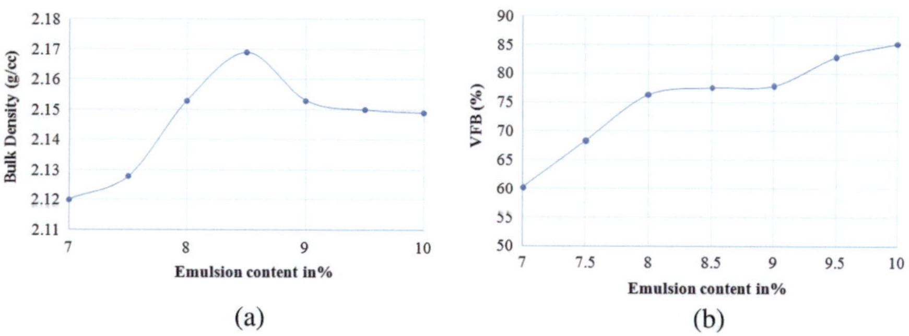

Fig. 4 Variations in bulk density, voids filled with bitumen (VFB) with emulsion content

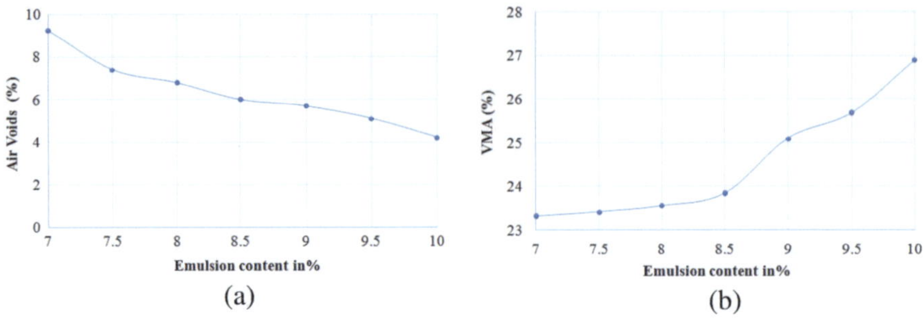

Fig. 5 Variations in percentage of air voids and VMA with emulsion

volumetric properties, establishing the significance of 8.5% BEC in enhancing the performance of CMSDBC mixture.

3.2 Mineral Filler Effect on Marshall Performance of CMSDBC Mix

The effect of GGBFS dosage as mineral filler on the Marshall Properties of CMSDBC mixture is presented in this section. The Marshall stability (MS) values demonstrated an increasing trend from 8.06 kN at 0% filler (GGBFS) to a peak of 13.76 kN at 3% filler, indicating consistent improvement in stability. Beyond 3% GGBFS, the MS slightly decreased. Marshall flow (MF) exhibited a declining trend from 7.92 mm at 0% filler to 5.95 mm at 3% filler, suggesting enhanced resistance to deformation. Beyond 3% GGBFS, the MF increased slightly. Bulk density increased from 2.169 to 2.25 g/cc up to 3% GGBFS and then stabilized.

VFB value decreased from 77.53 to 72.29% with increasing filler content, indicating improved binder filling within the mix. Percentage air voids (% AV) decreased from 6 to 6.8% up to 3% GGBFS, showcasing a reduction in void spaces and enhanced density, whereas the voids in mineral aggregates decreased from 23.85 to 22.47% up to 3% GGBFS, suggesting improved cohesion between aggregates.

From the overall result analysis, at 3% GGBFS, considered optimal, the CMA mix at the selected optimum emulsion content (OEC) demonstrated notable improvements in all properties. The % improvement for each property was calculated, emphasizing the positive impact of GGBFS on the CMA mix. Figure 6a, b visually represents the trends in major Marshall Properties with varying GGBFS mineral filler content.

Hence, in general, the incorporation of GGBFS as a filler in CMSDBC consistently enhanced many aspects of the mixture, including stability, resistance to deformation, compactness, binder filling, density, and cohesiveness between aggregates [7]. These findings provide confirmation of the beneficial impact of GGBFS on the properties of the mixture, particularly when used at a dosage of up to 3% GGBFS. Moreover,

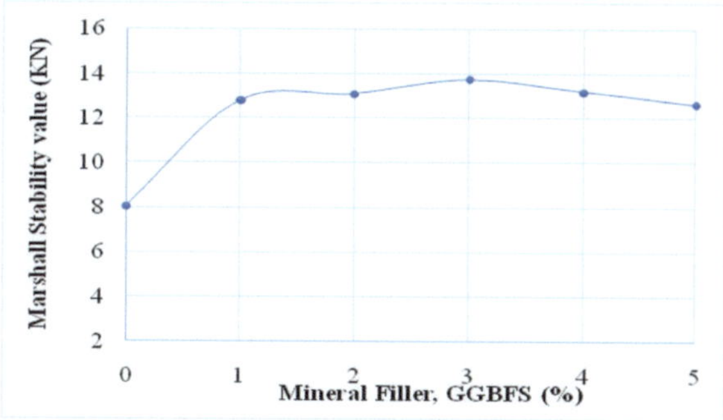

(a)

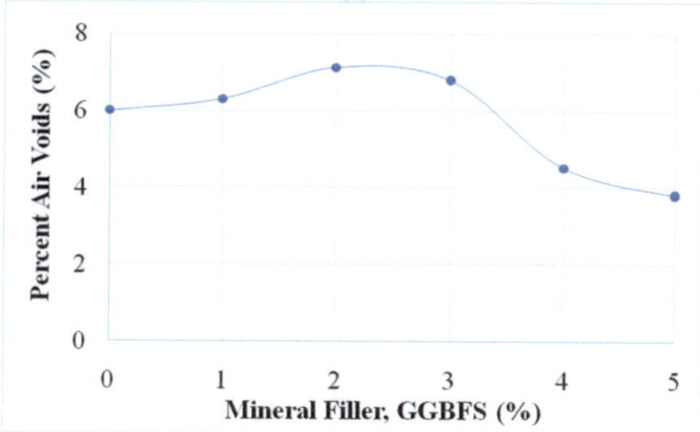

(b)

Fig. 6 Variations in Marshall stability and % air voids with GGBFS mineral filler

specific characteristics exhibited a slight decrease, indicating the significance of regulating the dosage of GGBFS in order to attain optimal performance in the mixture.

4 Conclusions and Future Recommendations

This research contributes to sustainable road development by exploring the eco-friendly attributes of bituminous emulsions in CMA technology. The use of CMA methodology not only reduces energy consumption and greenhouse gas emissions but also provides a promising and sustainable alternative for road construction. This research, encompassing the optimization of emulsion content and the incorporation of GGBFS as a mineral filler in CMSDBC mixtures, has yielded crucial insights into the material's suitability and sustainability for road construction. The Marshall performance results demonstrated a clear trend, with the Optimal Emulsion Content identified at 8.50%, emphasizing its pivotal role in determining CMA mix stability and structural integrity after satisfying the requirements as per IRC recommendations. In the second stage, GGBFS incorporation as a filler revealed significant improved impacts on Marshall Properties. The maximum stability and optimum properties were observed at 3% GGBFS content, emphasizing its role in enhancing mix characteristics. The key takeaway from these experimental investigations is the identification of 8.5% bituminous emulsion content and 3% GGBFS as the optimal conditions, striking a balance between stability, workability, and density in CMSDBC mixes. This comprehensive understanding contributes significantly to the sustainable deployment of CMA technology in road construction.

Future research scopes include exploring the impact of natural fiber additions, conducting comprehensive tests, such as indirect tensile strength and rutting, assessing CMA applicability for high-volume pavements, conducting wear and tear tests, and evaluating durability and permeability for a more holistic understanding of CMA's long-term performance. These directions aim to advance the knowledge and refine CMA technology in some extent for sustainable road infrastructural expectations.

Acknowledgements The experimental work was conducted at the Highway Engineering laboratory within the Civil Department of NMAMIT Nitte. Authors are appreciative to the following UG students: Mr. Shridhar Naik (4NM15CV096), Ms. Shriraksha (4NM15CV097), Mr. Madana (4NM16CV409), and Mr. Shridhara B Moger (4NM15CV095) for their aid in performing few of the fundamental laboratory experimentations. Authors wish to thank Dr. Anil Kumar, Former Assistant Professor of Civil Engineering, NMAMIT Nitte, for providing the valuable suggestions during the research work. Authors also wish to express their earnest thanks to the Prof. G. Udayakumar, former HoD of Civil Engineering, and Prof. Niranjan Chiplunkar, Principal of NMAMIT Nitte and Nitte Education Trust (R) for extending the timely supports for carrying out this laboratory-based research work.

References

1. AASHTO-T283 (2003) Resistance of compacted asphalt mixtures to moisture-induced damage
2. Akarsh PK et al (2022) Incorporation of sugarcane bagasse ash to investigate the mechanical behavior of stone mastic asphalt. Constr Build Mater 353(129089):1–13. https://doi.org/10.1016/J.CONBUILDMAT.2022.129089
3. Arya IR, Pundhir NKS (2001) Emulsified cold mixes: design, construction and field performance. IRC Indian Highw 29:3
4. Dash SS et al (2022) Design and performance of cold mix asphalt—a review. Constr Build Mater 315:125687. https://doi.org/10.1016/j.conbuildmat.2021.125687
5. Ellis C et al (2011) Properties of GGBS-Bitumen emulsion systems with recycled aggregates. Road Mater Pavement Des 5(3):373–383. https://doi.org/10.1080/14680629.2004.9689977
6. IRC:SP:100-2014 (2014) Use of cold mix technology in construction and maintenance of roads using Bitumen emulsion. http://archive.org/details/govlawircy2014sp100
7. Marathe S et al (2023) Investigations on stone matrix asphalt mixes by utilizing slag and cellulose fiber. In: Anjaneyulu MVLR et al (eds) Recent advances in transportation systems engineering and management. Springer Nature Singapore, Singapore, pp 253–259
8. Marathe S, Kotrabasappa AP, Bhat AK (2025) Properties of rejuvenated reclaimed asphalt pavement mixtures with waste glass powder and sisal fibers for pavements. J Mater Civ Eng 37(2). https://doi.org/10.1061/JMCEE7.MTENG-18775
9. Pundhir NKS (2012) Construction of rural road with cationic Bitumen emulsion based cold mix technology. In: Workshop on non-conventional materials/technologies. Central Road Research Institute, New Delhi, pp 48–55

10. Pundhir NKS, Nanda PK (2006) Development of bitumen emulsion based cold mix technology for construction of roads under different climatic conditions of India. J Sci Ind Res 65:729–743
11. Thanaya INA et al (2009) A laboratory study on cold-mix, cold-lay emulsion mixtures. Proc Inst Civ Eng Transp 162(1):47–55. https://doi.org/10.1680/tran.2009.162.1.47
12. Yadav RK (2018) Emulsified cold mix design for Bituminous Macadam (BM). Int J Innov Sci Res Technol 3(1):658–663

Fly Ash-Based Solutions for Treating Construction and Demolition Waste: An In-Depth Analysis

S. Usha, G. D. Shivaraju, and T. V. Mallesh

Abstract This study investigates the treatment of recycled aggregates (RA) using a fly ash solution with varying concentrations and analyzes its impact on mechanical and durable parameters. Recycled aggregates often exhibit undesirable characteristics, such as porosity, high water absorption, and low crush resistance, necessitating innovative approaches for enhancement. In this research, recycled aggregates were subjected to treatment with fly ash solutions of different concentrations to explore the influence on their properties. The study comprehensively examines the mechanical parameters, including compressive strength and other relevant mechanical attributes, before and after the treatment with varying concentrations of the fly ash solution. Additionally, the durable parameters of the treated recycled aggregates are rigorously analyzed to assess their long-term performance and resilience. By varying the concentration of the fly ash solution, the research aims to identify the optimal treatment condition that maximizes improvements in mechanical strength and durability. The findings of this study contribute valuable insights into tailoring the treatment process for recycled aggregates, addressing specific challenges associated with varying concentrations of fly ash solutions. This approach not only seeks to enhance the performance of recycled aggregates but also aligns with sustainable practices by repurposing industrial by-products. The research outcomes aim to advance the understanding of effective and environmentally friendly methods for optimizing the use of recycled aggregates in construction applications.

S. Usha
Civil Engineering Department, SSIT, SSAHE, Agalakote, Tumakuru, India

G. D. Shivaraju · T. V. Mallesh (✉)
Civil Engineering Department, SSIT, Tumkur, India
e-mail: ushas@ssit.edu.in

T. V. Mallesh
e-mail: drtvmallesh@gmail.com

G. D. Shivaraju
e-mail: shivarajugd@ssit.edu.in

Y. K. Guruprasad et al. (eds.), *Recent Advances in Infrastructure Development and Management—Volume 1*, Lecture Notes in Civil Engineering 570,
https://doi.org/10.1007/978-981-96-1554-4_17

Keywords Fly ash · Recycled aggregates · Stabilization · Pozzolonic reactions · Sustainability

1 Introduction

The utilization of recycled aggregates (RA) in construction has garnered significant attention due to its potential for sustainable development and resource conservation. However, despite its environmental benefits, the widespread adoption of RA is often hindered by inherent shortcomings, such as porosity, high water absorption, and inadequate crush resistance [1]. These limitations underscore the critical need for innovative strategies to enhance the properties of recycled aggregates, thereby improving their applicability in construction projects.

In response to this challenge, this study delves into the treatment of recycled aggregates using fly ash solutions with varying concentrations. Fly ash, a by-product of coal combustion in power plants, presents an opportunity for repurposing industrial waste while addressing the deficiencies of recycled aggregates [2]. By subjecting RA to treatment with fly ash solutions, the research aims to explore the potential enhancements in mechanical and durable parameters, thus bolstering the overall performance and longevity of recycled aggregate-based constructions [6].

The investigation is structured to comprehensively analyze the mechanical attributes of treated recycled aggregates, encompassing parameters, such as compressive strength and other relevant mechanical properties [3]. Moreover, the study rigorously evaluates the durable parameters of the treated aggregates to assess their resilience and long-term performance under varying environmental conditions.

A key aspect of the research involves varying the concentration of the fly ash solution to identify the optimal treatment condition that maximizes improvements in mechanical strength and durability [4]. This systematic exploration seeks to provide valuable insights into tailoring the treatment process for recycled aggregates, offering tailored solutions to mitigate the specific challenges associated with different concentrations of fly ash solutions.

Beyond the technical advancements, this study underscores the broader significance of aligning construction practices with sustainability goals. By repurposing industrial by-products like fly ash, the proposed approach not only enhances the performance of recycled aggregates but also promotes environmentally friendly practices within the construction industry [5].

In summary, the outcomes of this research endeavor are poised to advance the understanding of effective and sustainable methods for optimizing the use of recycled aggregates in construction applications. By bridging the gap between theoretical knowledge and practical implementation, this study contributes to the ongoing pursuit of greener and more efficient construction practices [7].

2 Treatment Approaches

In the treatment of recycled aggregates, a series of immersion processes was conducted using fly ash solutions at different water-to-fly ash ratios. The concentrations investigated included 1:0.5, 1:0.75, and 1:1 (water: fly ash). Each solution was meticulously prepared by blending the specified proportions of water and fly ash, creating distinct concentrations for the treatment process.

Subsequently, RCA was introduced into each slurry and left to soak for a period of 4 h. Following this soaking phase, the RCA was extracted from the slurry bath and air-dried at room temperature for 3 days. Lastly, the solidified RCA underwent sieving to eliminate excess materials before undergoing various laboratory tests.

3 Experimental Procedure

A comprehensive exploration of the macro properties of recycled concrete aggregates (RCA) was undertaken through a series of laboratory tests, both before and after employing the treatment methods. The aim was to evaluate the impact of these treatments on the physical and mechanical characteristics of RCA. The investigation encompassed a range of tests conducted in accordance with established standards:

(i) Specific gravity (IS: 2386-1963).
(ii) Water absorption (IS: 2386-1963).
(iii) Aggregate crushing value (IS: 2386 (Part 4)-1963).
(iv) Aggregate impact value (IS: 2386 (Part 4)-1963).
(v) Aggregate soundness test (IS 2386 (Part 5): 2016).

Later, based on the results of the investigation, an optimal ratio of water to fly ash concentration was identified. Additionally, the optimal replacement percentage of these treated aggregates in a concrete mix was identified.

4 Results and Interpretation

Based on the findings presented in Table 1 from the aforementioned tests, it has been observed that the optimal ratio for the water-to-fly ash solution can be set at 1:0.75.

4.1 Test on Concrete

Concrete testing was performed by replacing coarse aggregates with treated recycled aggregates at varying percentages, including 0, 25, 50, 75, and 100%, in M20 grade

Table 1 Test on aggregates

S. No.	Tests	Water-to-fly ash ratios					
		NCA	Untreated RCA	1:0.5	1:0.75	1:1	IS-recommended values
1	Specific gravity	2.65	2.23	2.32	2.48	2.41	2.5–3
2	Water absorption (%)	0.98	4.72	3.98	3.36	3.4	0.1–2
3	Crushing value (%)	28.21	31.74	32.45	31.78	34.52	< 45
4	Impact value (%)	25.54	32.5	31.8	31.45	31.67	< 50
5	Soundness test (%)	3	5.58	4.97	4.21	4.32	< 12

NCA—Natural Coarse Aggregates
RCA—Recycled Coarse Aggregates

Table 2 Concrete testing results with treated recycled aggregates in M20 grade concrete

Sl. No.	Percentage of replacement	Compressive strength in MPa	Compressive strength in MPa after subjecting to		
			Alternative wetting and drying test	Sulfate attack test	Chloride attack test
1	0	28.12	28.45	27.68	27.41
2	25	27.55	27.35	26.95	27.10
3	50	26.58	26.53	25.45	24.55
4	75	23.69	23.20	23.32	23.10
5	100	22.64	22.91	22.29	22.07

(1:1.5:3) concrete. The results are compiled and presented in Table 2. The objective was to examine the effects of different replacement levels on the characteristics and performance of the concrete mix.

The most effective ratio for replacing natural coarse aggregate with recycled aggregate was found to be 50% (See Fig. 1).

5 Conclusions

Optimal Water-to-Fly Ash Ratio

The study determined that the best concentration for the water-to-fly ash solution is 1:0.75.

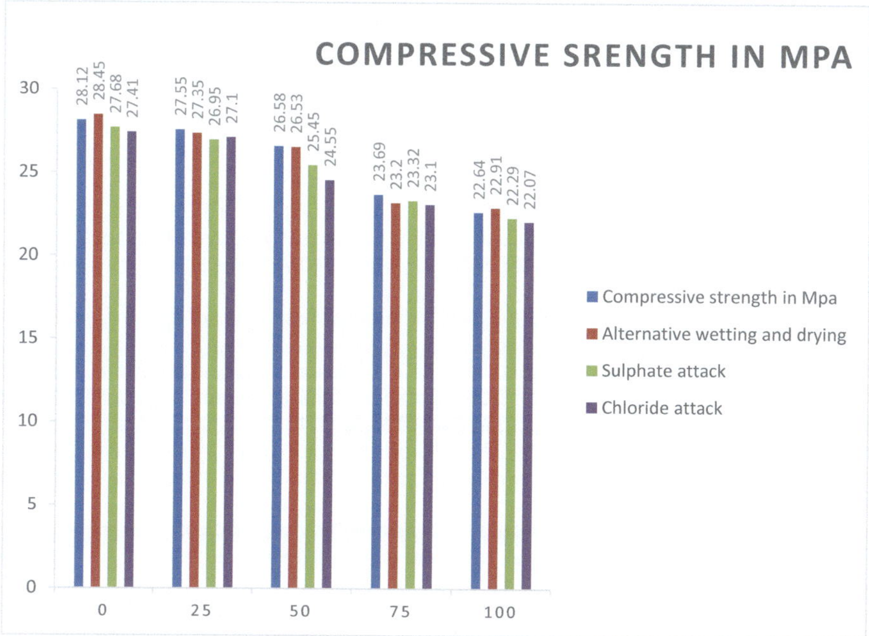

Fig. 1 Concrete testing results with treated recycled aggregates in M20 grade concrete

This ratio was found to significantly enhance the mix's performance.

Replacement of Coarse Aggregates

Various quantities of treated recycled aggregates were tested as replacements for coarse aggregates in M20 grade concrete.

The optimal replacement percentage was identified as 50%.

This substitution level provided the best balance between performance and material efficiency.

Performance Improvement

The combination of the optimal water-to-fly ash ratio and the ideal replacement percentage led to notable improvements in the concrete mix's characteristics and performance.

Treated recycled aggregates demonstrated enhanced properties when integrated into the concrete mix.

Sustainable Practices

The approach promotes sustainability by repurposing industrial by-products, such as fly ash.

This method addresses conventional concrete production challenges by reducing reliance on natural resources.

It aligns with the goals of environmentally conscious construction practices, advocating for a greener construction industry.

References

1. Usha S, Mallesh TV (2024) Strength and durability assessment of concrete bricks enhanced by construction & demolition waste integration. Appl Sci Eng J Adv Res 3(1):14–19. https://doi.org/10.54741/asejar.3.1.3
2. Usha S, Shivaraju GD, Mallesh TV, Reddy P. Performance assessment of fly-ash aggregates in concrete
3. Maratha K, Singh SP, Bibhuti Bhushan B (2023) Optimization of slurry impregnation technique for upcycling carbonated recycled concrete aggregates for paving concrete applications. J Mater Civ Eng. https://doi.org/10.1061/(asce)mt.1943-5533.0004712
4. Amit Kumar G, Singh SK, Kumar R (2023) Performance-based quality optimization approach for mechanically treated recycled concrete aggregates. J Mater Civ Eng. https://doi.org/10.1061/jmcee7.mteng-15284
5. Sandra M, Girija MK, Sreedevi A (2022) Strengthening of recycled coarse aggregate using different surface treatment techniques. https://doi.org/10.38208/acp.v1.490
6. Wafaa M, Shaban, Yang J, Su H, Liu Q-F, Daniel CW, Wang TL, Xie J, Li L (2019) Properties of recycled concrete aggregates strengthened by different types of pozzolan slurry. Constr Build Mater. https://doi.org/10.1016/J.CONBUILDMAT.2019.04.231
7. Madumita S, Katarina M, Agnes N (2020) Improving recycled aggregate quality by mechanical pre-processing. Materials. https://doi.org/10.3390/MA13194342

Effect of Steel Slag and Rap in Bituminous Concrete Grade-I Mix on Marshall Parameters and Fatigue Behavior

S. Poorna Chandra Thejaswi, K. Ganesh, Chethan Chandru, and S. V. Vidhath

Abstract The study explores the impact of reclaimed asphalt pavement (RAP) and steel slag on the Marshall parameters and fatigue behavior of bituminous concrete grade-1 mix. The study aims to determine the feasibility of utilizing these materials as sustainable alternatives in road construction. The research involves laboratory testing of bituminous concrete mixtures containing 5% steel slag and 10% RAP. The Marshall parameters, which evaluate the compatibility and mechanical properties of the mix, are assessed. Additionally, the fatigue behavior of the mixtures is analyzed to understand their resistance to repetitive loading. The results reveal that incorporating 5% steel slag and 10% RAP in the bituminous concrete grade-1 mix positively affects the Marshall parameters and fatigue behavior, indicating their potential as viable materials for improving the performance and sustainability of bituminous concrete grade-1 mix.

Keywords Bituminous concrete grade-1 mix · Reclaimed asphalt pavement (RAP) · Steel slag · Marshall stability · Optimum binder content (OBC) · Water sensitivity · Tensile strength ratio (TSR) · Indirect tensile strength · Fatigue characteristics

S. Poorna Chandra Thejaswi (✉) · C. Chandru · S. V. Vidhath (✉)
Department of Civil Engineering, Bangalore Institute of Technology, Bengaluru, India
e-mail: poornachandra.thejaswi@gmail.com

S. V. Vidhath
e-mail: vidhathsv@gmail.com

C. Chandru
e-mail: chethanchandru92@gmail.com

K. Ganesh
Department of Civil Engineering, BMS College of Engineering, Bengaluru, India
e-mail: gkiyer68.civ@bmsce.ac.in

Y. K. Guruprasad et al. (eds.), *Recent Advances in Infrastructure Development and Management—Volume 1*, Lecture Notes in Civil Engineering 570,
https://doi.org/10.1007/978-981-96-1554-4_18

1 Introduction

1.1 General

Nowadays, many countries try to fulfill the principles of sustainable development, which aims to reduce energy consumption and recycle waste materials [1, 2]. In this context, the incorporation of and recycled materials and industrial by-products as secondary and alternative resources in road construction has gained significant acceptance and is increasingly recognized as crucial. The demand for these materials is steadily growing each year. By utilizing industrial by-products in pavement engineering, not only we can potentially save on construction materials costs, but we can also alleviate the strain on natural resources. Disposing of steel slag in landfills not only results in the wastage of valuable resources but also poses serious environmental pollution concerns [3]. Nevertheless, directly utilizing untreated steel slag in construction materials presents significant risks to their mechanical properties and durability. In recent years, there has been remarkable progress in research on utilizing steel slag as environmentally friendly construction materials, especially after the discovery that accelerated carbonation can enhance its inferior properties (Fig. 1).

Bituminous pavements made of steel slag and reclaimed asphalt pavement (RAP) are gaining popularity due to their sustainable and environmental-friendly nature. RAP is a recycled material obtained from old asphalt pavements, and steel slag is a by-product of steel manufacturing processes [4–6]. By incorporating these materials into bituminous pavement, the construction industry is reducing waste and

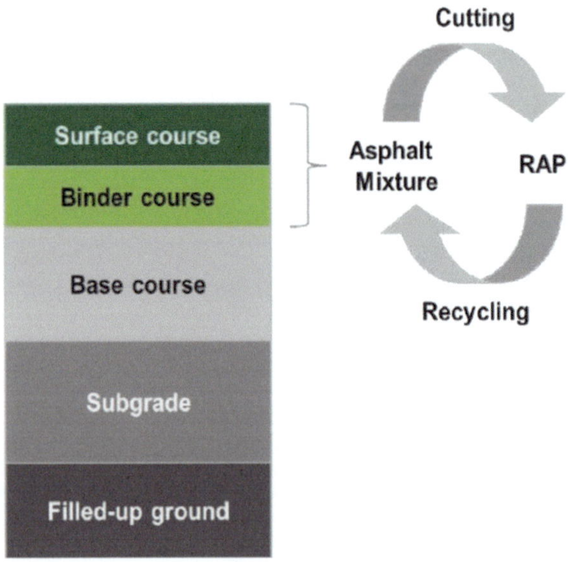

Fig. 1 Pictorial representation of RAP being used in bituminous pavement. *Source* ctc-n.org

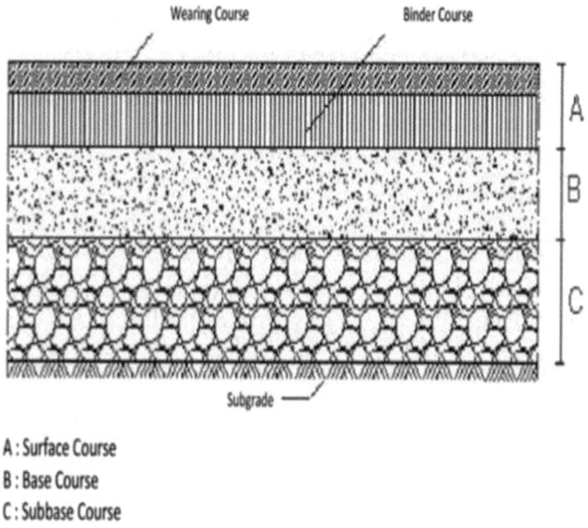

Fig. 2 Typical flexible pavement structure. *Source* researchgate.net

conserving natural resources. This type of pavement has shown improved properties, such as higher stability and durability, compared to traditional bituminous pavement. Furthermore, it has the potential to reduce construction costs and decrease the carbon footprint associated with road construction [7, 8] (Fig. 2).

The layers of flexible pavement typically include the following:

- Subgrade,
- Subbase,
- Base course,
- Binder course,
- Surface course.

1.2 Objectives

To determine the physical properties of virgin aggregate, RAP, bitumen and steel slag, optimum binder content (OBC), moisture susceptibility and fatigue characteristics of BC grade-I mix blended with RAP and steel slag.

1.3 Literature Review

Steel slag is a by-product of steel making process and is commonly generated during the production of steel, and it consists of various metallic oxides and non-metallic components. Due to its composition, the steel slag can be used as alternate for aggregates and fillers in road construction. The use of steel slag contributes to reduction of natural resources consumption minimizing waste disposal. It helps in mitigate greenhouse gas effect by capturing CO_2. It has been found to provide good stability, high resistance to rutting, and improved skid resistance in road pavements [9]. It also exhibits excellent load-bearing capacity, which is crucial for withstanding heavy traffic loads. It also enhances the durability and longevity of road surfaces due to its resistance to weathering and abrasion. The use of steel slag contributes to sustainable development but also offers favorable mechanical and environmental benefits. Mechanical characterizations, Marshall stability and indirect tensile stiffness modulus tests, were used to ascertain the laboratory performance of the mixtures [10]. The study highlights that steel slag possesses favorable properties for road construction and offers numerous benefits. Its physical, mechanical, and chemical composition make it preferable for use in road infrastructure projects, enhancing stability and durability. The utilization of steel slag can also contribute to sustainable development by reducing waste generation, conserving natural resources, and mitigating environmental impacts. Moreover, the incorporation of steel slag in road construction can lead to cost savings. Overall, the findings suggest that steel slag has significant potential as a road construction material, promoting sustainable and cost-effective infrastructure development [11]. This article provides a detailed overview of preventive maintenance strategies for flexible pavement and discusses the mechanical properties of steel slag asphalt. The authors emphasize the importance of proactive maintenance practices to extend pavement life and describe the potential benefits of incorporating steel slag into asphalt mixtures for improved performance and sustainability [12]. The study found that incorporating steel slag aggregates in WMA mixtures improved their resistance to moisture damage. This research contributes to the understanding of using steel slag as a sustainable alternative aggregate material in asphalt mixtures, particularly in WMA applications. The findings can potentially be applied in the design and construction of asphalt pavements to strengthen their durability and performance in humid and wet environments [13]. They have developed a method to assess the reactivity of aged asphalt binder in reclaimed asphalt pavement (RAP). The method involves sample preparation, extraction, purification, heating, and rheological measurements. The rheological properties are evaluated using a dynamic shear rheometer (DSR), which measures viscoelastic response and complex shear modulus. The data is then analyzed to quantify the degree of binder activity, using various rheological parameters and mathematical models. The results can be used to compare different RAP materials or evaluate the effectiveness of different binder modification techniques. The data can also predict the performance of RAP binders in asphalt mixtures, such as their resistance to rutting or cracking [14]. The study highlights the feasibility and benefits of recycling RAP and steel slag aggregates in WMA production. The findings support the use of these recycled materials in sustainable asphalt pavement construction, promoting resource conservation and reducing environmental impacts [15]. This study investigated the effects of incorporating steel slag aggregate and different types of bitumen emulsions on the performance of micro-surfacing mixtures. The results of this research can contribute to enhancing the durability, sustainability, and overall performance of micro-surfacing as a pavement maintenance technique [16]. The study highlights the potential of steel slag as a sustainable alternative in road construction materials. By utilizing steel slag, researchers and engineers can develop more durable and eco-friendly asphalt mixtures and aggregate bases, contributing to the overall sustainability and performance of infrastructure projects [17]. indicated use of bagasse fibers enhanced the Marshall properties and moisture susceptibility in stone mastic asphalt mixes which contributes to sustainability [3].

2 Methodology

The methodology for the present work is shown in Fig. 3.

2.1 *Material Characterizations*

Tests on Aggregates

Crushing test of the aggregates were obtained as **21.98%** which is less than **30%** as per the specifications of IRC and MORTH [18, 19], and which influences the compressive strength of the pavement. The toughness value of aggregates was found to be **22.1%** which is less than **24%** as per the specifications of IRC and MORTH, which influences the ability of the aggregates that resist shock or sudden impact load on it. The combined index of aggregates was found to be **15.78%** which is less than **35%** as per the specifications of IRC and MORTH. The particle shape regulates how aggregates were packed into a dense configuration and also regulates the internal

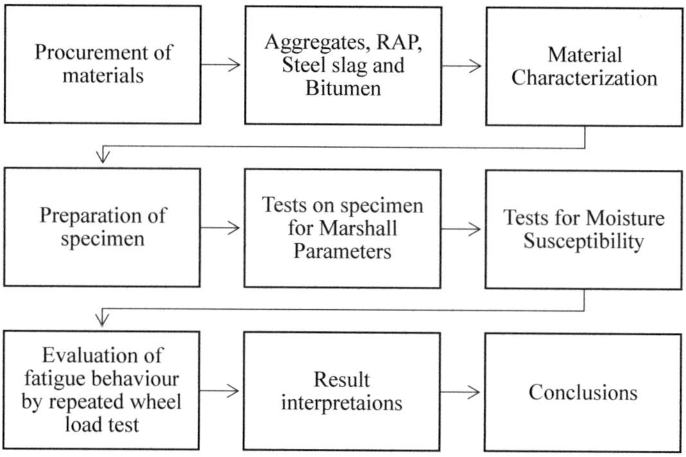

Fig. 3 Flow chart representing proposed methodology

resistance of a mix. The specific gravity of coarse aggregates were found to be **2.62,** and fine aggregates were found to be **2.17.** Materials which have higher specific gravity are considered as having higher strength.

Tests on Bitumen

Softening point test of bitumen was found to be **52.5 °C** which influence the material's resistance to deformation under high temperature. Grade of bitumen was found to be **62** which influences the stiffness and viscosity of asphalt mixture, a lower penetration grade indicates a harder bitumen, while a higher penetration grade implies a softer one. Ductile length was found to be **95 cm**, adequately ductile bitumen is more flexible and fatigue-resistant, which prolongs the life of road surfaces. Specific gravity was found to be **1.04**, which is heavier than water and influences the properties of bleeding, rutting, raveling, and fatigue of bituminous mixtures [18, 20].

Tests on Steel Slag

Crushing test of the steel slag was obtained as **45.43%.** The toughness value of steel slag was found to be **40.41%.** The specific gravity of steel slag was found to be **2.55.** These findings improve the strength and stiffness, rutting and fatigue resistance, and low moisture susceptibility in bituminous mixture.

Tests on RAP (Recycled Asphalt Pavement) Materials

Bitumen content of RAP was found to be **3.288%** which influences in the reduced bitumen content of bituminous mixtures. Crushing strength was found to be **22.33%** which is less than **30%** as per the specifications of IRC and MORTH and which influences the compressive strength of the pavement. Toughness was found to be **22.42%** which is less than **24%** as per the specifications of IRC and MORTH, which influences the ability of the aggregates that resist shock load or sudden impact on it. Specific gravity was found to be **2.42**, and materials which have higher specific gravity are considered as having higher strength.

3 Tests on Bituminous Mixes

The Marshall stability test is a widely used method for evaluating the optimum binder content of bituminous mixes, specifically hot mix asphalt (HMA). It measures the resistance of the mix to deformation and provides an indication of its strength and stability. The test helps determine the optimum bitumen content for a given mix design and assess the quality of the resulting asphalt mixture.

3.1 *Test on Conventional Bituminous Concrete Grade-1 Mix*

The details related to computation of OBC for conventional mix and marshall parameters are shown in Table 1.

Graphical representations of test on Conventional Bituminous Concrete Grade-1 mix.

See Figs. 4, 5, 6, 7, 8 and Table 2.

Calculations of Optimum Bitumen Content (OBC) of Conventional Bituminous Concrete Grade-1 Mix

Bitumen content with respect to maximum stability—B1.
Bitumen content with respect to maximum density—B2.

Table 1 OBC Calculations for conventional mix

BC (%)	Gm (g/cc)	Vv (%)	VFB (%)	Flow (mm)	Stability (kN)
4.5	2.34	5.33	65.54	1.86	19.05
5	2.35	4.11	73.34	1.87	20.06
5.5	2.36	2.85	81.45	2.215	17.15
6	2.33	3.42	79.76	3.37	11.65

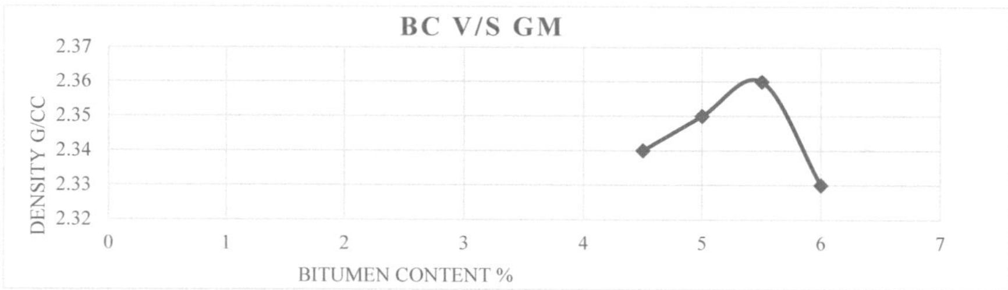

Fig. 4 Asphalt content versus bulk density for conventional BC mix

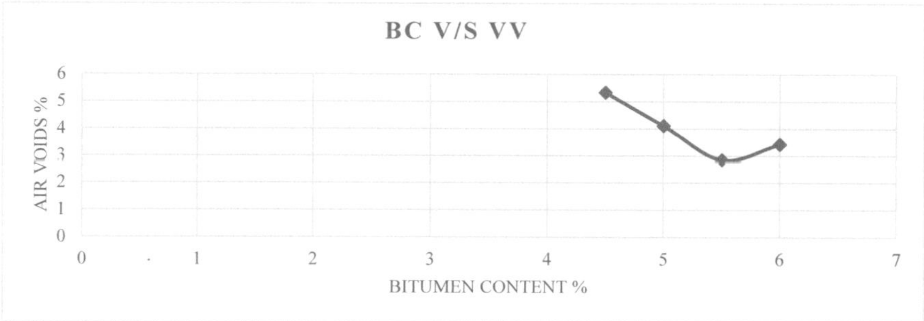

Fig. 5 Asphalt content versus air voids for conventional BC mix

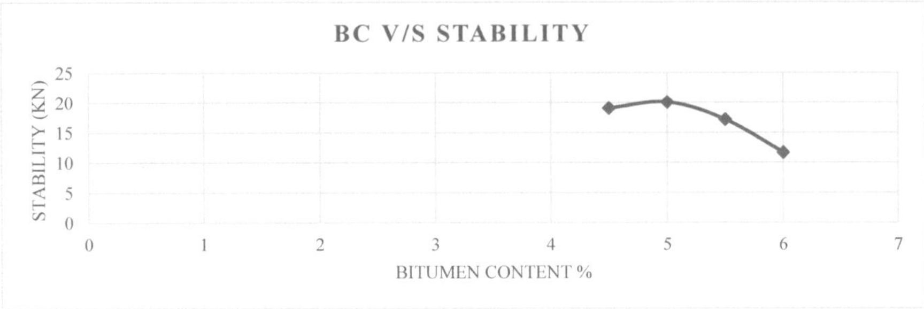

Fig. 6 Asphalt content versus stability for conventional BC mix

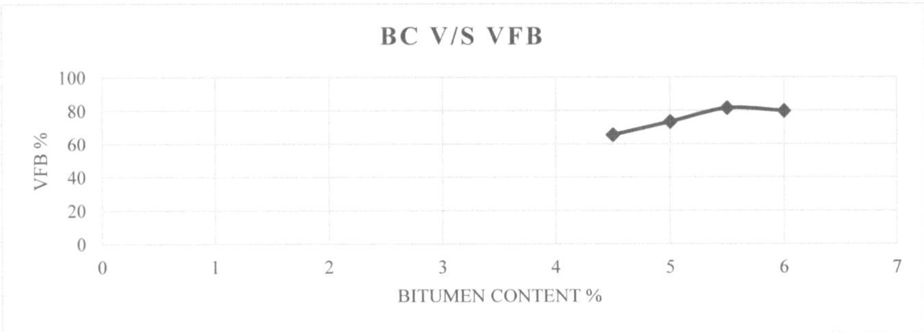

Fig. 7 Asphalt content versus VFB for conventional BC mix

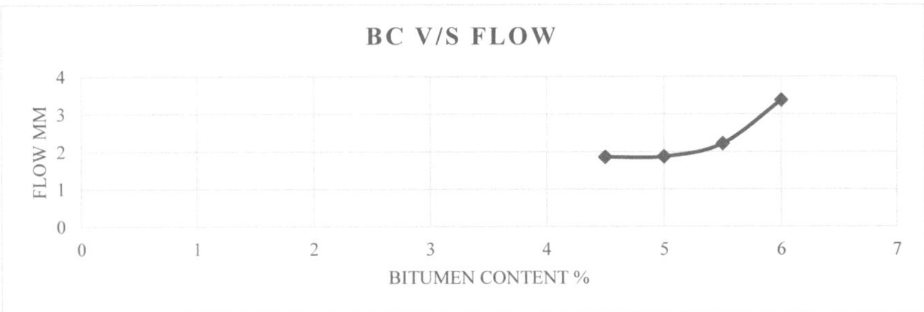

Fig. 8 Asphalt content versus flow for conventional BC mix

Table 2 OBC calculations for conventional mix

Marshall parameters	Percentage of bitumen
Maximum stability	5
Maximum density	5.5
4% voids	5.01
OBC	5.17

Bitumen content with respect to a value of 4% Vv—B3.
Hence, optimum bitumen content (OBC) = [B1 + B2 + B3]/3.

3.2 Test on Bituminous Concrete-1 Mix Blended with 10% RAP and 5% Steel Slag

See Table 3 and Figs. 9, 10, 11, 12 and 13.

Calculations of Optimum Bitumen Content (OBC) of Bituminous Concrete Grade-1 Mix Blended with RAP and Steel Slag

Bitumen content with respect to maximum stability—B1.
Bitumen content with respect to maximum density—B2.
Bitumen content with respect to a value of 4% Vv—B3.
Hence, optimum bitumen content (OBC) = [B1 + B2 + B3]/3.

See Table 4.

Table 3 Observations of test on BC-1 mix blended with 10% RAP and 5% steel slag

BC (%)	4.5	5	5.5	6
Gm (g/cc)	2.35	2.37	2.36	2.34
Gt (g/cc)	2.45	2.43	2.41	2.39
Vv (%)	4.10	2.58	2.03	2.19
Vb (%)	10.17	11.38	12.49	13.50
VMA	14.27	13.96	14.51	15.68
VFB	71.28	81.51	86.04	86.05
Stability (kN)	14.36	16.67	11.55	10.65

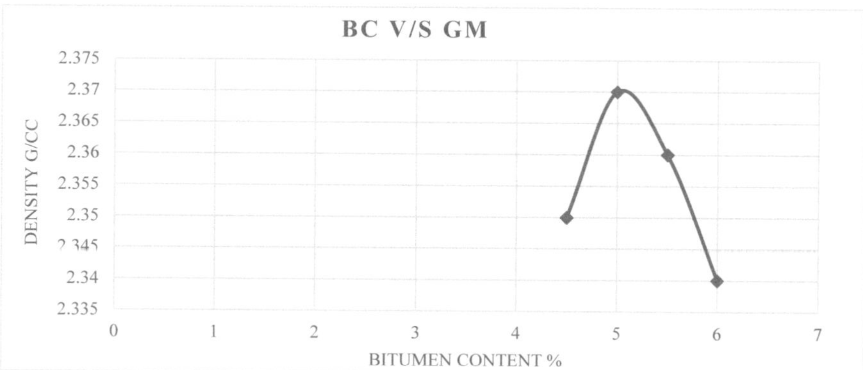

Fig. 9 Asphalt content versus bulk density for BC mix with steel slag and RAP

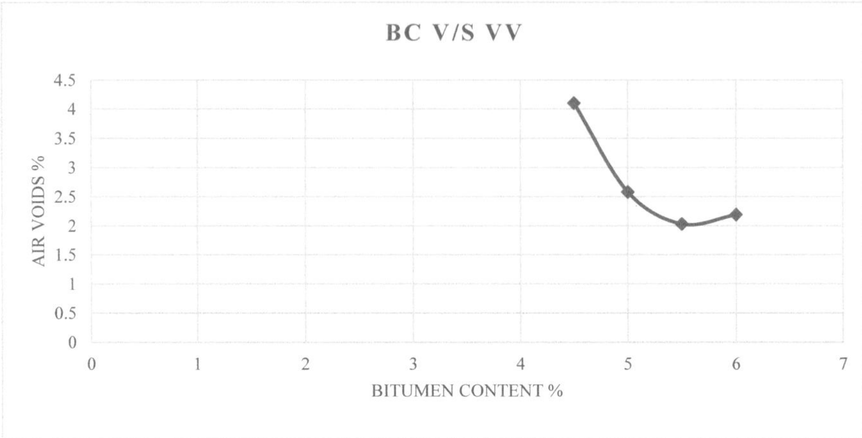

Fig. 10 Asphalt content versus air voids for BC mix with steel slag and RAP

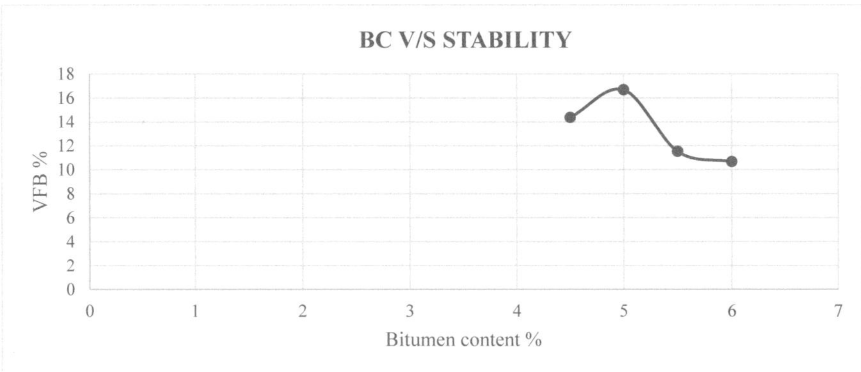

Fig. 11 Asphalt content versus stability for BC mix with steel slag and RAP

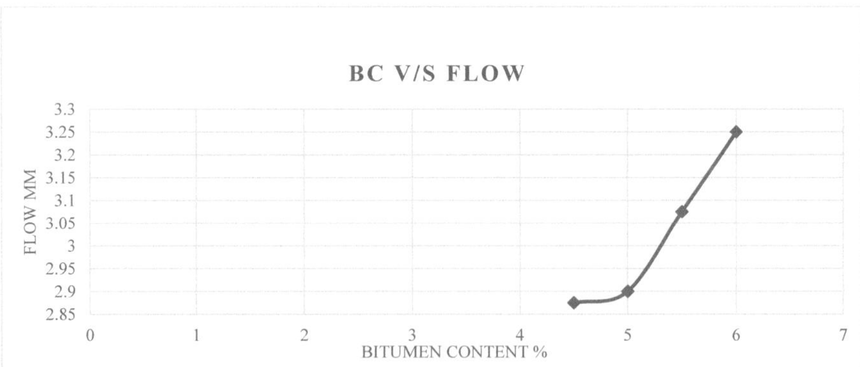

Fig. 12 Asphalt content versus flow for BC mix with steel slag and RAP

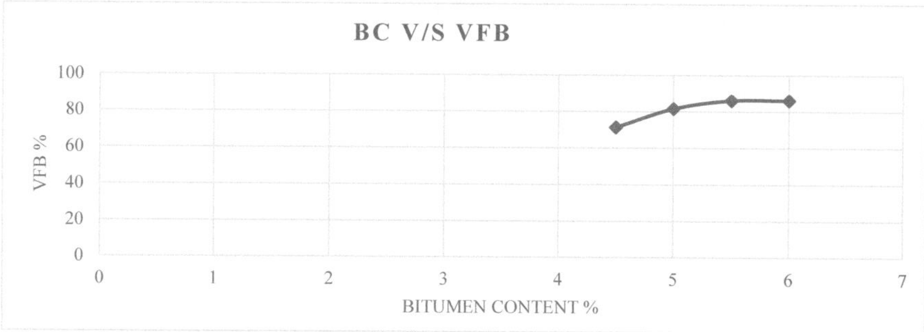

Fig. 13 Asphalt content versus VFB for BC mix with steel slag and RAP

Table 4 Calculation of OBC for BC-1 mix blended with Steel Slag and RAP

Marshall parameters	Percentage of bitumen
Maximum stability	5
Maximum density	5
4% voids	4.6
OBC	4.87

3.3 Comparison of Marshall Stability Test Results

See Table 5.

4 Water Sensitivity Test

The indirect tensile strength was determined for conventional BC grade-1 mix and mix blended with steel slag and RAP, and following results were obtained as shown (Tables 6 and 7; Figs. 14 and 15).

Table 5 Marshall Parameters for conventional BC grade-1 mix

Mix type	OBC (%)	Gm (g/cc)	Gt (g/cc)	Vv (%)	Vb (%)	VMA (%)	VFB (%)	Stability (kN)	Flow (mm)
Conventional	5.17	2.34	2.44	7.6	11.73	15.28	76.75	19.85	3.35
Blended with steel slag and RAP	4.87	2.36	2.46	5.00	10.97	15.97	68.68	21.86	2.8

Table 6 Water Sensitivity test results

OBC (%)	Unconditioned			Conditioned			Tensile strength (kPa)	
	P (kN)	D (mm)	t (mm)	P (kN)	D (mm)	t (mm)	Unconditioned sample(S1)	Conditioned sample(S2)
5.17	7.82	101.6	65.5	6.87	101.6	67	0.748	0.642
4.87	9.16	101.6	63.5	8.52	101.6	67	0.90	0.80

Table 7 TSR test results

Mix type	TSR	TSR (%)
Conventional	0.859	85.88
Blended with RAP and steel slag	0.882	88.16

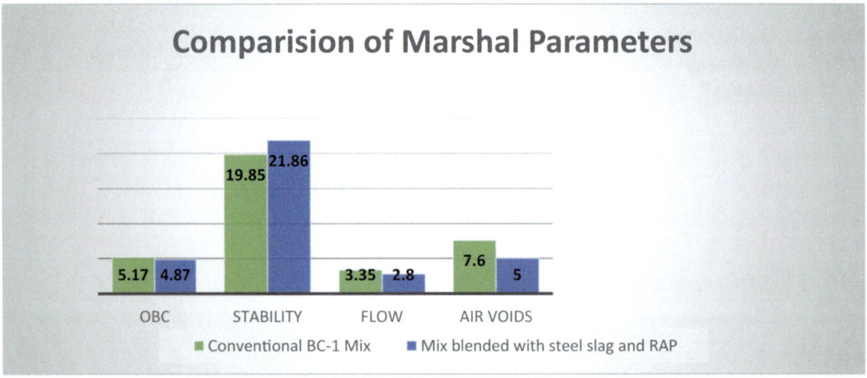

Fig. 14 Comparative analysis of Marshall parameters

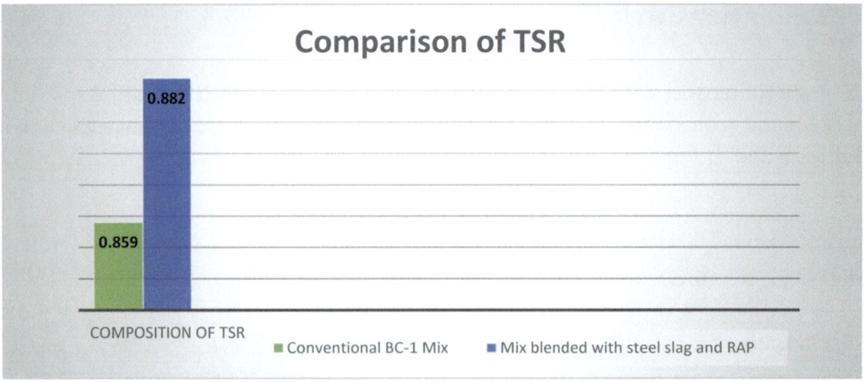

Fig. 15 Comparison among different types of mixes is shown

Table 8 Fatigue test results

Specimen	OBC	Maximum tensile load (kg)	Load applied (40% of max load) in kg	Number of cycles taken to attain failure
Unconditioned	5.17	782	312.8	432
	4.87	916	366.40	908

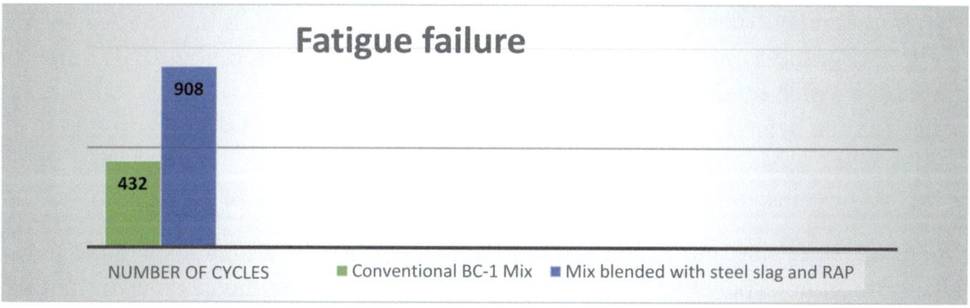

Fig. 16 Bar chart showing number of cycles taken by test specimens to attain fatigue failure

5 Determination of Fatigue Characteristics

5.1 Repeated Wheel Load Test

The details related to repeated wheel load test is shown in Table 8.

5.2 Results

The number of cycles taken by the bituminous concrete grade-1 mix blended with RAP and Steel slag is 908 and that taken by conventional bituminous concrete grade-1 mix without RAP and steel slag is 432. The greater difference is attributed to the fact that the mix blended with steel slag and RAP can resist repeated varying wheel load (Fig. 16).

6 Conclusion

The incorporation of steel slag and RAP in bituminous concrete mix showed significant impacts on Marshall parameters and fatigue behavior.

- The physical properties of RAP aggregates like crushing strength and impact value were found to be 22.095% and 22.42%, respectively, and that of steel slag was found to be 45.43% and 40.4%, respectively.
- The Optimum Bitumen Content (OBC) decreased from 5.17% to 4.87% when the mix was blended with steel slag and RAP, indicating reduced bitumen requirement and improved performance.
- Steel slag and RAP enhanced stability by 21.86% which contributed to increased strength and stiffness.
- The BC grade-1 mix with RAP and steel slag exhibited a higher TSR value of 88.16% which rendered mix to be moisture susceptible.
- The fatigue behavior of BC grade-1 mix with RAP and steel slag was found to be 908 cycles greater than fatigue behavior of conventional BC grade-1 mix which was 432 cycles and thereby depicts improved performance of pavement.

Overall, steel slag and RAP can be effective, eco-friendly, and sustainable materials in Bituminous pavement construction. This innovation enhances road durability, stability, and load-bearing capacity and reduces environmental impact, providing a cost-effective approach to paving.

References

1. Vengala J, Dharek MS, Pramod K, Thejaswi P, Poudel A (2024) Effect of sugarcane bagasse ash and lime on physico-mechanical properties of clayey soil. In: Usman M (ed) Advances in civil engineering, vol 2024 Issue 1. Wiley. https://doi.org/10.1155/adce/3516016
2. Dharek MS, Sunagar P, Sreekeshava KS, Nagashree B, Thejaswi P, Kilabanur P, Nruthya K, Satish Chandra C (2021) Experimental investigations on strength performance of the brick produced by blending demolished waste with pozzolanic materials. In: Lecture notes in civil engineering. Springer, Singapore, pp 573–583. https://doi.org/10.1007/978-981-33-4590-4_54
3. Thejaswi P, Vengala J, Dharek MS, Manjunatha M, Poudel A (2023) Sugarcane bagasse fibers for enhancing moisture susceptibility properties in stone mastic asphalt. In Nawab Y (ed) Advances in materials science and engineering, vol 2023. Hindawi Limited, pp 1–13. https://doi.org/10.1155/2023/5378738
4. Ochieng N, Owino PCO, Bhattarai SK (2017) Effects of reclaimed asphalt pavement (RAP) on mechanical properties of asphalt concrete mixtures. Constr Build Mater 132:366–375
5. Shu J, Wu Z, Li H (2016) Performance evaluation of steel slag asphalt mixtures: laboratory and field investigations. Constr Build Mater 105:99–106
6. Al-Khateeb BKA, Bani-Hani MS (2015) Evaluating the properties of asphalt mixtures containing reclaimed asphalt pavement (RAP) using accelerated loading tests. Constr Build Mater 101:543–551
7. Khanna SK, Justo CEG, Veeraragavan A, Highway materials testing laboratory manual. Nem Chand Bros, Roorkee
8. Kadiyali LR,Highway engineering. Khanna Publishers, New Delhi
9. Huang B, Chen C (2014) Performance evaluation of polymer-modified Bituminous concrete mixtures. Constr Build Mater 52:148–154
10. Dondi G, Mazzotta F, Lantieri C, Cuppi F, Vignali V, Sangiovanni C (2021) Use of steel slag as an alternative to aggregate and filler in road pavements. Materials (Basel) 14(2):345. PMID: 33445570; PMCID: PMC7826912. https://doi.org/10.3390/ma14020345

11. Hainin MR, Aziz A, MdM, Ali Z, Putra Jaya R, El-Sergany MM, Yaacob H (2015) Steel slag as a road construction material. Jurnal Teknologi 73(4). Penerbit UTM Press. https://doi.org/10.11113/jt.v73.4282
12. Hassan Z, Khabiri MM. Preventive maintenance of flexible pavement and mechanical properties of steel slag asphalt. J Environ Eng Landsc Manage. 1648-6897. https://doi.org/10.1080/16486897.2007.9636928
13. Hesami S, Ameri M, Goli H, Akbari A (2014) Laboratory investigation of moisture susceptibility of warm-mix asphalt mixtures containing steel slag aggregates. Int J Pavement Eng 16(8):745–759
14. Sobieski T, Arámbula-Mercado E, Epps Martin A (2021) A method for characterizing RAP degree of binder activity. In: Road materials and pavement design, vol 23, issue 10. Informa UK Limited, pp 2250–2270
15. Fakhri M, Ahmadi A et al (2017) Recycling of RAP and steel slag aggregates into the warm mix asphalt: a performance evaluation. Constr Build Mater 147:630638. ISSN 09500618. https://doi.org/10.1016/j.conbuildmat.2017.04.117. https://www.sciencedirect.com/science/article/pii/S0950061817307687
16. Zalnezhad M, Zalnezhad M, Hesami E (2018) Effect of steel slag aggregate and bitumen emulsion types on the performance of microsurfacing mixture. J Traff Transp Eng (English Ed). https://doi.org/10.1016/j.jtte.2018.12.005
17. Shiha M, El-Badawy S, Gabr A (2020) Modelling and performance evaluation of asphalt mixtures and aggregate bases containing steel slag. Constr Build Mater 248:118710. ISSN 0950-0618. https://doi.org/10.1016/j.conbuildmat.2020.118710
18. Specifications for roads and bridge work (Fifth Revision). MORTH & Indian Road Congress, New Delhi
19. IS: 2386 (Part–I) (1963) Code of practice for methods of tests for aggregate for concrete. BIS, New Delhi, India
20. IS 1201-1220 (1978) Code of practice for methods for testing tar and bituminous materials. BIS, New Delhi, India

A Study on Embodied Energy of GGBFS-Based Mortar Mixes

E. Chaithra, Mangala Keshava, and B. N. Varsha

Abstract The study looks into the idea of employing ground granulated blast-furnace slag (GGBFS) as a partial replacement in cement mortar to reduce environmental issues associated with waste disposal. This study examines cement mortar mixes 1:3 and 1:6, replacing cement with GGBFS in partial replacements of 0, 10, 20, 30, 40, and 50%. The engineering properties of materials, such as flow and compressive strength of each mix at the 7- and 28-day curing ages, were investigated. The embodied energy (EE) for all the mixes of mortar has been computed in order to check and compare EE of conventional mortar with GGBFS incorporated mortar. This study shows that GGBFS can be used in mortar application as partial substitute for cement. The present study thus aims to determine the mortar mix which has the least EE. The study also reveals that EE of GGBFS-replaced mortar is a step toward minimal energy consumption when compared with the control mix.

Keywords GGBFS · Compressive strength · Embodied energy · Mortar

1 Introduction

Waste generation has significantly increased as a result of rising urbanization and population. Solid wastes are produced by industrial, agriculture, mining, domestic, and C&D activities [1, 2]. The CPCB reported that India produces more than 55 million ton (MT) of solid waste annually, out of which 14 MT of waste is disposed of in dumps. The industrial revolution is crucial to the growth of any nation, and it

E. Chaithra (✉) · M. Keshava (✉) · B. N. Varsha
Department of Civil Engineering, BMS College of Engineering, Bengaluru, India
e-mail: chaithrae@bmsce.ac.in

M. Keshava
e-mail: mk.civ@bmsce.ac.in

B. N. Varsha
e-mail: varshabn.civ@bmsce.ac.in

Y. K. Guruprasad et al. (eds.), *Recent Advances in Infrastructure Development and Management—Volume 1*, Lecture Notes in Civil Engineering 570,
https://doi.org/10.1007/978-981-96-1554-4_19

is essential. The main downside of industrial development is the production of an excessive amount of waste as by-products, which pollute the environment once they are disposed of. The usage of industrial by-products in construction is adopted by developing countries for sustainable development [3]. The use of industrial residues in making of construction materials has been one of the most effective avenues in the field of civil engineering [2]. The primary contributors of solid wastes generation from the industries are fly ash, pond ash, and bottom ash which are the residues producing from thermal power plants, the Iron and Steel mills producing blast-furnace slag and steel melting slag, mines producing red mud and tailings, C & D activities generating brick waste, concrete waste, tiles waste, etc., cement factories releasing cement kiln dust, silica fume, etc. [4].

Ground granulated blast-furnace slag (GGBFS) by-product is one of the most common and largest producer in terms of quantity and proved to be one of the best cementitious material. GGBFS can be used as a partial or full replacement for cement in order to reduce environmental imbalances and disposal issues. In India, it is estimated that more than 10 MT of iron and steel slag is being produced per annum by existing iron and steel plants [1]. There is possibility of significant energy savings and a drop in CO_2 emissions when traditional material consumption is minimized through the use of alternative materials, methods, and technique. Embodied energy (EE) is the total energy used by building materials during the raw material extraction, processing, manufacturing, and distribution stages [5]. The primary objective of this study is to partially replace GGBFS in cement mortar with cement as a binder. The compressive strength of various mortar mixes is determined. It emerged that the EE of the alternate waste material database is limited. Thus, an assessment of EE of the GGBFS material and mortar mixes with GGBFS has been performed.

2 Earlier Studies

The study [6] investigated the rheological properties of cement mortar in a 1:3 mixture, with fly ash and GGBS replacing 10% to 40% and 10% to 70% of the cement content, respectively. The results indicated that the strength of mixtures containing GGBS was superior to those containing fly ash. Additionally, the rheological behavior of mortars was found to be similar across the two types of mixes.

This paper [7] explored the effect of pre-dissolving ground granulated blast-furnace slag (GGBS) in water prior to incorporating it into mortar. Two mixing methods were compared: conventional and an new method where GGBS was dissolved in water for 1, 3, 6, and 12 h before being mixed with cement and sand. Various GGBS replacement levels of 0%, 2.5%, 5%, 10%, and 20% were examined. The new method produced higher compressive strength compared to the conventional method, with the highest strength observed at 2.5% and 5.0% GGBS after 1 h of immersion in water. This improvement was attributed to the hydrolysis of GGBS in water and increased ion mobility in the 1:3 mortar mix.

The study [8] was carried out to improve early-age strength by adding cement and GGBS to mortar mixes. Cement replacement with 0, 30, 50, and 70% of GGBS in a 1:3 mortar mix with a 0.5 water-to-binder ratio was examined. Notably, mortar mix compressive strength results replaced with 30% and 50% GGBS were nearly equal, showing 47% improvements in percentage and 32%, respectively. This highlights that 30% replacement level offers the greatest value in compressive strength.

This paper [5] examined the energy required and carbon dioxide (CO_2) emissions from construction materials, specifically aggregates and masonry units, in Bengaluru, India. It used a "cradle-to-gate" method to assess the environmental effects of these materials. The research emphasizes the need for region-specific data and standardized methods for sustainable development. The embodied energy (EE) of sand obtained is 0.119 MJ/Kg.

The study [9] analyzed EE in building materials from manufacturers in Bengaluru City. It covers the entire lifecycle of a material, from production to transportation. The research provides insights into energy dynamics, enabling more sustainable construction practices. They conclude that the embodied energy of sand for a distance of 50 km is 76.85 MJ/m^3, and for cement, it is 3.6 MJ/Kg.

This paper [10] evaluates the embodied energy (EE) of common building materials in India, focusing on the extraction, transportation, and manufacturing processes. The building industry, a significant consumer of materials, contributes to 20–25% of the country's total energy demand. The study proposes alternatives with lower embodied energy, promoting sustainable construction practices. The embodied energy of M-sand 0.105 MJ/kg and that of cement 7.36 MJ/Kg.

The study [11] highlighted the need for energy conservation in the production and transportation of building materials, particularly in India. They examined the energy consumption in manufacturing building materials like cement and steel, as well as transporting these materials. The findings suggest that adopting energy-efficient and alternative materials can reduce reliance on conventional materials and reduce greenhouse gas emissions. The EE of cement obtained is 5.85 MJ/Kg.

The earlier studies have been carried out on 1:3 mortar mixes with GGBFS as partial replacement to cement in various proportions. GGBFS replacement up to 30% has yielded good compressive strength, thus indicating the effective utilization of GGBFS. Studies on computing EE of common available building materials, such as cement (C), fine aggregate (FA), and coarse aggregate (CA), have been reported. Embodied energy (EE) of alternate materials is limited. EE of mortar mixes incorporating the alternate material is also limited. Hence, there is scope for computing EE of marginal materials and EE of mortar mixes utilizing these by-products. This is the main findings of the present study.

Hence, following are the objectives drawn:

- To determine the EE of GGBFS material available in Karnataka.
- To calculate EE for standard and alternate mixes of mortar mixes of proportion 1:3 and 1:6.

Scope of the Present Study

- The purpose of the study is to identify appropriate mortar mixtures that meet both the strength parameter and energy criteria when the GGBFS is utilized as binder replacement.
- IS: 1905-1987 lists seven different categories of mortar mixes varying from high, medium, and lean mixtures. Commonly used mix in the category of high mixture is H2-1:3 and in the medium mixture is M2-1:6 for masonry purposes. Thus, the same mixes have been chosen for further study.

3 Materials and Methodology

1. **Cement**—The ordinary Portland cement (OPC)-53 grade was utilized as the main binder collected from nearby market for this investigation. The tests were carried out to examine physical properties, such as specific gravity, initial and final setting times, and consistency, were found to comply with IS: 12269-1987.
2. **GGBFS**—This is made by reducing melting iron slag from a blast furnace in steam or water, resulting in a grainy, smooth residue that is quickly cooled and powdered into small particles. This finely ground substance has properties equivalent to the properties of cement [12]. The GGBFS is obtained from Jindal South West (JSW) local dealer in Bengaluru. The basic property testing was done as per IS: 16714-2018. The standard consistency test of GGBFS done by using Vicat Apparatus confirming IS code. Cement and GGBFS characteristics are displayed in Table 1.
3. **Fine aggregate (FA)**—Manufactured sand (M-sand) is utilized as fine aggregate. The physical characteristics of fine aggregate conducted include specific gravity, water absorption, and fineness modulus. Table 2 represents properties of M-sand.

Table 1 Properties of cement and GGBFS

Properties	Cement	GGBFS
Specific gravity	3.12	2.87
Standard consistency (%)	31	35
Initial/final setting time	120 min/300 min	–
Silicon dioxide (SiO_2)	21.08*	34.81*
Aluminum oxide (Al_2O_3)	4.42*	17.92*
Ferric oxide (Fe_2O_3)	3.71*	0.66*
Calcium oxide (CaO)	63.78	37.63*

* Values are from Literature [12]

Table 2 Properties of fine aggregate

Properties	Values
Specific gravity	2.87
Water absorption (%)	2.79
Fineness modulus	3.58

3.1 Methodology

Figure 1 indicates the brief procedure adopted in the current investigation.

4 Details of Experimental Work

The effects of using GGBFS to partially replace cement have been investigated in this study. One control mix (CM) and five various replacements 10, 20, 30, 40, and 50% of GGBFS have been used in place of cement. An initial amount of cement was taken, and water content was determined through a flow table test and designed to have a flow of $110 \pm 5\%$. The water/binder (w/b) ratio was thus obtained. Laboratory experiments were carried out, such as flow test and compressive strength based on Indian Standards (IS). The compressive strength of substituted mortar mixes has been compared with control mix (CM) at curing ages 7 and 28 days. Table 3 shows the mix proportion of mortar mixes. The proportions have been obtained considering the densities of cement and fine aggregate separately. The standard size of 50 mm $\times$ 50 mm $\times$ 50 mm cubes was cast. Each combination has six cubes cured under normal water condition, and a total of seventy-six specimens were cast, demolded, cured, and tested.

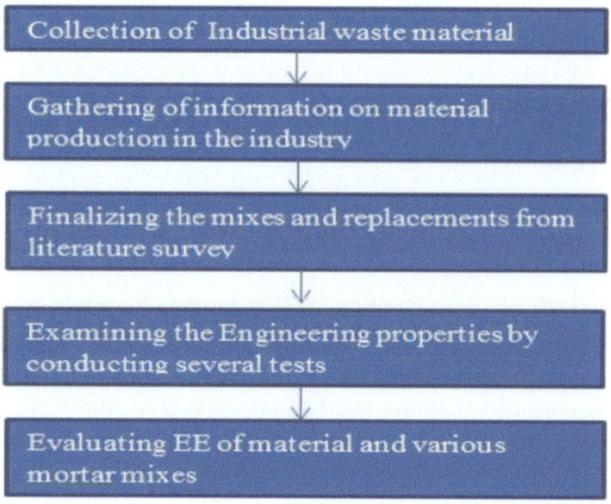

Fig. 1 Brief methodology

Table 3 Mix proportion for 1 m^3 of mortar

Grade	Cement (Kg/m$^{3)}$	Fine aggregate (Kg/m$^{3)}$
1:3	378	1432
1:6	216	1640

Table 4 EE of common materials in MJ/Kg

Materials	Min EE	Max EE	Comb1	Comb2	Comb3	Comb4
Cement	3.6	5.85	3.6	3.6	5.85	5.85
FA	0.08	0.12	0.08	0.12	0.08	0.12

The consistency of fresh mortar is evaluated in order to calculate the necessary flow for masonry mortar. With the use of this test, it is possible to ascertain the necessary flow of 110 ± 5% in accordance with IS: 2250-1981. Cubes were tested for compression in compliance with IS 516-1959. Three cubes each were examined and tested on the 7th and 28th days.

4.1 Embodied Energy of Common Materials

The EE of common materials were obtained from the literature, for water, fine aggregate, and cement. According to published research, EE of water is zero, fine aggregate varies from 0.08 to 0.12 MJ/Kg, and cement ranges from 3.6 to 5.85 MJ/Kg [5, 9, 11]. Therefore, in order to calculate the EE of mortar mixes, the minimum and maximum ranges of EE of common materials are taken and their four possible combinations are considered. Materials with a lower embodied energy are more environmentally friendly than those with a greater energy [13]. Combinations 1 to 4 are when the minimum and maximum EE values for cement and fine aggregate are used in mortar shown in Table 4. Four combinations were adopted to determine the EE considering all possibilities of the minimum and maximum values of EE of the materials. Comb1 represents combination 1.

4.2 EE of Alternate Material (GGBFS)

The EE of alternate material is computed based on "cradle-to-gate" boundary condition. This includes energy utilized for the production of GGBFS. Hence, the GGBFS is a by-product from iron manufacturing unit. Thus, only the processing activities like grinding, drying has been done before reuse. In this study, GGBFS is obtained from JSW and the information about the energy consumption utilized for processing of GGBFS is collected from JSW Office, Bengaluru [15], for computation of EE. The details are as follows.

Data from JSW, Bengaluru
Energy consumption for production of one ton of GGBFS.

Type of energy	Energy consumption (kWh/T)
Electricity	30
Thermal energy	0.075
Total energy	30.075

$$\text{Embodied Energy} = \text{Total Energy} \times \text{P}$$

where P—conversion factor.

Conversion factor: 1 kWh of power requires 0.68 kg of coal + 20% transmission loss [5].

$$1\ \text{unit/kwh of power} = 0.68 + (0.2 \times 0.68) = 0.816\ \text{Kg of coal} \tag{1}$$

$$\text{Energy density of 1 Kg of coal} = 24\ \text{MJ} \tag{2}$$

$$\text{P} = (1) \times (2) = 0.816 \times 24 = 19.584$$

$$\text{Embodied energy} = (30.075 \times 19.584)/1000 = \mathbf{0.58\ MJ/kg}$$

4.3 *Embodied Energy in Mortar Mixes*

Embodied energy of mortar mixtures is obtained by multiplying the quantity of materials by their embodied energy [14]. For one cubic meter of mortar, the embodied energy of various mixes was computed. Using the following formula, the EE of control and GGBFS-substituted mortar mixes was calculated and tabulated in Table 5.

$$\text{EE} = Q_{\text{B}} \times \text{EE}_{\text{B}} + Q_{\text{FA}} \times \text{EE}_{\text{FA}} + Q_{\text{W}} \times \text{EE}_{\text{W}}.$$

where

EE—Embodied energy in MJ.

Q_{B}, Q_{FA}, and Q_{W}—Quantity of binder, fine aggregate and water in Kg.

EE_{B}, EE_{FA}, and EE_{W}—Embodied energy of binder, fine aggregate, and water in MJ/Kg.

Table 5 EE of various mortar mixes per kg

Replacement (%)	EE of mortar mix 1:3 (MJ)				EE of mortar mix 1:6 (MJ)			
	Comb1	Comb2	Comb3	Comb4	Comb1	Comb2	Comb3	Comb4
CM	1475.36	1532.64	2325.86	2383.14	908.80	974.40	1394.80	1460.40
10	1361.20	1418.48	2126.65	2183.93	843.57	909.17	1280.97	1346.57
20	1247.05	1304.33	1927.45	1984.73	778.34	843.94	1167.14	1232.74
30	1132.89	1190.17	1728.24	1785.52	713.40	778.70	1053.30	1118.9
40	1018.75	1076.02	1529.04	1586.32	647.88	713.48	939.47	1005.07
50	904.58	961.86	1329.83	1387.11	582.64	648.24	825.64	891.24

EE of marginal and replaced mortar mixes have been computed using the above formula. EE of cement varies from 3.36 to 5.85 MJ/Kg, fine aggregate ranges 0.08–0.12 MJ/Kg, and water is zero. Hence, in this study the EE is computed using the various combinations as listed in Table 5.

5 Result and Discussion

The characteristics of cement and GGBFS were examined, which are as follows:

1. **Consistency Test**—Flow table test is done to determine the water content with the flow rate maintained between 110 and 115% as per IS code provision in mortar for all the mixes. Figure 2 represents the variation in the water/binder ratio for 1:3 and 1:6 mixes.

According to the current study, when GGBFS was used in cement as binder replacement in a higher percentage, there is a fluctuation in water/binder ratio when flow is maintained within the range of 110 to 115% and it has observed that the w/b ratio is more fluctuating in 1:6 grade of mortar.

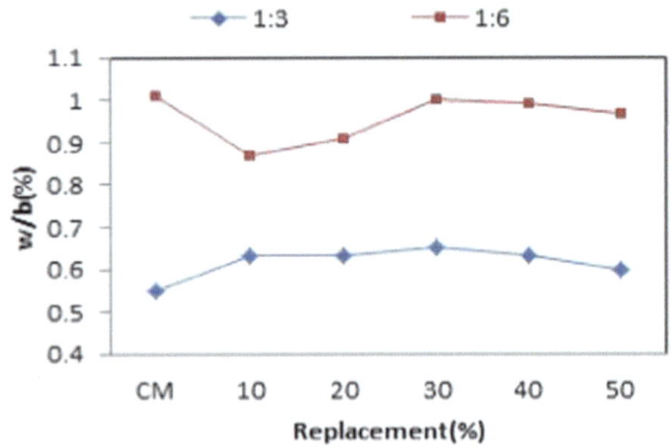

Fig. 2 Consistency test of 1:3 and 1:6 mixes

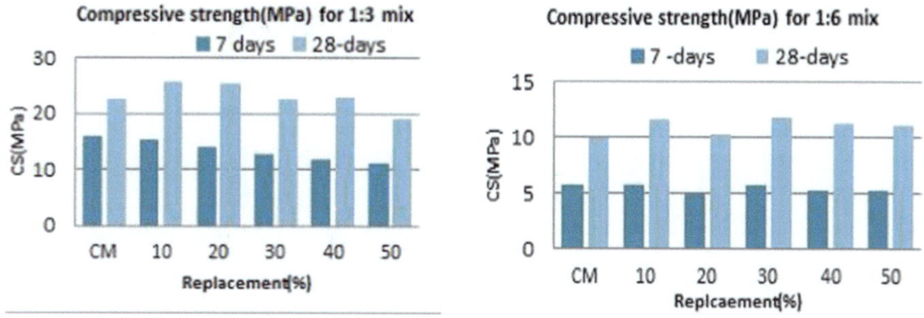

Fig. 3 Compressive strength (MPa) for 1:3 and 1:6 mortar mixes

2. **Compressive Strength**—Compressive test was examined for 7-day and 28-day ages for 1:3 and 1:6 mixes. Figure 3 exhibits the compressive strength results thus obtained.

As per IS specification, the minimum strength for H1 mix is 10 MPa, and for M2 its 5 MPa, the compressive strength properties following 7 days and 28 days of curing demonstrate the results showing that among the replacement percentages, the compressive strength for GGBFS mortar mixes is satisfied. The GGBFS-substituted mortar mixes up to the 50% replacement achieving minimum compressive strength as per code. Thus, the GGBFS, an industrial by-product, can be utilized as binder replacement up to 50% replacements without compromising strength in both the mortars. The maximum compressive strength was found for 10% replacement in 1:3 mortars and 30% replacement in 1:6 mortars.

Embodied Energy (EE)—The EE of GGBFS material was computed for the cradle-to-gate boundary condition, and the result was 0.58 MJ/kg. Cement has a greater EE than GGBFS does. As a result, the GGBFS material is more eco-friendly and can replace for cement as an alternative. Moreover, in order to calculate the EE of mortar mixes, the same EE of GGBFS and common material combinations are adopted. Table 5 results have been presented pictorially in Fig. 4. The EE for the CM and GGBFS-replaced mixes have been computed.

The EE value for various combinations of mix 1:3 has been computed. It is observed that there is a reduction in the EE value when GGBFS material is replaced to cement. The reduction observed for all the four combinations are 39%, 37%, 42.82%, and 41.79%, respectively.

The EE value for various combinations of mix 1:6 has been computed. It is observed that there is a reduction in the EE value when GGBFS material is replaced to cement. The reduction seen in all the four combinations is 35.9%, 33.47%, 40%, and 38.90%, respectively.

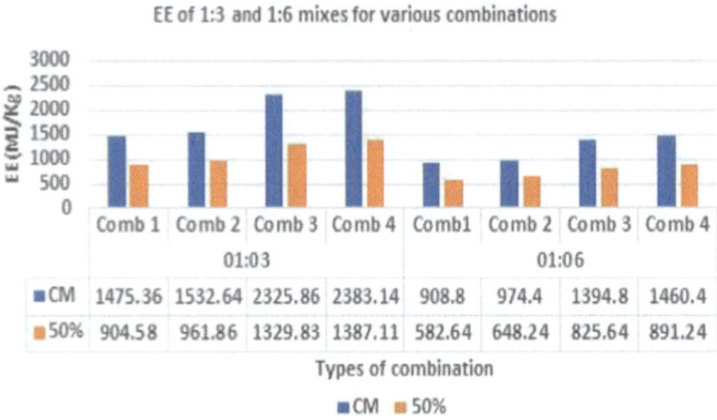

Fig. 4 Variation of EE of 1:3 and 1:6 mortar mixes

6 Conclusion

The investigation yields the following conclusions.

1. There is a fluctuation in w/b ratio when flow is maintained with range of 110–115% for both the mixes, and more fluctuation can be seen in 1:6 mortar mixes.
2. From the study, it has observed GGBFS can be substituted as a binder in place of cement and has adequate strength in both the mortar grades.
3. The GGBFS industrial by-product can be utilized as binder up to 50% replacements without compromising strength in 1:3 and 1:6 mortar mixes.
4. The maximum compressive strength was found for 10% replacement in 1:3 mortars and 30% replacement in 1:6 mortars.

1. The EE of GGBFS locally available was computed to be 0.58 MJ/Kg. The literature reports the EE of GGBFS as 1.58 MJ/Kg which is as per international findings.
2. Percentage savings in EE of 1:3 mixes are observed to be in the range of 37% to 42.82%.
3. Percentage savings in EE of 1:6 mixes are observed to be in the range of 33.47% to 40%.
4. Finally, it was also observed that the reduction in EE in 1:3 mixes is higher than that of 1:6 mortar mixes.
5. Both the mortar mixes have satisfied the strength parameter as specified in the code and show considerable reduction in EE, thus paving the path toward sustainability.

References

1. James J, Pandian P (2016) Soil stabilization as an avenue for reuse of solid wastes: a review. Acta Tech Napocensis Civ Eng Archit 58:50–76
2. Madurwar M, Ralegaonkar R, Mandavgane S (2012) Application of agro-waste for sustainable construction materials: a review. Constr Build Mater 38:872–878
3. Chandra S, Reddy NG, Rao B (2016) Assessment of industrial wastes as a road construction material: a review, 12
4. Siddique R (2014) Utilization of industrial by-products in concrete. Procedia Eng 95:335–347. The 2nd International conference on sustainable civil engineering structures and construction materials
5. Varsha BN, Saranya SJKR, Jayanth KR (2018) Embodied energy of aggregates and masonry units produced around Bengaluru, India. Int J Adv Sci Eng Technol 6:42–45
6. Arularasi V, Thamilselvi P, Avudaiappan S, Saavedra Flores EI, Amran M, Fediuk R, Vatin N, Karelina M (2021) Rheological behavior and strength characteristics of cement paste and mortar with fly ash and ggbs admixtures. Sustainability 13(17):9600
7. Malipeddi R, Adiseshu S (2021) Study of dissolution parameter of ground granulated blast furnace slag as cement replacement on mechanical properties of mortar. Mater Today Proc 44:642–650. International conference on materials, processing characterization
8. Korde C, Cruickshank M, West RP, Pellegrino C (2019) Activated slag as partial replacement of cement mortars: effect of temperature and a novel admixture. Constr Build Mater 216:506–524
9. Surekha B, Hegde MN, Jagadish KS (2016) Energy and building materials. Int J Civ Eng 5:13–24
10. Mithra P, Kuriakose LT, Unnithan A (2015) Embodied energy assessment for building materials. Int J Res Eng Technol 4(3):27–28
11. Reddy B, Jagadish K (2003) Embodied energy of common and alternative building materials and technologies. Energy Build 35:129–137
12. Paul A, Hussain M (2020) Sustainable use of ggbs and rha as a partial replacement of cement in the stabilization of Indian peat. Int J Geosynth Ground Eng 6:03
13. Ramachandra Murthy A, Iyer NR (2014) Assessment of embodied energy in the production of ultra high performance concrete (UHPC). Int J Students Res Technol Manage 2(3):113–120
14. Ramesh T, Prakash R, Shukla KK (2013) Life cycle energy analysis of a multifamily residential house: a case study in Indian context. Open J Energy Effic 2:34–41
15. https://www.jsw.in/groups/reports-jsw-energy

Experimental Analysis of Effect of Water Quality on Strength of Concrete

G. S. Brunda, Ambreshwar, Peddireddy Sreekanth Reddy, Kavya Gurumath, and Bijayananda Mohanthy

Abstract Rapid migration of population to cities forced to utilize the available lands, which may be near water bodies leads to declination in water quality due to contamination. In such conditions, maintaining water quality in the construction activities is expensive due to transportation. In view to mark these issues, the present study aimed to assess the impact of water quality on compressive strength of concrete. For the same, the M25 concrete mix having a water cement proportion of 0.5 was employed and tested. Water samples, such as potable water, treated and untreated sewage water from Chikkabanavara Lake and Hesarghatta Lake, were gathered from various sources and were used for casting the cubes and cylinders. Further, the casted samples were subjected to curing times of 7, 14, 21, and 28 days, compressive and split tensile strength may be achieved. The outcomes revealed that the compressive strength and split tensile strength of the test specimens increased with potable water and treated sewage water from Hesaraghatta lake with curing periods. It is suggestible to utilize the water from Hesaraghatta lake for the nearby construction activities when compared to the prior.

Keywords Water quality · Compressive strength · Tensile strength

G. S. Brunda (✉) · P. S. Reddy · K. Gurumath
Department of Civil Engineering, Acharya Institute of Technology, Soladevanahalli, Bengaluru, Karnataka, India
e-mail: brundags@gmail.com

Ambreshwar
Department of Civil Engineering, Guru Nanak Dev Engineering College, Bidar, Karnataka, India

G. S. Brunda · Ambreshwar · B. Mohanthy
Department of Civil Engineering, National Institute of Technology Mizoram, Aizawl, Mizoram, India

P. S. Reddy
Department of Civil and Infrastructure Engineering, Indian Institute of Technology Dharwad, Dharwad, India

Y. K. Guruprasad et al. (eds.), *Recent Advances in Infrastructure Development and Management—Volume 1*, Lecture Notes in Civil Engineering 570,
https://doi.org/10.1007/978-981-96-1554-4_20

1 Introduction

During building, the construction profession uses a large amount of water for the production of mortar, blending of cement concrete, and cure operations, among other things. It's very difficult to get good quality water or potable water in a huge amount nowadays. The durability of mortar and cement-based concrete used in building work is greatly influenced by the quality and amount of water. The availability of good quality water for construction is very difficult as we have a scarcity of water for the daily needs. Water quality influences concrete qualities such as strength at compression, setting time, and hardening. Water is required in concrete for the hydrating of the cement and its ability to work. Water elements may influence chemical reactions/hydration and consequently the fresh and hardened characteristics of concrete. The existence of different impurities in the concrete-making water influences the compressive capacity of the concrete. The quantity of chemicals or contaminants in concrete also has a substantial influence on its resilience. The use of low-quality water in concrete causes deterioration which leads to collapse, because of this it is very difficult to use good quality water all the time in concrete when the availability of good quality water is difficult.

In this context, Nikhil et al. [1] conducted an experimental analysis on the impact of water quality on the setting, hardening and strength of the concrete, and the findings showed that drinking water resulted in excellent strength attributes in concrete, with 33.34% rise in compressive property while versus wastewater for seven, fourteen, and the entire 28-day strength. Olugbenga [2] evaluated the influence of several impurities like copper, manganese, tin, sodium, zinc, lead, and salts which are present in water on the compressive property of concrete. According to the findings, the origin of water utilized when mixing concrete has a considerable influence on the strength of the concrete when tested. Ikpa et al. [3] the effect of mixing water on the compressive property parameters of concrete is investigated utilizing two types of wastewater treatment sources, groundwater, and distilled water sources. Before utilizing, the setting time required characteristics were determined, with SWWT and TWWT. SWWT and TWWT have longer setup times due to the presence of a higher degree of COD than DW and BW. According to the findings, purified water had the highest mean compressible strength of 17.96 N/mm^2, while secondary wastewater had the lowest average compressive capacity of 12.09 N/mm^2. Ramkar et al. [4] this research investigates the influence of several types of wastewater that have been treated on the durability of concrete qualities such as tensile strength, compressive strength, and flexural strength in comparison with potable water. The results show that the first and last setting times of cement were identical for drinkable water and STWW but dropped for PTWW, compressive strength improved in STWW and home wastewater over an extended period of time, and the tensile and flexural strength tests yielded identical outcomes, indicating that there was no enhancement in tensile and flexural resilience when employing STWW. Saha et al. [5] this paper investigates the effects of different quality of water from different sources on the properties of concrete. The study determines that the impact of deionized water, alkali material, and existence of algae in water affects the setting time of concrete and compressive strength, and the concentration of alkali substances like Na_2CO_3 and $NaHCO_3$ more than 4 g/L extensively affects the mortar setting time. Moses et al. [6] the study focuses on examining the compressive strength of concrete when the different quality of water is used to mix. Tap water, rainwater, and river water are used to cast the concrete, and the cubes are tested for 7 and 28 days. The pH and hardness are varied by adding hydrochloric acid and sodium hydroxide. The study reveals that more pH value water shows more strength than less pH value water, and it is preferable to utilize water with more pH while mixing with concrete.

2 Materials and Methodology

Aggregates are carried from the nearby crusher. Ordinary Portland cement of 43 grade is collected from the local suppliers, and the water samples were collected from Hesaraghatta lake, Chikkabanavara lake. Potable water, Sewage water prior, and later treatment from Acharya sewage water treatment plant. The water samples have been collected by following the guidelines of IS 3025. Water quality analysis has been done for all collected samples. The mix proportions of 1:1:2 have been made by considering the target characteristic compressive strength of 25 MPa and W/C ratio of 0.5. The strength characteristics such as compressive and split tensile strength were conducted on all the prepared specimens in corresponding with the guideline of ASTM C109 and ASTM C498, respectively, with a curing period of 7, 14, 21, and 28 days. The typical photograph of mix and curing can be seen in Figs. 1 and 2.

Fig. 1 Process of mixing in laboratory

Fig. 2 Process of curing

Table 1 Basic test on fine and coarse aggregates

Specific gravity	
Fine aggregate	2.64
Coarse aggregate	2.79
Cement	3.15

3 Results and Discussion

3.1 Basic Tests on Fine and Coarse Aggregates

See Table 1.

3.2 Water Quality Analysis

Physical, chemical, and biological tests were conducted for samples collected from different sources. The test results are shown in Table 2.

Based on the test results shown in Table 2, most of the parameters of Chikkabanavara lake and sewage before treatment samples show not recommendable values for the usage of water in construction. Because of more pH, hardness, alkalinity, and other parameters which are responsible in strength of concrete, the concrete may deteriorate and weaken the concrete. Potable water, sewage water after treatment, and Hesaraghatta lake water show recommendable values for the usage of water in construction. The below clauses explain whether or not we can use the water for building based in the strength characteristics check.

Table 2 Physical, chemical, and biological test results for different water samples

S. No.	Tests	Potable water	Sewage before treated	Sewage after treated	Chikkabanavara lake	Hesarghatta lake
1	pH	6.9	7.6	7.9	8.8	7.7
2	Total hardness (mg/l)	320	856	576	448	425
3	Chloride (mg/l)	179	340	222	344	230
4	Nitrate (mg/l)	9	90	46	26	17
5	Total alkalinity (mg/l)	42	200	140	150	76
6	Dissolved oxygen (mg/l)	24	< 1	6.13	12	16
7	Acidity	Nil	Nil	Nil	Nil	Nil
8	COD (mg/l)	26	464	175	128	48
9	BOD (mg/l)	12	150	56	110	82

3.3 Strength Characteristics

Compressive and split tensile tests are the fundamental tests to test the relevance of concrete in distinct building activities. The same has been adopted to access the overall performance of concrete prepared from exclusive water samples with a curing period of 7, 14, and 28-day periods.

The comparison of concrete's compressive property for different water samples is shown in Fig. 3. Observation shows that for cured periods of 7, 14, 21 and 28, the compressive property of concrete increases. For all the specimens, cast from different water samples and obtained values are shown in Table 2. Before treatment, the Chikkabanvara lake and sewage water show a less compressive strength than other specimens. The increasing pattern of split tensile strength can be observed in Fig. 4. pH values of water increases, the strength of the concrete decreases, and the pH of Chikkabanavara lake and sewage water before treatment shows high value with a decreased values in compressive strength. In addition to compression test, split tensile strength test was also conducted and the outcomes are shown in Tables 3 and 4. With respect to curing days for different water samples, strength of concrete specimens with untreated sewage water and Chikkabanavara lake show less value and the specimen's breaks early for the less load application for tension.

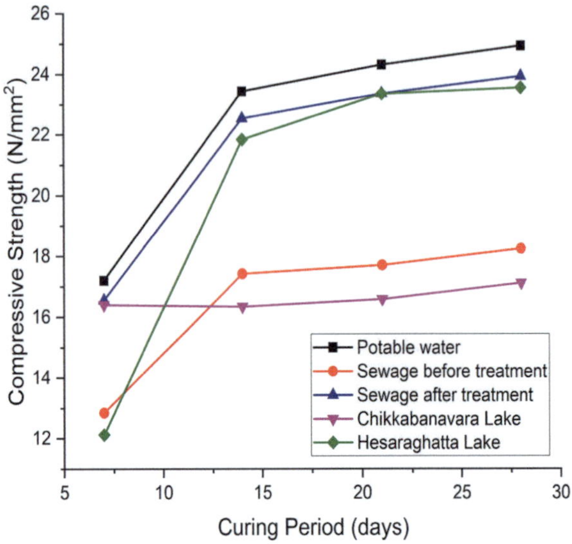

Fig. 3 Concrete strength in compression analysis for several water sample collections

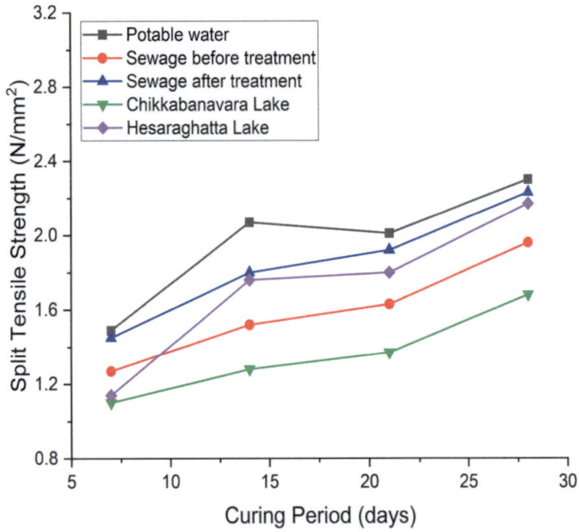

Fig. 4 Concrete strength in tension analysis for several water sample collections

Table 3 Compressive strength of concrete

Source of water sample	Compressive strength (N/mm^2)			
	7 days	14 days	21 days	28 days
Potable water	17.20	23.44	24.31	24.93
Sewage before treated	12.85	17.42	17.70	18.24
Sewage after treated	16.55	22.54	23.36	23.93
Chikkabanavara lake	16.04	16.34	16.58	17.11
Hesaraghatta lake	12.12	21.83	23.35	23.55

Table 4 Split tensile strength of concrete

Source of water sample	Split tensile strength (N/mm^2)			
	7 days	14 days	21 days	28 days
Potable water	1.49	2.07	2.01	2.30
Sewage before treated	1.27	1.52	1.63	1.96
Sewage after treated	1.45	1.80	1.92	2.23
Chikkabanavara lake	1.10	1.28	1.37	1.68
Hesaraghatta lake	1.14	1.76	1.80	2.17

4 Conclusions

The concrete strength characteristics of mix proportion have been evaluated by varying the type of water collected from different locations with a curing time of 7, 14, 21, and 28 days are available. The mix proportions were subjected to experimental investigations, and the following conclusions were derived.

(1) Chikkabanavara lake water and sewage water before treatment show high concentrations in some of the parameters that prohibit the development of strength of concrete.
(2) The concrete compressive strength has been achieved 90% in all the cubes casted using different water samples except Chikkabanavara lake sample for 28 days.
(3) When compared to cubes formed with water from Hesarghatta lake and potable water, the compressive strength of concrete has been shown to a higher strength.
(4) Concrete combined with treated sewage water has a compressive strength that is only slightly lower than that of Hesarghatta lake water after 28 days.
(5) Concrete specimens with untreated sewage water and Chikkabanavara lake water have 30% less compressive strength than potable water and treated sewage water.
(6) Split tensile strength of concrete specimens with untreated sewage water and Chikkabanavara lake showing less value, the specimen's breaks early for the less load application for tension.

References

1. Nikhil TR et al (2011) Impact of water quality on strength properties of concrete. Indian J Appl Res 4(7):197–199
2. Olugbenga ATA (2014) Effects of different sources of water on concrete stress: a case study of Ile-Ife. Civ Environ Res 6:39–43
3. Ikpa CC et al (2021) Evaluation of water quality impact on the compressive strength of concrete. Jurnal Kejuruteraan 33(3):539–550
4. Ramkar AP et al (2016) Effect of treated waste water on strength of concrete. J Mech Civ Eng (IOSR-JMCE) 13(6 Ver. II):41–45. e-ISSN: 2278-1684, p-ISSN: 2320-334X

5. Saha S, Saha P, Talluri N (2022) Effects of quality of water on the setting times and compressive strength of concrete, vol 60, part 1. Science Direct, pp 378–383
6. Moses MK, Nsahlai BL, Penka BJ, Fru CP (2024) Analysis of the influence of water qualities on the strength of concrete. J Eng Appl Sci 110

Sustainability Assessment of a Building Construction Project

Jagdish and S. P Sreenivas Padala

Abstract Sustainability standards for monitoring and evaluation practices currently available around the world are implemented with the long-term impact ensuring resources support and capacity to continue by producing positive outcomes of construction materials by reducing their environmental footprints and conserving resources. Comprehensive sustainable assessment for designed building project employs diverse approaches such as including green building certification programmes, lifecycle assessment, material selection, and embodied carbon analysis. Comparison and validation of building assessment of calculated carbon footprint by manual calculation in comparison with tools such as OneClick LCA and EC3 tool for the sustainability assessment based on the conventional database and sustainable database used for the designed building. The change of material has led to decrease in the carbon footprint throughout its life cycle. The percentage difference for the designed building using conventional practices and sustainable practices of the building comparison wise for the conventional design and sustainable design was found out to be 20.56, 16.86, 19.51% for manual calculation, OneClick LCA tool, and EC3 tool, respectively, for the reduction in carbon footprint upon replacement of materials using sustainable practices.

Keywords Building materials · Carbon emission · Life cycle analysis

Theme Futuristic building materials · Recycled and sustainable materials adopted in construction and · Management practices

---

Jagdish · S. P Sreenivas Padala (✉)
Department of Civil Engineering, M S Ramaiah University of Applied Sciences, Bengaluru, Karnataka, India
e-mail: sreenivasiit@gmail.com

Y. K. Guruprasad et al. (eds.), *Recent Advances in Infrastructure Development and Management—Volume 1*, Lecture Notes in Civil Engineering 570,
https://doi.org/10.1007/978-981-96-1554-4_21

1 Introduction

The pursuit of sustainability has become a paramount concern in modern construction and building projects. As environmental awareness grows, there is an increasing demand for structural materials that not only meet human needs but also minimize their ecological footprint. To address this challenge, a comprehensive approach is required, encompassing various methodologies and strategies that ensure a holistic assessment of a building's sustainability. In this research study, the selected approaches are applied to assess the sustainability of building projects, considering their environmental, social, and economic dimensions. Through rigorous analysis and data collection, the aim is to develop a comprehensive understanding of how these approaches interact and contribute to the overall sustainability of a building from its inception to its end-of-life cycle. This holistic assessment serves as a guide for future construction practices that prioritize both human needs and the health of the planet.

2 Literature Review

LCA is not only just about quantifying emissions but it's also about using the results to guide decisions and improvements. [4] states the life cycle Assessment of materials to the operation stage of the building. Kurian et al. [8] states that for calculation of life cycle assessment, the information required is the amount of material at construction stages and machinery electricity consumption, the number of hours and fuel consumption of machines at transportation stages, and energy consumption during operational stage. Considering a typical building, the results and calculations for greenhouse emissions emitted from the building by Carvalho et al. [3], using manual calculation and software calculation. Various tools have also been developed such as Sbp tool and SBAT [1, 2, 9], discuss building assessment approaches and the importance of multi-criterion system in assessment. Salah et al. [10] propose a framework for assessing the sustainability of construction projects.

3 Objective

The objective of the research is to design a building project and assess its sustainability across its entire life cycle by employing a diverse range of assessment tools and methods, life cycle assessment and embodied carbon analysis with a focus on minimizing the negative impacts on the environment, society, and economy. Comparison and validation of the sustainability results is carried out by the available software such as OneClick LCA and EC3 tool. The comparison of the results will ensure that the chosen sustainable approach effectively reduces the environmental impact, enhances social well-being, and optimizes the economic considerations thereby providing a valuable insight for stakeholders into the construction industry.

4 Problem Statement

The overall challenge is to encapsulate the complexity of conducting a holistic sustainability assessment that addresses environmental, social, and economic considerations. This statement should convey the need to balance design choices with sustainability objectives and provide insights into reducing carbon emissions, optimizing material selection, and integrating renewable energy sources. The problem involves conducting a multi-faceted sustainability assessment using diverse approaches to evaluate a building project's impact on the environment, society, and economy [10]. The objective is to create a comprehensive action plan that fosters sustainability and reduces negative impacts throughout the building's life cycle. The aim is to evaluate the environmental impact of the building's entire life cycle, from raw material extraction to end-of-life disposal, while promoting sustainability and minimizing negative impacts on the environment, society, and economy.

5 Methodology

The sustainability assessment can be carried out by the methods of cradle to gate, cradle to handover, cradle to end of life, cradle to end of grave, and cradle to cradle. The cradle to cradle analysis has been considered in this project. The cradle to cradle (C2C) analysis is an innovative approach to sustainable building design that goes beyond minimizing negative names to create positive contributions to the environment society and economy the framework reimagines the concept of waste and encouraging encourages the design of products materials and buildings that are regenerative and beneficial throughout their entire life cycle (see Fig. 1).

6 Proposed Solution

6.1 Framework

The environmental impact of different material is considered. The life cycle phases of the building are determined by blending of two databases. To begin with, initially manual calculations are carried out by utilizing inventory of carbon and energy database (ICE) and software computations are carried out using OneClick LCA and EC3 tool. OneClick LCA tool is trailed by EPDs given in the ISO 14040 and EN

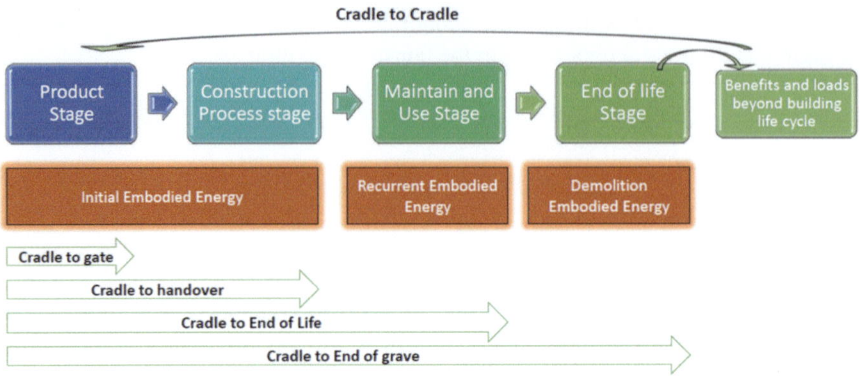

Fig. 1 Flowchart of cradle to cradle stages for life cycle of building

15,804 standards. This study investigates the carbon footprint of different materials used in the building with the four phases of LCA. For the research, building life cycle stages are considered particularly, incorporation of construction stage (that includes material production, material transportation and construction on site), transportation stage (that includes the fuel consumption, number of vehicles and quantity of materials transported), operational stage (that includes HVAC, lighting, water supply and equipment use), and demolition stage (destruction and renovation). For an alternate extent of LCA, different life cycle stages are either compulsory or optional. In any case, various accreditations and estimation frameworks may restrict the modules determined. The manual material mapping entirely has been carried out using available database from Kurian et al. [8], and Chukka et al. [4] as shown in Tables 1, 2, 3 and 4. The entire computation of LCA and EC3 tool with available dataset and calculation is carried out for dataset that is present in the software. Due to the lack of quality of data in tallyCat EPD database (EC3 tool), for the research OneClick LCA is also adopted which enables the process of automated material mapping by user defined values (Figs. 2 and 3).

6.2 *Proposed Design for Analysis*

A residential building apartment of G+8 is designed and location considered for the Project is Yeshwanthpur, Bangalore. Design is prepared of venture appearing in the (Fig. 4). The building has a constructed area of 4400 m^2. The flat roof of building is of 150 mm thickness. The design is carried out in AutoCAD Software as shown in (Figs. 2 and 3) and then exported to Autodesk Revit 2020. Autodesk Revit 2020 enables 3D drawing and assign specifications as software allows a multidisciplinary BIM for the project realizing floor plans of building. Creating a database that integrates data from AutoCAD, and Revit BIM (Building Information Modeling)

Table 1 Calculation of total construction equipment energy

Model	Equipment type	Cost/hr	H	FC (l/hr)	γ_{fuel}	$H * FC * \gamma_{fuel}$
LS108.BSS	Crawler Crane/ Sumitomo Link belt/ 50 TON 30 M Boom	1200	120	15	875	1,575,000
EX60	Excavator/ Tata Hitachi/EX60	1200	45	5	875	196,875
BP250HDD	Concrete Pump / Greaves/Schwing/ BP250HDD	1800	90	6	875	472,500
FM 7	Tipper/Volvo/ FM-7/ 30 T	1000	36	12	875	378,000
6M3	Concrete mixer / Transit/Shirke/06	1700	114	4	875	399,000

Table 2 Construction stage database for proposed building

Construction material	Carbon emission coefficient (S_{mp})	Density (kg/m^3)	Quantity (W_{mp}) (m^3)
Concrete (1:1:2)	0.148	2400	1065
Steel	1.77	7800	14.04
Fine aggregate (CM)	0.036	1670	–
Coarse aggregate (CM)	0.036	1450	–
Red brick (CM)	0.327	1920	1700
Aerated block (SM)	0.23	540	1700
Cement (CM)	0.73	1440	–
Cement with 30% flyash replacement (SM)	0.62	1120	–
PVC window (CM)	28.1	1380	6.4
Stainless steel	5	7850	0.006
Ceramic tiles (CM)	0.74	2250	148
Plywood door (CM)	1.07	650	35.12
Timber door and window (SM)	0.46	500	35.12
Recycled fine aggregate (SM)	0.0024	1700	–
Recycled coarse aggregate (SM)	0.0012	2300	–
Glass	0.85	2500	0.015
Granite (SM)	0.107	2650	148

Table 3 Fuel consumed by transportation vehicle

Vehicle Used	Type	Load capacity	Fuel efficiency	*F.C* (l)	
				Conventional	Sustainable
Tata Insigna 2823 k	Dumper	8 m^3	6 km/l	5120	5120
Eicher 2114	Loader	12 tonnes	7 km/l	118,396	87,881

Table 4 Electrical appliances used during operational stage

Appliances	Load watt (Non energy rated appliances) (a)	Load watt (Energy rated appliances) (a)	No. of appliances (b)
LED		18 (LED Batten)	114
Tube light	40		114
Ceiling fan	80	40	148
Colour television (32 inch)	70 (LCD)	40 (LED)	32
Refrigerator	200	200	32
Washing machine	600	400	32
HVAC	3000	2500	32
Pump motor	3700	3700	1
Water heater	2500	–	1
Fire-fighting centrifugal pump	200	200	1
Solar water heater	–	–	1

can be powerful tool for construction and project management. Such a database can help streamline information sharing, and enhance collaboration.

6.3 *Lifecycle Assessment of the Building*

The life cycle assessment is executed in a study stage to determine the carbon emissions effect on the environment during the construction, transportation, material use/ operational use and demolition phase [4].

The formula to calculate the life cycle assessment of the building is

$$U_{\text{tot}} = U_{\text{con}} + U_{\text{mt}} + U_{\text{op}} + U_{\text{dem}} \tag{1}$$

where U_{tot} represents total carbon emission; U_{con} represents construction stage; U_{mt} represents transportation stage; U_{op} represents operational stage; U_{dem} represents end of life stage.

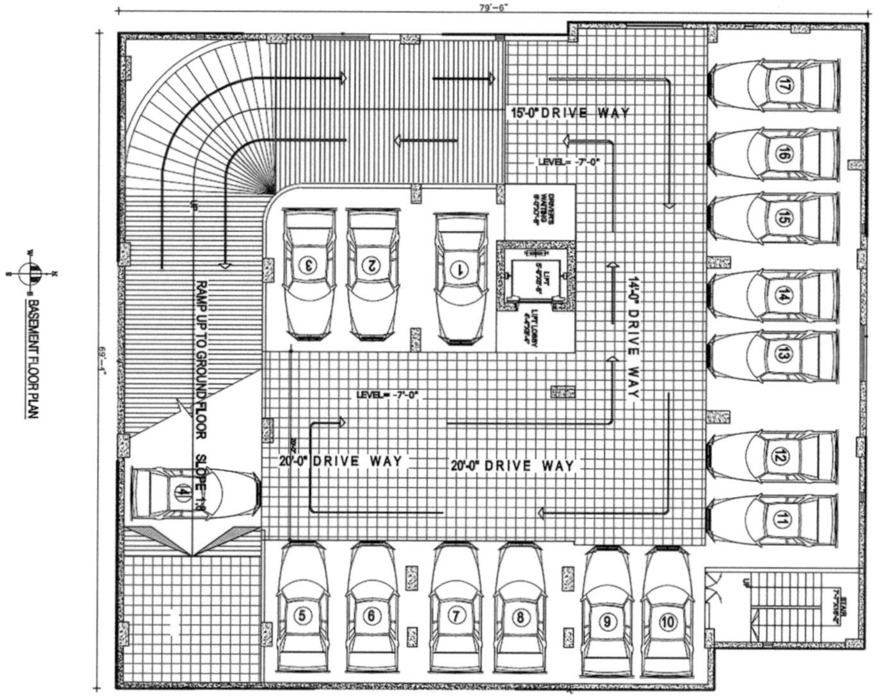

Fig. 2 2D view of basement

7 Calculation

A comparative study for the building using conventional methods and sustainable materials has been carried out. Carbon emission of a residential building constructed with conventional materials [4]. Use of recycled materials have also been used [5, 7]. The calculations for carbon emission are carried by data obtained from REVIT BIM. The database is fed for calculating the carbon emissions using the OneClick LCA and EC3 tool (software analysis). The entire life cycle assessment of the building is calculated for the design period of time of 75 years, and calculation is carried out as following Eq. (1).

7.1 Construction Stage

The construction stage of the building involves the construction material along with the construction equipment that are used during the building construction. The materials along with their carbon emission coefficients are considered for calculation of

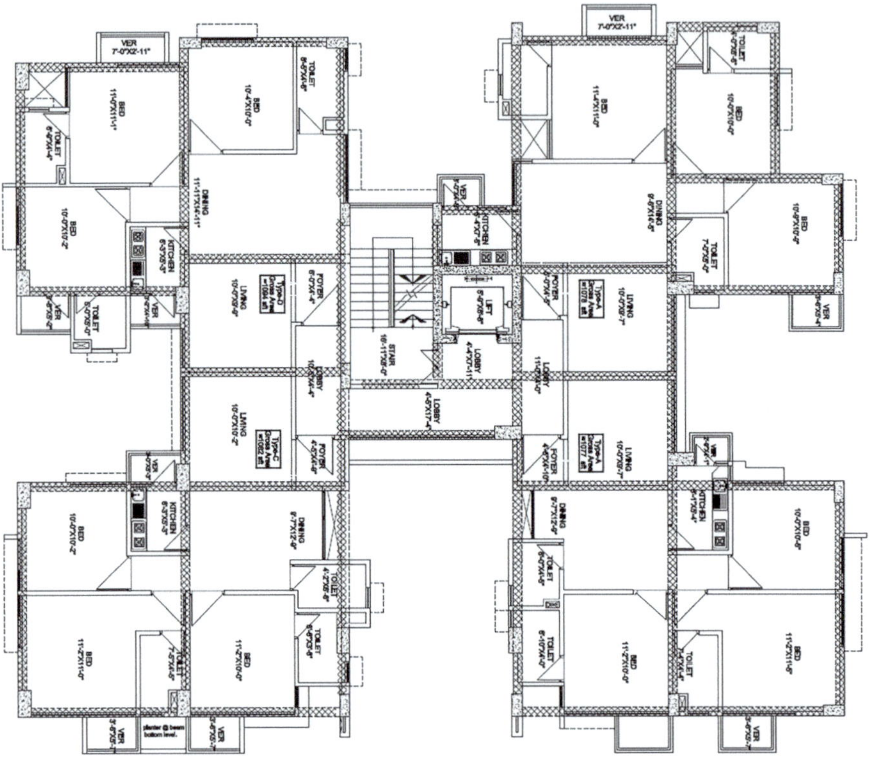

Fig. 3 Typical floor plan

Fig. 4 3D model in Revit

the construction materials. The materials that are used regularly without any replacements are considered for the conventional project materials (CM), and the materials that are environment friendly are termed as the sustainable materials (SM). The equation and data required for the calculation of the carbon emissions have been derived from Kurian et al. [8], and Chukka et al. [4] for the calculation of the building.

$$U_{con} = \sum_{i=1}^{n} \left(W_{mp}, i \cdot S_{mp}, I + C_{ce} \right) \tag{2}$$

where U_{con} represents total emission during construction stage; W_{mp} represents quantity of material; S_{mp} represent CEC; C_{ce} represents emission from the construction equipment.

$$C_{ce} = \sum_{i=1}^{n} (H * \mathrm{FC} * \gamma_{fuel} * \mathrm{EF}) \tag{3}$$

where H = total operating hours of equipment; FC = average fuel consumption for equipment per unit time (m^3/hour); γfuel= Density of fuel (kg/m^3); EF = Energy factor for certain fuel type.

Embodied carbon emission for construction energy, $C_{pu} = 0.25$ kgCO_2/kg

$$C_{ce} = C_{pu} \times V_{eq} \tag{4}$$

where C_{ce} = construction equipment energy; C_{pu} = embodied carbon emission; V_{eq} = total equipment energy

Embodied carbon emission for construction energy, $C_{pu} = 0.25$ kgCO_2/kg.

Total equipment energy (V_{eq}) = 3,021,375 kgCO_2/kg.

Total carbon emission (C_{ce}) = 755,344 kgCO_2/kg.

7.2 Transportation Stage

$$U_{mt} = Smt \times F.C \tag{5}$$

where U_{mt} represents carbon emissions generated by material during transportation; S_{mt} represents the CEC of construction material hauling; $F.C$ represents fuel consumption in litre.

The carbon emission coefficient (CEC) (S_{mt}) for a diesel vehicle is 2.64 kg CO_2e/l.

7.3 Operational Stage

The only carbon emissions produced by electricity consumption are the essential measure to be calculated. Operational CEC for electricity usage

$$U_{\mathrm{op}} = C_{\mathrm{emi}} \times \mathrm{TP}_{cons} \tag{6}$$

where C_{emi} represents the carbon emission per unit power; $\mathrm{TP}_{\mathrm{cons}}$ represents the total power consumption of all devices throughout life cycle.

The operational lifecycle of the building is considered to be around 60 years, and operational energy is calculated based on the life of the building.

7.4 End of Life or Destruction Stage

Destruction stage of the building includes demolition of the entire building using necessary available tools and machinery available. The destruction stage is assumed to be about 20% of the construction stage due to unavailability of demolition data.

Equipment that shall be used are concrete breakers, plumb bob machinery, steel cutters, and demolition crawlers.

$$U_{\mathrm{dem}} = U_{\mathrm{con}} * 20\% \tag{7}$$

where U_{dem} = carbon emission at destruction stage; U_{con} = carbon emission due to construction materials.

8 Result

The breakdown of database into the sustainability aspects is represented in Tables 5 and 6. The analysis is carried out using the Eqs. (2, 4–7) for each activity and presented by comparison of the conventional and sustainable design database used in the project as shown in Fig. 5.

The database for the conventional design with the set of values for each activity is shown in Table 5.

The database for sustainable design with set of values for each activity as shown in Table 6.

For the conventional material design, database states the total emission for the OneClick LCA expressed in $KgCO_2e/kg$ values was 5.45% lower than manual calculation and EC3 tool values was 7.93% lower than manual calculation. For the sustainable material design, database states the total emission for the OneClick LCA

Table 5 Comparison of conventional database using OneClick LCA tool, EC3 tool, and manual calculation

Activities	Manual calculation	OneClick LCA	EC3 tool
Construction materials	2,159,034	2,706,100	2,150,086
Construction equipment	755,344	30,000	45,125
Transportation stage	312,566	310,000	1,990,000
Operational stage	39,232,939	36,269,854	34,698,972
End of life	439,953	305,300	742,000
Total emission ($KgCO_2/kg$)	42,899,836	40,621,254	39,626,183

Table 6 Comparison of sustainable database using OneClick LCA tool, EC3 tool, manual calculation

Activities	Manual calculation	OneClick LCA	EC3 tool
Construction materials	744,455	925,919	806,065
Construction equipment	755,344	30,000	45,125
Transportation stage	232,006	190,000	1,010,000
Operational stage	31,502,204	36,269,854	32,397,288
End of life	159,110	92,270	45,900
Total Emission ($KgCO_2/kg$)	33,393,119	34,337,043	34,304,378

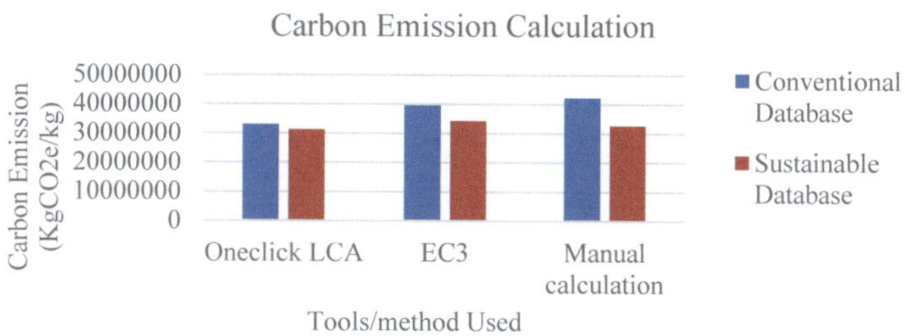

Fig. 5 Comparison of total carbon emission of conventional building design and sustainable building design by sustainability assessment using OneClick LCA tool

expressed in KgCO2e/kg values was 2.78% higher than manual calculation and EC3 tool values was 2.69% higher than manual calculation.

9 Conclusion

The reduction in carbon emissions by 24.92% for the conventional database and sustainable database from manual calculation highlights the positive impact of sustainable design strategies on the environment. The difference in value was mainly due to change in EPDs of different tools used for analysis. This reduction signifies a significant contribution towards mitigating climate change, conserving resources, and promoting a more responsible approach to construction and development. The results underline the potential of sustainable building design to significantly reduce carbon emissions and contribute to a more sustainable future. By incorporating environmentally conscious practices from construction to end of life, the construction industry can play a pivotal role in mitigating climate change and promoting responsible resource use.

References

1. Berardi U (2011) Sustainability assessment in the construction sector: rating systems and rated buildings. Sustain Dev 10(1002):1–14
2. Bragança L, Mateus R, Koukkari AH (2010) Building sustainability assessment. Sustainability 2(7):1–14
3. Carvalho JP, Braganca L, Mateus R (2019) Optimising building sustainability assessment using BIM. Autom Constr 102:170–182
4. Chukka NDKR et al (2022) Environmental impact and carbon footprint assessment of sustainable buildings: an experimental investigation. Adsorpt Sci Technol 2022:1–8
5. Devi SV, Gausikan R, Chithambaranathan S, Jeffrey JW (2020) Utilization of recycled aggregate of construction and demolition waste as a sustainable material. Mater Today: Proc 1–6
6. Hossain MU et al (2017) Development of social sustainability assessment method and a comparative case study on assessing recycled construction materials. Int J Life Cycle Assess 1–18
7. Iodice S, Garbarino E, Cerreta M, Tonini D (2021) Sustainability assessment of construction and demolition waste management applied to an Italian case. Waste Manage 128:83–98
8. Kurian R et al (2021) Estimation of carbon footprint of residential building in warm humid climate of India through BIM. Energies 14:1–16
9. Reddy AS, Raj PA, Kumar PR (2019) Developing a sustainable building assessment tool (SBAT) for urbanization challenges in emerging economies, pp 136–148
10. Salah M, Elmasry M, Mashhour IM, Amer N (2023) A framework for assessing sustainability of construction projects. Clean Eng Technol 13:1–13
11. Santos R, Costa AA, Silvestre JD, Pyl L (2019) Integration of LCA and LCC analysis within a BIM-based environment. Autom Constr 103:127–149
12. Singh RK, Murty H, Gupta S, Dikshit A (2012) An overview of sustainability assessment methodologies. Ecol Indic 15:281–289

Static Analysis of FGM Beam Using Polynomial Beam Theory

Anuja S. Jape, Sayali S. Nehe, Shraddha D. Arane, and Aniket S. Gaikwad

Abstract Polynomial beam model is used in the present research study to determine the deflections of functionally graded beams under uniformly distributed loading. This hypothesis is formulated using the principle of virtual work. A polynomial beam theory was used to compare the deformations of functionally graded (FG) beam loaded uniformly. This polynomial beam theory ascertains the traction-free conditions on the both top and bottom ends of the beam. The FG beam is treated as a simple supported beam. Hence, Navier's solution is employed to develop governing equations. The numerical results were then compared with existing theories in the literature and found to be consistent with the current theory.

Keywords Static-bending analysis · FGM · Polynomial beam theory

1 Introduction

Functionally graded material (FGM) refers to a material type that exhibits a gradual and controlled variation in its composition, microstructure, or properties over its volume. The idea of functionally graded materials was originated in the year 1984 through the work of Japanese scientists. Stress concentration, often a concern in materials with abrupt changes in properties, is mitigated in FGMs due to their gradual transitions. This property is particularly valuable in applications where mechanical stresses vary, as FGMs can distribute these stresses more evenly throughout the material. The advantages of functionally graded materials include improved resistance to issues like delamination and stress concentration, enhanced thermal and mechanical properties, and increased bond strength. FGMs find applications in aerospace, civil engineering, nuclear reactors, and other areas where materials with specific and tailored properties are crucial. Static analysis provides information about the

A. S. Jape (✉) · S. S. Nehe · S. D. Arane · A. S. Gaikwad
Department of Structural Engineering, Sanjivani College of Engineering, Savitribai Phule Pune University, Kopargaon, Maharashtra, India
e-mail: japeanujast@sanjivani.org.in

Y. K. Guruprasad et al. (eds.), *Recent Advances in Infrastructure Development and Management—Volume 1*, Lecture Notes in Civil Engineering 570,
https://doi.org/10.1007/978-981-96-1554-4_22

distribution of deformations and stresses within the FGM beam. This knowledge is crucial for ensuring that the material does not experience localized failures or excessive deformations. Classical beam theory (CBT) [1, 2] is the most fundamental theory in the literature for analyzing both isotropic and anisotropic beams. However, it is inadequate for analyzing thick composite beams made of FGM, as it neglects transverse shear deformation, rendering it inaccurate for such materials. Timoshenko [3] introduced the first‾order shear deformation theory (FSDT) to tackle this issue; however, it suffers from drawbacks like constant shear strain across the thickness and the necessity of a shear correction factor. To address these shortcomings, advanced shear deformation theories have emerged, delivering improved accuracy by appropriately accommodating shear deformation. Reddy [4] introduced a higher-order shear deformation theory (HSDT) specifically for analyzing laminated composite plates. This theory relaxes the assumption of straightness of normal to the mid-surface often made in classical plate theories. Chen [5] improved beam element of composite laminates and higher-order shear deformation theory. Behera [6] used boundary characteristic orthogonal polynomials to create a free vibration of Euler and Timoshenko nanobeams. Filippi et al. [7] developed functionally graded material beam static analyses using different theories and finite elements. Ghumare and Sayyad [8] advanced the study of static and buckling behavior of P-FGM beam by developing a novel fifth-order shear deformation theory including normal deformation. Shariq Beg [9] enhanced the analysis of laminated and FGM beams, through the use of several theories. Sayyad [10, 13] used modified exponential and inverse trigonometric shape function in the displacement field for the static bending, buckling and vibration study of FG-beam with several boundary value cases. Hebbar et al. [11] presented analytical modeling approach for static bending, instability analysis, and vibration behavior of FG beam. HSDT is utilized to account for transverse shear deformation effects. This is useful for understanding and predicting load-deflection and modal responses in functionally graded beam structures. Fogang [12] presented buckling as well as vibration analysis using the finite difference approach. The finite difference method is applied to conduct numerical analysis of Timoshenko beam. Jape and Sayyad [14, 18] presented bending behavior of functionally graded nanoscale beams based on hyperbolic nonlocal continuum theory. Valencia [15] presented a linear-static-thermal analysis of FG beam. Kolapkar and Sayyad [16] developed the refined-shell theory, which accounts for higher-order transverse shear strain as well as normal rotational effects. Tamnar and Sayyad [17] used trigonometric shape function applied for the static response of FG porous shell.

In the present research paper, central transverse deflection and axial deflection of functionally graded beam are calculated using polynomial beam theory. This polynomial shape function accounts for the realistic behavior of transverse shear stress at top and bottom of rectangular cross section of the FG beam. The characteristics of the FG beam are gradually varying across the thickness of beam as per the power-law distribution. Mathematical formulation for the static analysis of FGM beam using present theory is done based on principal of virtual work. In the current study, Navier's approach is implemented for the simply supported FG beam. The

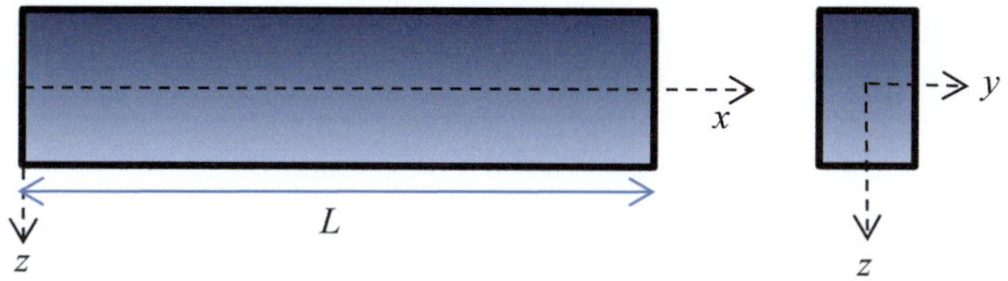

Fig. 1 FGM beam under consideration

numerical analysis for the central transverse deflection and axial deflection of the functionally graded beam is then compared with existing findings in the literature.

2 FGM Beam Under Consideration

For the mathematical formulation, FGM beam of length L and cross-sectional dimension $b \times d$, as illustrated in Fig. 1, is considered with simply supported end condition. Downward z-axis is considered as positive. The FG beam is composed of Al/Al_2O_3, characterized by gradation in material properties along the thickness.

This variation in material properties of the FG beam follows power-law gradation as given in Eq. (1).

$$E(z) = E_M + (E_C - E_M)\left(0.5 + \frac{z}{h}\right)^p \tag{1}$$

where E represents Young's modulus of elasticity; E_M, E_C are the elastic properties metal and ceramic material, respectively. p is the power-law coefficient which ranges from zero to infinity. z/h is the thickness ordinate. Zero value of power law coefficient represents a beam fully made of ceramic material and power law coefficient value of infinity represents a beam of metal material.

3 Mathematical Formulation

Equation (2) establishes the displacement field for the current theoretical framework. The traction-free boundary conditions on both the top and bottom surfaces of the FG beam are derived utilizing the parabolic shape function inherent in the present beam model.

$$\begin{aligned} \Delta_x(x, z) &= \Delta_{x0}(x) - z\frac{\partial \Delta_{z0}}{\partial x} + \left[z - \frac{4}{3}\left(\frac{z}{h}\right)^2 z\right]\theta(x) \\ \Delta_z(x) &= \Delta_{z0}(x) \end{aligned} \tag{2}$$

where, Δ_x is the axial deflection along x-axis, Δ_z is the transverse deflection along z-axis, respectively, and θ is the rotation. Parabolic shape function accounts for the impacts of transverse shear deformation, in the axial displacement. The non-zero strains, as outlined in Eq. (3), are derived from elasticity theory.

$$\begin{aligned} \varepsilon &= \frac{\partial \Delta_x}{\partial x} - z\frac{\partial^2 \Delta_y}{\partial x^2} + \left[z - \frac{4}{3}\left(\frac{z}{h}\right)^2 z\right]\frac{\partial \theta}{\partial x} \\ \gamma &= \left[1 - \frac{4z^2}{h^2}\right]\theta \end{aligned} \tag{3}$$

Stresses in FGM beam are calculated from Hooke's law stated in Eq. (4).

$$\sigma = E(z)\,\varepsilon \quad \text{and} \quad \tau = \frac{E(z)}{2(1+\mu)}\,\gamma \tag{4}$$

Differential equations are developed using the principle of work done as articulated in Eq. (5).

$$b\int_0^L \int_{-h/2}^{h/2} (\sigma\,\delta\varepsilon + \tau\,\delta\gamma)\,dzdx - \int_0^L q(x)\,\delta\Delta_z\,dx = 0 \tag{5}$$

Where $q(x)$ is the transverse load acting on FGM beam. The differential equation stated in Eq. (6) is governed by integration of Eq. (5) by partwise and gathering the differential factors of unknowns, and then equating the equation with zero. Equation (6) states the governing equations of the present theory.

$$\begin{aligned} &\mathrm{AA}\frac{\partial^2 \Delta_{x0}}{\partial x^2} - \mathrm{BB}\frac{\partial^3 \Delta_{z0}}{\partial x^3} + \mathrm{CC}\frac{\partial^2 \theta}{\partial x^2} = 0 \\ &\mathrm{BB}\frac{\partial^3 \Delta_{x0}}{\partial x^3} - \mathrm{DD}\frac{\partial^4 \Delta_{z0}}{\partial x^4} + \mathrm{FF}\frac{\partial^3 \theta}{\partial x^3} = -q(x) \\ &\mathrm{CC}\frac{\partial^2 \Delta_{x0}}{\partial x^2} - \mathrm{FF}\frac{\partial^3 \Delta_{z0}}{\partial x^3} + \mathrm{HH}\frac{\partial^2 \theta}{\partial x^2} - \mathrm{JJ}\theta = 0 \end{aligned} \tag{6}$$

where

$$(\mathrm{AA},\ \mathrm{BB},\ \mathrm{DD},\ \mathrm{CC},\ \mathrm{FF},\ \mathrm{HH},\ \mathrm{JJ}) = b\int_{-\frac{h}{2}}^{\frac{h}{2}} E(z)\left[1,\ z,\ z^2,\ f(z),\ \mathrm{zf}(z),\ f^2(z),\ (f\prime(z))^2\right]dz \tag{7}$$

The transverse and axial deflections in a simply supported FG nanobeam are computed using Navier's solution scheme, which precisely fulfills the conditions of simple support, i.e., $N_x = \Delta_z = M_x^b = M_x^s = 0$. The unknown deformations of the present problem as well as the transverse loading acting on the beam are presumed in the single Fourier series stated in Eq. (8).

$$\Delta_{x0} = \sum_{m=1}^{\infty} \Delta_{\text{xm}} \cos \text{rx}, \ \Delta_{z0} = \sum_{m=1}^{\infty} \Delta_{zm} \sin \text{rx}, \ \theta = \sum_{m=1}^{\infty} \theta_m \cos \text{rx}, \ q(x) = \sum_{m=1}^{\infty} \frac{4q_0}{m\pi} \sin \text{rx} \tag{8}$$

where $r = m\pi/L$, $(\Delta_{\text{xm}}, \Delta_{\text{zm}}, \theta_m)$ are unknown deformation, q_0 represents the peak intensity of uniform loading acting on the FG-beam. Equation (8) is substituted in Eq. (6), and the final equations are written in matrix form as given in Eq. (9). Solution of Eq. (9) leads to determination of unknown deformation.

$$\begin{bmatrix} -\text{AAr}^2 & \text{BBr}^3 & -\text{CCr}^2 \\ \text{BBr}^3 & -\text{DDr}^4 & \text{FFr}^3 \\ -\text{CCr}^2 & \text{FFr}^3 & -(\text{HHr}^2 + \text{JJ}) \end{bmatrix} \begin{Bmatrix} \Delta_{\text{xm}} \\ \Delta_{\text{zm}} \\ \theta_m \end{Bmatrix} = \begin{Bmatrix} 0 \\ -\frac{4q_0}{m\pi} \\ 0 \end{Bmatrix} \tag{9}$$

4 Numerical Problems and Discussions

The static problem of the FG beam is covered in this section. The FGM beam made up of Al/Al_2O_3 metal/ceramic material is considered for generating the results. Properties of metal/ceramic are presented in Eq. (10).

$$E_C = 380\text{GPa}, \ E_M = 70\text{GPa}, \ \mu_c = \mu_m = 0.3 \tag{10}$$

Equation (11) shows the nondimensional form for which all numerical results are generated.

$$\overline{\Delta_x} = \frac{100 E_M h^3}{q_0 L^4} \Delta_x, \quad \overline{\Delta_z} = \frac{100 E_M h^3}{q_0 L^4} \Delta_z \tag{11}$$

The comparison of nondimensional axial displacement $\left(\overline{\Delta_x}\right)$ of FG beam for $(p = 0, \ 1, \ 2, \ 5, \ 10, \ \infty)$ is calculated using the present theory and summarized in Table 1. All values of nondimensional axial deflection in Table 1 are calculated for $L/h = 5$. From Table 1, it is observed that axial displacement increases with the power law. The axial displacement value for power law $p = 0$ is $\overline{\Delta_x} = 0.9361$, whereas for $p = \infty$ is $\overline{\Delta_x} = 5.0815$. Hence, it can be stated that for functionally graded beam, the nondimensional axial deflection increases as the volume of ceramic material decreases and volume metal material increases. Further the numerical results are compared with nondimensional axial deflection of FSDT and CBT that shows close match with the present results.

Table 1 Comparison of axial displacement $\left(\overline{\Delta_x}\right)$ of FGM beam with varying power law $(L/h = 5,\ m = 100)$

	p					
Theory	0	1	2	5	10	∞
Present	0.9361	2.2976	3.1044	3.6947	3.8674	5.0815
FSDT [3]	0.921	2.2722	3.0739	3.6496	3.8096	5.000
CBT [1]	0.921	2.2722	3.0739	3.6496	3.8096	5.000

Table 2 Comparison of transverse displacement $\left(\overline{\Delta_z}\right)$ of FGM beam with varying power law $(L/h = 5,\ m = 100)$

	p					
Theory	0	1	2	5	10	∞
Present	3.1080	6.1625	7.9198	9.5513	10.605	16.872
FSDT [3]	3.1635	6.2563	8.0666	9.8414	10.940	17.173
CBT [1]	3.1654	6.2594	8.0677	9.8281	10.938	17.183

The numerical results of transverse displacement $\left(\overline{\Delta_z}\right)$ of the FG beam are as shown in Table 2, for L/h=5 and subjected to uniform load for $(p = 0\ ,\ 1,\ 2,\ 5,\ 10,\ \infty)$. The nondimensional transverse deflection increases in a similar manner as the power law index rises. As shown in Table 2, the transverse deflection in the FG beam closely matches the results from CBT and FSDT, demonstrating the accuracy of the present theory.

5 Conclusion

The present research study employs parabolic beam theory to predict axial and transverse deflections in functionally graded beams under uniform load. The findings show that both deflections increase with the power law index. The accuracy of the suggested theory is confirmed by the results, which align with current hypotheses. This work is significant because it improves the accuracy of deflection predictions for FG beams, which is important when constructing sophisticated structural elements with customized mechanical properties.

References

1. Bernoulli J (1694) Curvatura laminae elasticae. Acta Eruditorum Lipsiae 3(6):262–276
2. Euler L (1744) Methodus inveniendi lineas curvas maximi minimive proprietate gaudentes. Lausanne and Geneva: Apud Marcum-Michaelem Bousquet and Socio 1–322

3. Timoshenko SP (1921) On the correction for shear of the differential equation for transverse vibrations of prismatic bars. Phil Mag 41:742–746
4. Reddy JN (1984) A simple higher order theory for laminated composite plates. ASME J Appl Mech 745–752
5. Chen W, Wu Z (2005) Higher-order shear deformation theory and refined beam element of composite laminates. Acta Mechanica Sinica/Lixue Xuebao 21(1):65–69
6. Behera L, Chakraverty S (2014) Free vibration of Euler and Timoshenko nanobeams using boundary characteristic orthogonal polynomials. Appl Nanosci (Switz) 4(3):347–358
7. Filippi M, Carrera E, Zenkour A (2015) Static analyses of FGM beams by various theories and finite elements. Compos B Eng 72:1–9
8. Ghumare SM, Sayyad AS (2017) A new fifth-order shear and normal deformation theory for static bending and elastic buckling of P-FGM beams. Lat Am J Solids Struct 14:1–19
9. Shariq BM, Yaqoob Y, Khalid H (2018) Analysis of laminated and FGM beams using various theories. IOP Conf Ser Mater Sci Eng 404(1)
10. Sayyad AS, Ghugal YM (2018) Analytical solutions for bending, buckling, and vibration analyses of exponential functionally graded higher order beams. Asian J Civ Eng 19(5):607–623
11. Hebbar N, Hebbar I, Ouinas D, Bourada M (2020) Numerical modeling of bending, buckling, and vibration of functionally graded beams by using a higher-order shear deformation theory. Frattura Ed Integrita Strutturale 14(52):230–246
12. Fogang V (2021) Timoshenko beam theory: first-order analysis, second-order analysis, stability, and vibration analysis using the finite difference method
13. Sayyad AS, Ghugal YM (2021) Bending, buckling, and vibration analysis of functionally graded nanobeams using an inverse trigonometric beam theory. Int J Nano Dimens 12(2)
14. Jape AS, Sayyad AS (2022) Bending of functionally graded nanobeams via hyperbolic nonlocal theory. IOP Conf Ser Mater Sci Eng 1236(1)
15. Valencia M, Gutierrez R, Celaya G (2023) Thermal–structural linear static analysis of functionally graded beams using reddy beam theory. Math Comput Appl 28(4)
16. Kolapkar SS, Sayyad AS (2023) Static response of FGM shell using refined higher-order shear and normal deformation theory. Mater Res Proc 31:117–127
17. Tamnar RB, Sayyad AS (2023) Static response of functionally graded porous spherical shells using trigonometric shear deformation theory. Mater Res Proc 31:76–85
18. Jape AS, Sayyad AS (2023) A hyperbolic theory for the analysis of laminated shallow shells with double curvature. Forces Mech 13:100246

A Case Study on Structural Audit of Commercial Building

Bharti M.Shinde, Priyanka A. Rakshe, Snehal V. Waghchaure, Rutika S. Jadhav, and Rutuja A. Gavhane

Abstract A structural audit is like a doctor checking a patient—it's like a detailed health checkup of a structure. Its main goal is to ensure the building's and its surroundings' safety and risk-free existence. A structural audit evaluates the overall health and function of the structure, much like a medical checkup. In particular, the building, constructed in 1984, is a two-storied load-bearing structure with commercial (education) usage on black cotton soil. Under the guidance of a qualified and experienced structural expert, this assessment explores the strength of the structure, including the collection of data, visual inspection, NDT testing, and suggesting necessary maintenance and retrofitting options to improve efficiency and safety. The caliber of construction work is frequently correlated with the degree of damage or degradation in a structure. Wear and tear on the structure is caused by a number of sources, including fire damage, cracking, chemical attack, and steel erosion over time. An overall structural audit takes these things into account and provides suggestions on how to make the structure more suitable and long-lasting. This audit was performed to assess the strength, structural integrity, and feasibility of building for further use.

Keywords Structural audit · Safety and risk · Retrofitting · Wear and tear

1 Introduction

Maintaining the lifetime of architectural assets and guaranteeing the safety of residents are dependent upon the structural integrity of buildings. In this case, a complete structural audit functions as an essential diagnostic tool, similar to a patient's physical examination, identifying the weak points and any dangers that could put at risk the stability and operation of the structure. The subject of our attention is a two-story

B. M.Shinde (✉) · P. A. Rakshe · S. V. Waghchaure · R. S. Jadhav · R. A. Gavhane
Department of Structural Engineering, Sanjivani College of Engineering, SPPU Pune, Kopargaon, Maharashtra, India
e-mail: shindebhartist@sanjivani.org.in

Y. K. Guruprasad et al. (eds.), *Recent Advances in Infrastructure Development and Management—Volume 1*, Lecture Notes in Civil Engineering 570,
https://doi.org/10.1007/978-981-96-1554-4_23

load-bearing building built on black cotton soil in 1984 for commercial (educational) use.

The structural audit's observations, which range from decay and corrosion in wooden parts to cracks in masonry walls and cracking cement connections, provide a clear picture of the difficulties the structure experiences. The need for intervention is further highlighted by uneven flooring settling, deteriorating old reinforced concrete parts, damaged doors and windows, weakened bitumen roof covering sheets, and problems with insects. Due to its age and exposure to the elements, the structure has a number of problems that need to be fixed right away in order to maintain safety regulations and stop more deterioration.

In addition to documenting the building's existing condition, this study report aims to offer practical suggestions for improvement and repair. The goal of the present paper is to assist stakeholders in making decisions that will contribute to the sustainable future of the structure by exploring the details of observed problems, suggesting solutions like the removal or retrofitting of cantilever-reinforced concrete elements, filling in flooring gaps, and claiming for the implementation of an extensive maintenance plan. This study presents itself as a comprehensive guide for maintaining the integrity and durability of the structure under examination since it integrates structural knowledge, data-gathering approaches, and the search for esthetic improvements.

Abuturab and Tande [1] have presented that maintenance is crucial for extending the lifespan of a structure and preventing damage, thus preventing failures caused by such issues like cracking, decomposition, etc. They perform primary inspection of RCC structure by NDT test and also using ETAB's Software. Mahadik and Jaiswal [2] have raised awareness among civil engineers, residents, and building owners about the importance of conducting health examinations on existing concrete buildings. Karnavat and Singh [3] have studied specific temple specimens which are assumed to be representative of the entire temple. Sonawane and Dhawale [4] deal with methods of estimating the audit of residential building whose life has crossed the age of years. Kondalkar and Pitale [5] have presented an audit report of old structure which is situated in Nagpur. The structural audit report includes the detailed investigation based on NDT consisting of ultrasonic pulse velocity test, rebound hammer test, half-cell test.

1.1 Structural Audit

(a) Definition

A structural audit is a detailed study of a building by experts to identify potential issues affecting its stability and safety. It helps determine the general state of the structure, identifies failure, cracks, or defects and suggests necessary maintenance or modifications to ensure its safety and functionality. It is a crucial tool for maintaining a building's quality.

(b) Factors Affecting Health of Structure

There are a number of reasons why a structure can fail, including

(1) Design and construction processes.
(2) The following factors contribute to corrosion-related loss: displacement, internal cracking, cracking of the concrete, cracks parallel to the reinforcement, cracking at the edges, and a reduction in the area of steel reinforcement.
(3) Modification of the loading conditions.

(c) Purpose of Structural Audit

(1) Understanding current building condition:

- Objective: Complete examination and evaluation of the building's current state.
- Importance: Identifying existing issues, weaknesses, and potential risks of building.

(2) Identification of critical areas and remediation:

- Objective: Identify specific susceptibility and critical areas.
- Importance: Enables targeted efficiencies and deal with issues to preserve structural integrity.

(3) Ensuring safety of human life and buildings:

- Objective: Assess and enhance safety for occupants and the overall structure.
- Importance: Reduce a potential risk, creating a safe environment for residents.

(4) Recommendation of necessary repairs and modifications:

- Objective: Provide concrete suggestions for repairs and modifications.
- Importance: Guides stakeholders in implementing corrective measures and increase structural strength.

(5) Improving lifespan and efficiency of buildings:

- Objective: Purpose to measures to enhance duration and efficiency.
- Importance: Aims to extend the structure's lifespan and corrective actions.

(6) Preventing sudden collapse or disaster:

- Objective: Carefully identify and address potential risks.
- Importance: Acts as a preventative measure, protect against unexpected structural failures and disasters.

2 Structural Examination Framework

2.1 Key Feature of the Structure

- Year of Construction: 1984
- Type of Structure: Two-storied load bearing structure
- Type of Building: Commercial (Education)
- Nature of Soil: Black Cotton Soil
- Location: Kopargaon, Dist. Ahmednagar, Maharashtra, India.

The structural characteristics of the building were examined, and the results indicated problems with the state of the building's components and structural parts, including information on their health and condition was summarized.

2.2 Visual Observations

The concerns with the current status of building components and structural elements were described, along with their health and condition, during the structural examination of the masonry structure.

(a) Cracks in Masonry Walls

Numerous cracks appear in the brick walls, both inside and outside. The size and direction of these cracks vary, suggesting possible defects in the building's structural stability. The observed cracks are shown in Fig. 1.

(b) Deterioration in Mortar Joints

There are indications of loss, cracking, and deterioration in the mortar joints that attach to the masonry parts over time. These places of the walls, general stability and ability to support weight, are at risk. Figure 2.

(c) Deterioration and Decomposition of Structural Components

Wooden components, such as joints and supports, showed easy to see signs of decomposition, deterioration, and damage from insects. This was especially noticeable in the cantilever balcony on the north facing entry. The impacted wooden elements' ability to support weight and maintain their structural integrity are under risk because of these conditions. Figure 3.

(d) Irregular Settlement in the Flooring and Degradation of Joints

There were multiple areas where the flooring showed signs of irregular settling, which resulted in sloped floors and an irregular flooring surface as shown in Fig. 4. It is most likely the result of moving soil or inadequate foundation support over the plinth region that is causing this settling.

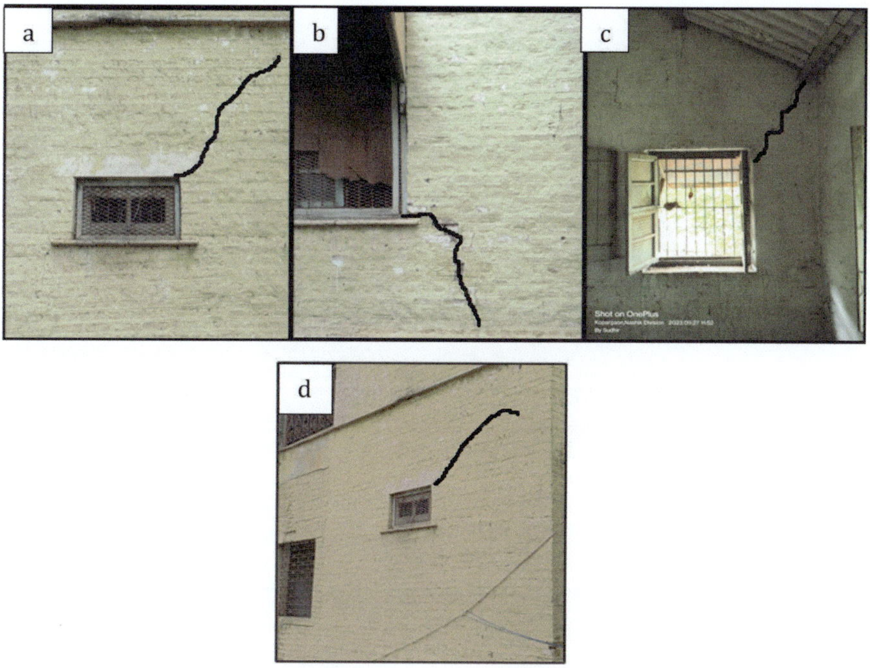

Fig. 1 Observed masonry cracks at location; **a** North face entry above lintel **b** East face below sill (**c**) First floor east face inner wall **d** South back side wall

Fig. 2 Deterioration of mortar joints; (**a**) East face wall outer side **b** South face wall backside

(e) Fractures and Structural Susceptibility in the Asbestos Roofing Sheets

The AC sheets showed signs of lifespan, cracking, and weakness as a result of increased exposure to environment. The damaged sheets reduce their ability to provide sufficient protection from rain and other weather. In addition, there were understandable holes between the AC sheets, which allowed water to leak into the building's floor as shown in Fig. 5.

Fig. 3 Deterioration and decomposition of structural component; **a** North side front balcony covering **b** North side front balcony parapet post

Fig. 4 Deterioration of flooring joints and irregularity

(f) Damaged Doors and Windows, Coupled with Pest Infiltration

The old building is highly infected with insects and spiders, which is a serious risk to safety. Many doors and windows that are essential to protection and safety are broken or in decay, which calls for immediate repairs and maintenance as found in Fig. 6. The problem increases by a build-up of waste, which results in a dirty and unclean environment. To solve these issues and guarantee that residents are living in a clean, safe environment, immediate action is required.

(g) Deterioration of Old Reinforced Concrete Elements

Throughout the audit, it was clear that column members, window chajjas, and reinforced concrete (RC) cantilever parts seen in Fig. 7 were showing symptoms of

Fig. 5 Roof covering AC sheet condition

Fig. 6 Observed doors and windows condition

degradation cracking in the concrete indicating material loss and surface degradation. In addition, there was corrosion on the reinforcing bars, which presented a serious structural risk. To ensure the stability and safety of the structure, these members must be reinforced and responded across immediately.

2.3 Nondestructive Testing

A building's structural soundness is first determined by visual examination, but concrete's long-lasting strength and durability are essential to both its lifespan and safety. When assessing concrete's strength, resistance to chemical attacks, corrosion

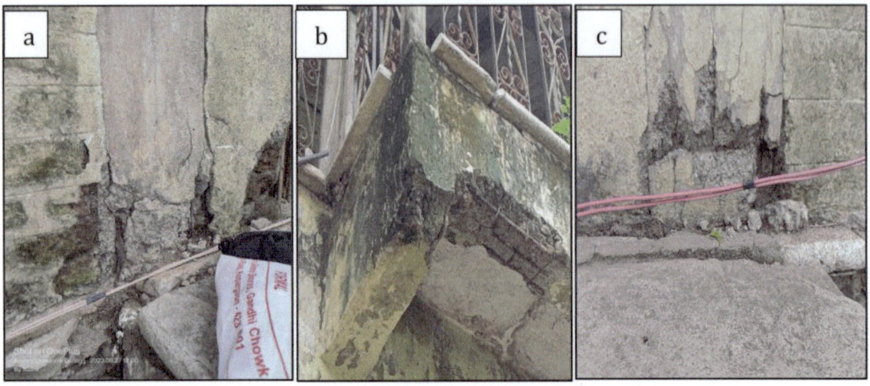

Fig.7 Deterioration of old reinforced concrete element; **a** Base of RC column, **b** Cantilever balcony on north front face, **c** North face entry door column

evaluation, and other properties, nondestructive testing (NDT) must be used. The ultrasonic pulse velocity test for crack inspection and concrete quality, the rebound hammer test for compressive strength, and core testing for strength determination are common methods. Table 1 shows the observations for rebound hammer test at various locations. Based on the type of audit, lifespans of the building, kind of construction, and environmental factors, several nondestructive testing tools are used. These tests help determine how to handle structural parts so that their strength and quality are maintained without causing harm.

Table 1 Rebound hammer test readings

Sr. No	Location	Avg. rebound No	Compressive strength (MPa)	Remark [As per IS 516: (Part 5/ Sect. 4) 2020]
1	Column on North face entry	41.50	20.35	Quality of concrete is Good
2	Column on South rear entry	40.50	20.10	Quality of concrete is good
3	Brick masonry load bearing walls outside building	11.62	5.70	Quality of load bearing masonry wall is satisfactory subjected to repairs and maintenance
4	Brick masonry load bearing walls inside building	14.10	6.90	Quality of load bearing masonry wall is satisfactory subjected to repairs and maintenance

2.4 Overall Observations

The building was found to be a load-bearing structure with satisfactory wall bearing strength. It was determined that undertaking timely and appropriate maintenance measures during the initial stages could potentially extend the buildings lifespan. Specific recommendations are made for structural improvements. The cantilever-reinforced concrete (RC) elements were identified for either removal or retrofitting, with a proposed reinforcement on the outer side using RC elements. Additionally, wooden balconies were recommended for removal to enhance structural stability. Regarding the flooring, it was suggested to either fill gaps or completely replace the existing flooring to achieve a more functional and safe flooring solution. Furthermore, urgent action was advised for cleaning and termite proofing to preserve a favorable indoor environment. To enhance esthetics and overall appeal, a plan for beautification is suggested, which included coloring all relevant structural elements, walls, wooden truss elements, doors, and windows.

3 Recommendations

Based on the structural audit conducted, several critical issues have been identified that require urgent attention and remediation to ensure the safety, stability, and lifespan of the building. Here are the recommended actions based on the identified problems to be undertaken in near future;

Masonry Walls Cracks: Appropriate structural repairs and reinforcements to restore the structural integrity of the masonry walls are needed. Injection grouting with high strength material for masonry repairs is recommended to stabilize the affected areas.

Deterioration of mortar joints: Perform thorough repointing with high-quality mortar to enhance load-bearing capacity and stability of walls. Apply compatible and durable mortar to ensure a strong bond. Consider waterproofing treatments to prevent future erosion and degradation of mortar joints.

Wood element rot and decay: Replace severely rotted or decayed wood elements like joists and supports to restore structural strength. Treat all wood elements with appropriate preservatives to prevent future decay and insect infestation.

Uneven flooring settlement and joint deterioration: Implement necessary repairs and leveling to address the flooring settlement issues. Repair or even replace flooring and reinforce deteriorated joints to ensure a smooth and safe flooring surface for usage.

Asbestos roof covering sheets weakness: Replace the weak asbestos sheets with new durable roofing materials to ensure effective protection from weather conditions.

Broken doors and windows, and pest infestation: Replace or repair damaged doors and windows to ensure proper security, insulation, and closure. Implement pest control measures to eradicate the infestation and maintain a healthy living or working environment.

Deterioration of old reinforced concrete elements: Undertake immediate repair for reinforcement of deteriorated RC members using appropriate materials and techniques. Implement corrosion protection measures to prevent further degradation of reinforcement bars and concrete surfaces.

4 Conclusions

1. A structural audit finds important problems that affect lifespan, safety, and demands quick fixes. The replacement of wood elements, masonry repairs, and improvements to the roof and flooring are the main points of advice.
2. Accurate evaluations and strengthening plans depend on tools like NDT. NDT testing finds weak points and reduced quality in different types of structural elements.
3. The observations highlight the need for quick action by addressing problems including irregular flooring, cracking concrete, broken doors and windows, and insect infestation.
4. Some useful recommendations are to remove or reinstall components, to fill up voids left by flooring, and to advise a detailed maintenance schedule.
5. Recommendations provide a detailed approach to maintaining the quality of the structure and vary from masonry repairs to visual improvements.
6. The conclusion focuses on what kind of careful structural checks are in identifying problems and promoting safe surroundings.
7. All things considered, rapid maintenance is stressed for improved structural integrity and a sustainable future.

References

1. Abuturab S, Tande SN (2023) Structural diagnosis, repair and retrofitting of RCC Structure. Int Res J Eng Technol 10(5):1181–1186
2. Mahadik AB, Jaiswal MH (2014) Structural audit of buildings. Int J Civ Eng Res 5(4):411–416
3. Karnavat R, Singh V (2017) Structural audit of a heritage structure: a case study. Res India Publ 47(3):170–175
4. Sonawane KR, Dhawale AW (2015) Structural audit: a case study of nasik residential building, Maharashtra, India. World Wide J 4:123–126
5. Kondalkar R, Pitale NH (2021) Structural audit of an old building: a case study. In: Proceedings of the international conference on advances in civil engineering (ICACE 2021). IOP Publishing, pp 1–13

Performance of Alkali-Activated Fly Ash-GGBS Concrete Under Acid, Sulphate and Carbonated Environment

T. Srividya, P. R. Kannan Rajkumar, R. Jeyalakshmi, J. Baskara Sundararaj, M. Jegan, and M. Sivasakthi

Abstract The physico-chemical evaluation of alkali-activated materials under aggressive conditions is a critical factor in determining their suitability and applications in construction. This paper presents the investigation on the performance of alkali excited blend of fly ash (FA) and ground granulated blast furnace slag (GGBS) concrete after exposure to aggressive chemicals such as sulphuric acid, sodium sulphate and carbon dioxide atmosphere. The main investigating factor is the weight percentage of GGBS in FA, and the concrete specimen casts in ambient temperature are GPC1 (FA 80%: GGBS 20%), GPC2 (FA70%: GGBS 30%) and GPC3 (FA 60%: GGBS 40%). The durability performance was verified under 5% concentrated sulphuric acid, 5% sodium sulphate solution and controlled carbon dioxide environment (3% CO_2, 65% relative humidity (RH), and temperature of 25 °C) as well as duration of exposure independently. The compressive strength, mass loss and physical changes were monitored before and after exposure. The result shows that increasing content of GGBS improved the strength in the as cured samples and attained the grade of M40. The strength retention is slightly decreased

T. Srividya
Department of Civil Engineering, School of Infrastructure, B.S. Abdur Rahman Crescent Institute of Science & Technology, Chennai, Tamil Nadu, India
e-mail: Srividya@crescent.education

P. R. K. Rajkumar (✉)
Department of Civil Engineering, College of Engineering and Technology, SRM Institute of Science and Technology, Kattankulathur, Tamil Nadu, India
e-mail: kannanrp@srmist.edu.in

R. Jeyalakshmi
Department of Chemistry, College of Engineering and Technology, SRM Institute of Science and Technology, Kattankulathur, Tamil Nadu, India

J. B. Sundararaj · M. Jegan
Centre for Advanced Concrete Research, Department of Civil Engineering, SRM Institute of Science & Technology, Kattankulathur, Tamil Nadu, India

M. Sivasakthi
Centre for Waste Management, International Research Centre, Sathyabama Institute of Science and Technology, Chennai, India

Y. K. Guruprasad et al. (eds.), *Recent Advances in Infrastructure Development and Management—Volume 1*, Lecture Notes in Civil Engineering 570,
https://doi.org/10.1007/978-981-96-1554-4_24

after exposing to aggressive acid, whereas retention of the strength after sulphate immersion at longer duration of 120 days. The carbonation depth of GPC specimens varied from 2 to 8 mm that are within the limits. The mineral phases were identified from powder *X*-ray diffraction and chemical structural changes from the IR frequency number of Si-O-Si and Si-O-Al bond in geopolymers. A high and prolonged strength capacity was accomplished proving that blended geopolymeric concrete has the potential to be applied as an alternate building material to traditional Portland cement concrete.

Keywords Carbonation · Acid attack · Sulphate attack · Geopolymers · Durability

1 Introduction

Geopolymer concrete (GPC) is an eco-friendly eliminates the utilisation of ordinary Portland cement (OPC). Cement-free concrete, i.e., geopolymers are now gaining more attention as sustainable construction practice as it utilises industrial waste materials for their production [1–4]. Geopolymer is produced from the alkalination reaction of aluminosilicates either by hydroxides or combination of alkali hydroxides with silicates of sodium or potassium. Geopolymers are the connectivity of silica and aluminium through shared oxygen atom forming tetrahedral unit Si-O-T (T = Si / Al). French Chemist Professor Davidovits named these condensed units as monomeric units of sialate, sialate-siloxo and sialate-disiloxo [5].

In designing concrete mixes, strength and durability-related qualities are the two most important factors. Generally, the water-to-cement ratio has a significant impact on the attributes of OPC concrete [6, 7]. On the other hand, a number of critical factors, including the alkali activator (AA) to binder ratio, the concentration of the AA, the silicate modulus (Ms) and the silica content affect the properties of the geopolymer binder system [8–15]. Therefore, a detailed study on durability is important because it will lead to the creation of GPC that is reusable, energy efficient and ecologically beneficial. Research investigation on durability dates back from Rangan et al. (2006) [16] on the heat-cured low calcium FA-blend GPC cylinder specimens were immersed in 5% Na2SO4 solution for a duration of 365 days. The specimens showed no evidence of surface erosion, cracking or spalling. The FA-based GPC exhibits great resistance to sulphate attack and experiences minimal creep and drying shrinkage. Role of acids, sulphates and chlorides on the durable properties of FA-based GPC performed better in terms of mass, strength and visual inspection than OPCC. Review studies relating to the durability and performance of geopolymers under elevated temperature and their superior behaviour over OPCC have been explored in detail [17–25].

Concrete carbonation is a key phenomenon for figuring out how durable concrete is since it can produce changes in the microstructure and moisture content. As a result, the permeability and diffusion coefficients of concrete may be affected [26].

Carbonatation happens naturally when CO_2 from the surrounding environment enters the material directly and is absorbed by the concrete surface [27]. After penetration, calcium silicate hydrate (C-S-H) and calcium hydroxide ($Ca(OH)_2$) combine with CO_2 to form H_2O and calcium carbonate ($CaCO_3$) [28]. From the literature studies, it was inferred that one of the current limitations in adopting geopolymer technology in structural applications is energy consumption for heat curing. To overcome this issue, binary combination of FA and GGBS can be adopted to achieve ambient curing (AC) in site, thereby avoiding energy required for curing. Both early strength gain and setting time can be improved with this FA-GGBS blend GPC system [29]. These factors favour adopting this binary system in structural applications [4, 30, 31]. There is a significant gap in comprehensive research concerning the service life of these materials for adopting this technology, migrating from traditional materials to this cutting-edge technology. Long-term behavioural studies of this binary system along with their mineralogical changes in different environments not yet been explored. Also, few research has been published analysing the carbonation of alkali-activated materials, and even fewer have focused on low calcium content in geopolymers [32]. Keeping these points, this study examines the durability properties of a binary system of FA-GGBS synthesised GPC along with their mineralogical changes using analytical techniques XRD and FT-IR. Durability studies such as resistance to concentrated sulphuric acid, resistance to sodium sulphate solution and carbonation test were conducted on GPC specimens up to a period of 120 days discussed.

2 Materials and Methods

Class F Fly ash and Ground granulated blast furnace slag are used as precursors for geopolymerisation in this research. FA conforms to ASTM C618 [33] with specific gravity (SG) of 2.22. GGBS with SG of 2.85 satisfying ASTM C989 [34]. Mineral oxides present in the precursors are explored using *X*-Ray fluorescence and found to have, silica and alumina forms as major components along with other oxides in FA, whereas in GGBS, comprising CaO forms nearly 50% of total mineral oxides. XRD of FA and GGBS are shown in Fig. 1. Coarse aggregate (CA) of size 12.5 mm with SG of 2.72 is used. M-sand and CA are subjected to testing as per ASTM C33 [35], ASTM C136/C136M-14 [36] and IS 2386 [37, 38]. Sodium hydroxide solution (SHS) of molarity 18.25M and sodium silicate solution (SSS) of Ms 2.0 is used to prepare alkaline activators [39].

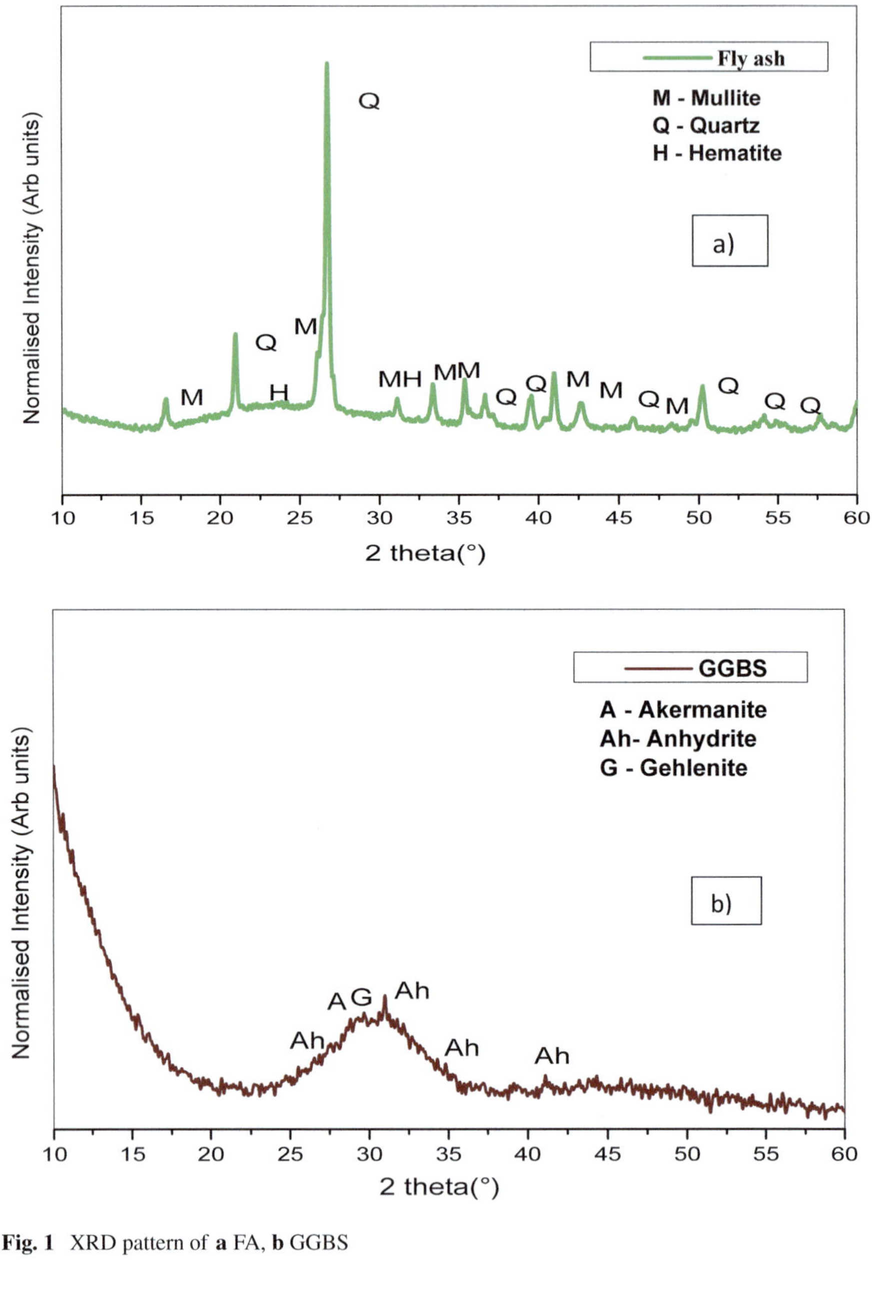

Fig. 1 XRD pattern of **a** FA, **b** GGBS

Table 1 Mix Proportion of GPC

Material (kg/m^3)	GPC 1	GPC 2	GPC 3
FA	326	284	242
GGBS	74	126	158
SSS	80	80	80
SHS	80	80	80
Water	40	40	40
M-sand	600	600	600
CA	1000	1000	1000
Slump (mm)	120	105	90
SSS / (SHS + water)		0.67	

2.1 Material Mix Proportion and Fabrication of Geopolymer Concrete

The mix composition of M40 grade FA-GGBS-based AC GPC is developed after various laboratory trials by varying the proportions of precursors and the ratio of alkaline liquid. In this study, FA and GGBS are taken in three different combinations as GPC 1(80%FA:20%GGBS), GPC 2(70%FA:30%GGBS) and GPC 3(60%FA:40%GGBS) with 0.67 as activator ratio (SSS / (SHS + water)). The quantities of GPC mixes per m^3 along with the slump value are tabulated in Table 1. Fabrication of concrete specimens is carried out by homogenising the ingredients as dry mixture in 100 L pan mixer for 3 min after that alkaline liquid is added. Activators are prepared one day prior to casting. The homogenous concrete mix is cast into cubes of dimensions 100 × 100 × 100 mm and prism specimens of size 100 × 100 × 500 mm. After demoulding, the specimens are kept in atmospheric conditions for 28 days. Before immersion, the compressive strength of specimens was determined. The compressive strength (CS) of cubes was found to be 44.07 MPa (GPC 1), 48.82 MPa (GPC 2) and 56.3 MPa (GPC 3).

2.2 Experimental Program Acid Resistance

Geopolymer concrete cube specimens were immersed in 5% concentrated H_2SO_4 acid to evaluate their resistance to acid attack conforming to ASTMC1898-20 [40].

Mass loss was calculated from the difference in mass before and after exposure to acid. The percentage of mass loss is continuously monitored over a period of 120 days with a time interval of 28 days. Visual appearance of concrete specimens followed and was shown in Fig. 2a, b and c.

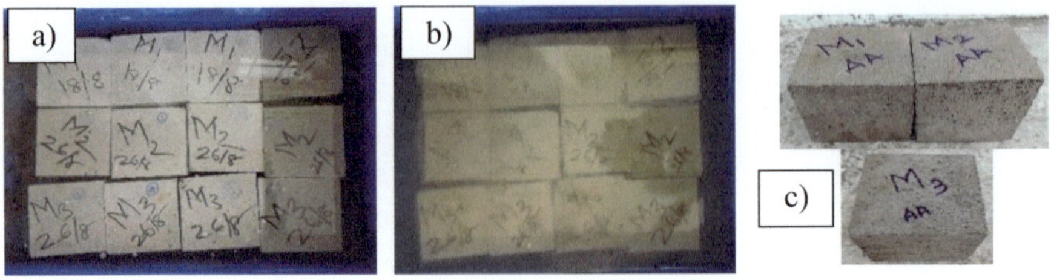

Fig. 2 Visual appearance of GPC samples **a** During immersion, **b** Immersed specimens after 120 days, **c** Dried GPC specimens exposed up to 120 days

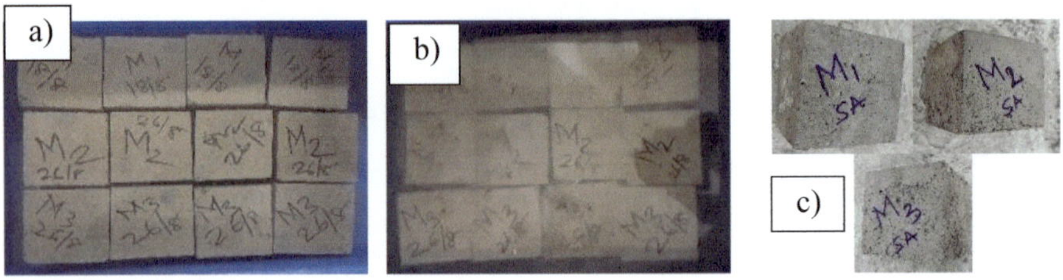

Fig. 3 Typical photographs of GPC specimens **a** During immersion, **b** Immersed specimens after 120 days, **c** Dried specimens exposed up to 120 days

Sulphate Resistance

The sulphate attack test conducted as per ASTM C1012-04 [41] standards, GPC cube specimens immersed in Na_2SO_4 solution shown in Fig. 3a, b and c

Accelerated Carbonation Test

The accelerated carbonation test (ACT) was conducted in an accelerated carbonation chamber using GPC prisms of size 100 mm × 100 mm × 500 mm cured at standard days of 28 days at AC conditions. Wax was applied as an anti-carbonation coating on two sides and both the ends of the prisms as in Fig. 4. Subsequently, these were placed in the accelerated carbonation chamber as shown in Fig. 5. The test conditions adopted were CO_2 concentration level of 3%, temperature of 25 °C and 65% of RH. After exposure for 28, 56, 90, and 120 days, the carbonation depth was determined by spraying 1% phenolphthalein indicator on the newly cut surface from the prism in accordance with EN 13295:2004 [42].

Fig. 4 Photographs of specimens coated on both sides and ends

Fig. 5 Accerelated chamber with wax coated specimens

3 Results and Discussion

3.1 Acid and Sulphate Resistance

The CS and mass loss values before and after exposure in acid and sulphate environments were plotted in Fig. 6a, 6b and 6c. From the visual observation, it was inferred that GPC remains intact without any changes in the surface as well as found to be structurally intact as shown in Fig. 2c and 3c.

Erosion of the surface was not observed in GPC specimens even after 120 days of exposure in 5% concentrated H_2SO_4 and 5% Na_2SO_4 solution and did not exhibit any noticeable colour change. Figure 6a represents the performance of GPC specimens in terms of loss in weight under sulphate and acid environment. Under sulphuric

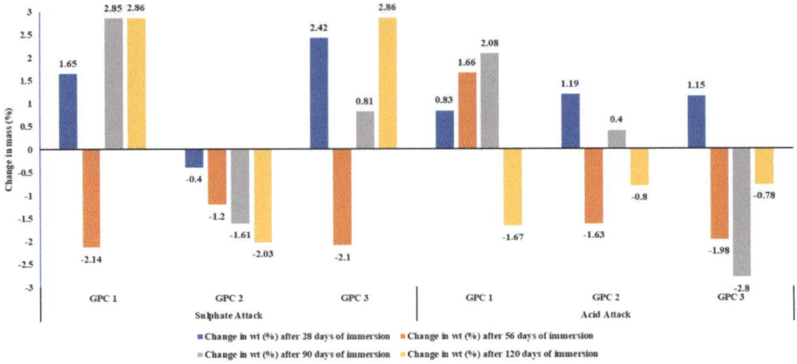

a Change in weight (%) of GPC mixes immersed in sulpahte and sulphuric acid

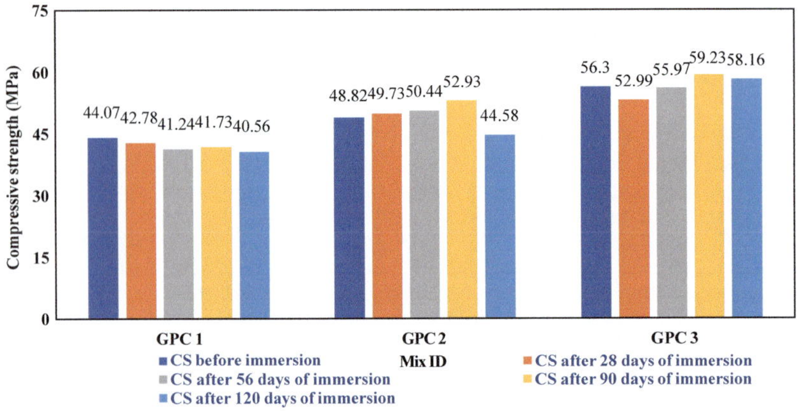

b CS of GPC mixes immersed in sulphuric acid

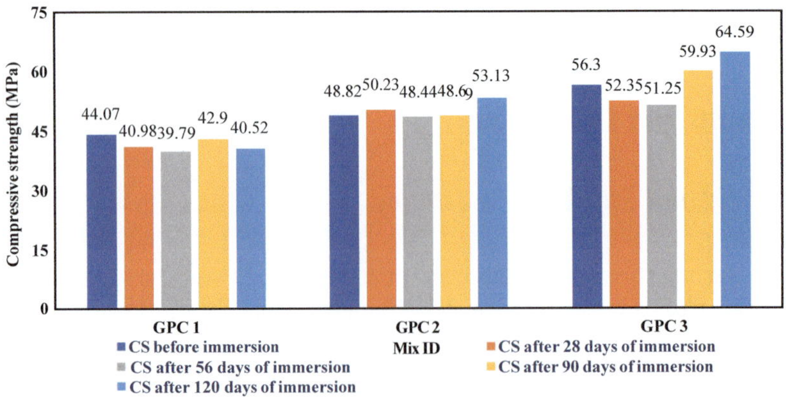

c CS of GPC mixes immersed in sodium sulphate solution

Fig. 6 **a** change in weight (%) of GPC mixes immersed in sulphate and sulphuric acid, **b** CS of GPC mixes immersed in sulphuric acid, **c** CS of GPC mixes immersed in sodium sulphate solution

acid, all the GPC specimens retained their weight under differing proportions of FA and GGBS as well as with the immersion ages. The GPC samples had weight loss of less than 2% at 120 days. The GPC 1 specimen performed well with a weight gain of 3% at 90 days and was found to have 1.67% weight loss at 120 days. GPC 1 and GPC 3 specimens had weight gain of 2.8% at 120 days, whereas a loss in weight of 2% was observed in GPC 2 samples in a sulphate environment. Both acid and sulphate-immersed specimens performed well with less than 2% weight loss at 120 days.

Figure 7 depicts the strength evolution of GPC samples on acid exposure. GPC 1 specimens had significant strength loss of 10.23% at 120 days. Mix with higher slag content showed percentage reduction in loss in strength with ages, i.e., iIn GPC 3 samples, initially there was a decrease in strength of 5.8% after 28 days followed by 4.15% strength gain over a period of immersion time extended after 120 days.

Figure 6c depicts the strength evolution of GPC samples on sulphate exposure. Under sulphate exposure, GPC 1 samples had 9.71% decline in strength after 56 days but recovered slightly to 2.65%. In GPC 2 and GPC 3 specimens, a similar trend of decrement in strength along with significant strength increments of 8.82% (GPC 2) and 12.08% (GPC 3). Similar findings on the performance of geopolymers on acid and sulphate resistance were reported by few others [43–46]. From the observations, exposure of activated GPC specimens under acid and sulphate environment was found to be performing better than OPCC as per literature studies.

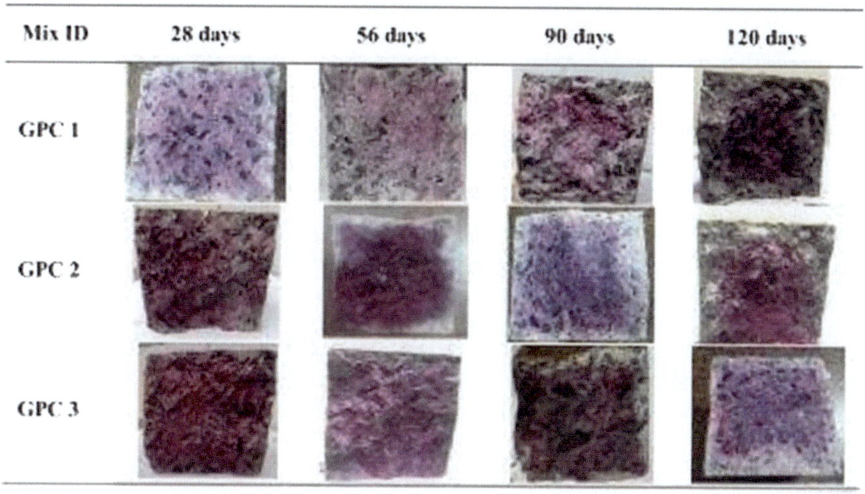

Fig. 7 Pink colour ranges of GPC specimens at every 28 days interval upto 120 days after the accelerated carbonation

Table 2 Carbonation depth measurement of GPC specimens at 28, 56, 90 and 120 days of accelerated curing

Mix ID	28 days	56 days	90 days	120 days
GPC 1	Fc: Nil Pc:7.5 mm	Fc: Nil Pc:11.25 mm	Fc:4.5 mm Pc:17.5 mm	Fc: 8 mm Pc:17.5 mm
GPC 2	Fc: Nil Pc:10 mm	Fc: 4.5 Pc: up to 12.5 mm	Fc: 4.5 Pc: up to 17.5 mm	Fc: Nil Pc: upto 17.5 mm
GPC 3	Fc: Nil Pc: less than 1 mm	Fc: 2 mm Pc: up to 5 mm	Fc: 5.5 mm Pc: Nil	Fc: 2 mm Pc: 10 mm

Fc: Full carbonation (concrete colour); Pc: partial carbonation (light purple red colour)

3.2 Accelarated Carbonation Test

Carbonation depth of specimens under accelerated curing at 28, 56, 90 and 120 days of GPC specimens is presented in Table 2. Colour changes of GPC specimens after spraying phenolphthalein solution at every 28-day interval are presented in Figure 7. Initially, in all the GPC samples, partial carbonation was observed with minimum of less than 1 mm (GPC 3) and maximum range of 10 mm (GPC 2) at 28 days. After 56 days of accelerated curing, specimens showed both partial and full carbonation.

The depth of complete carbonation ranged from 2 mm (GPC 3) to 8 mm (GPC 1), whereas the depth of partial carbonation varied from 1 mm (GPC 3) to 17.5 mm (GPC 1). Studies conducted on longer duration exposure at specific composition of mixes revealed that increasing slag reduced the carbonation depth but longer exposure enlarged the carbonation depth. Resistance to carbonation of FA-slag blended system increases with decrease in porosity, i.e., better pore structure leading from strong geopolymeric gel, slag ratio and alkaline concentration.

3.3 Mineralogical Characterisation XRD Analysis

X-ray diffractograms of alkali-activated FA-GGBS blend mixes, before and after exposure to 5% sodium sulphate solution upto 120 days shown in Fig. 8a. The main phase detected is a poorly crystalline C–A–S–H phase, from the reflection ranging between 28° and 30° 2θ [47, 48].

Hydrotalcite, a product of alkali-activated slags [49–51], is observed. Vaterite, metastable calcium carbonate polymorph is also formed as sulphate is known to favour stabilisation for longer periods [52]. The reaction product of sodium sulphate identified in all the samples different peaks. Similarly, from Fig. 8b, specimens immersed in H_2SO_4 acid showed the presence of crystalline phases of Quartz, mullite and hematite [43, 53, 54]. In addition to that, gypsum identified at different intensity which enhances the strength factor. Figure 8c shows the XRD patterns of accelerated carbonated mixes wherein the main crystalline calcium carbonate phases: calcite,

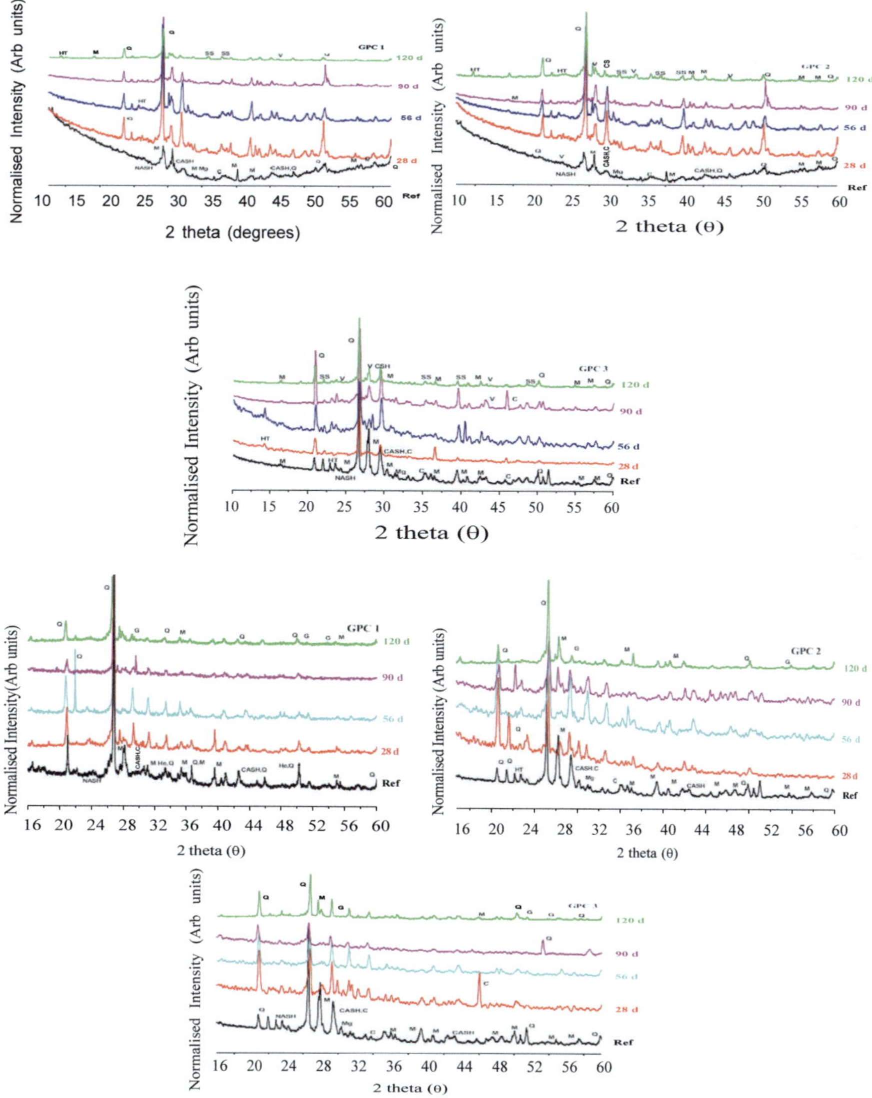

Fig. 8 **a** XRD patterns of GPC before and after sulphate attack. Mg, maghemite, M, mullite, CASH, calcium alumino silicate hydrate, HT, hydrotalcite, C, calcite, NASH, sodium alumino silicate hydrate, Q-Quartz. **b** XRD patterns of GPC before and after acid exposure. Mg, maghemite, M, mullite, CASH, calcium alumino silicate hydrate, SS, sodium sulphate, HT, hydrotalcite, C, calcite, NASH, sodium alumino silicate hydrate, CSH, calcium silicate hydrate Q-Quartz, G, gypsum. **c** XRD patterns of GPC before and after carbonation. V, Vaterite, A, Aragonite, H, Huntite, Nh, Nahcolite, C, Calcite, C-A-S-H, calcium alumino silicate hydrate, N-A-S-H, sodium alumino silicate hydrate, Q, Quartz, M, Mullite, He, Hematite

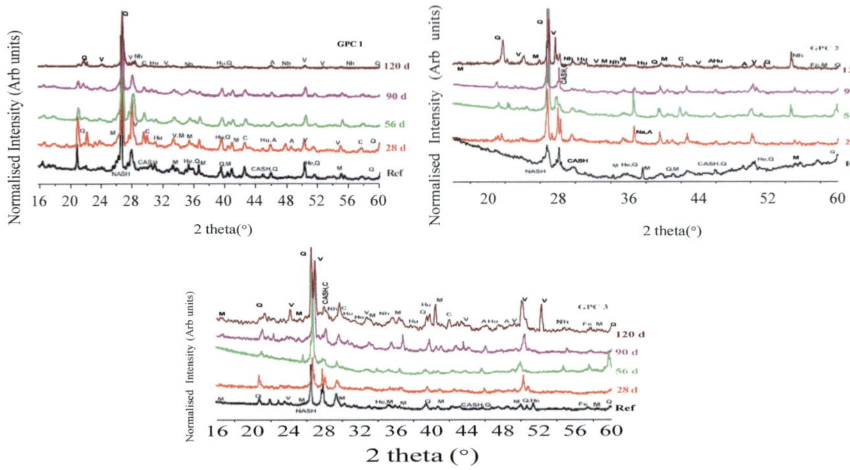

Fig. 8 (continued)

vaterite and aragonite are identified [55]. In addition to these, Mg-Ca carbonate phase Huntite emerged in the blended systems and their peaks are less evident in more matured gel, i.e., 120 days of curing since the inclusion of Mg in the calcite structure resulted in contraction of unit cell as described in the literature [56].

Na-rich carbonation product, Nahcolite in all samples is a prominent product exposed at higher CO_2 partial pressures [57]. Natron, yet another Na-rich product usually prominent in alkali-activated slag system, is identified at the intensity of 37° 2θ in the blended system sample with higher content of slag (GPC 3).

ATR FT-IR Analysis

The FT-IR analysis was used to understand the chemical bonds present in the different geopolymers GPC 1, GPC 2 and GPC 3 before and after exposure to different environments and presented in Fig. 9a, b and c. The interpretation of the IR spectral bands was referred from the previously published literature [58]. Before the exposure of control geopolymer materials, the fingerprint bond of the geopolymer asymmetric stretching vibrations of Si-O-Al/Si were found in the range of 975–1025 cm^{-1}. The C-O stretching vibrational frequency was observed near to 1400 cm^{-1} due to the formation of carbonates from the reaction of unreacted alkali and CO_2 from the environment present in all the samples. The bands at 565, 545 and 565 cm^{-1} for GPC 1, GPC 2 and GPC 3 were due to bending vibrations of Si- O-. The symmetric stretching vibrational frequency of Si-O-Al/Si was found in the region of 650–680 cm^{-1}. The network stretching vibrational bands observed at 765 and 772 cm^{-1} were observed in the GPC 3 may be from the enhanced reaction by GGBS phase.

After the carbonation, all bands present in the control samples were observed except for the change in intensity of the carbonated products. GPC 1 does not experience the changes in peak position in the fingerprint region however Si-O bending vibrational peaks shifted to lower end from 565 to 530 cm^{-1}. In GPC 2, the new

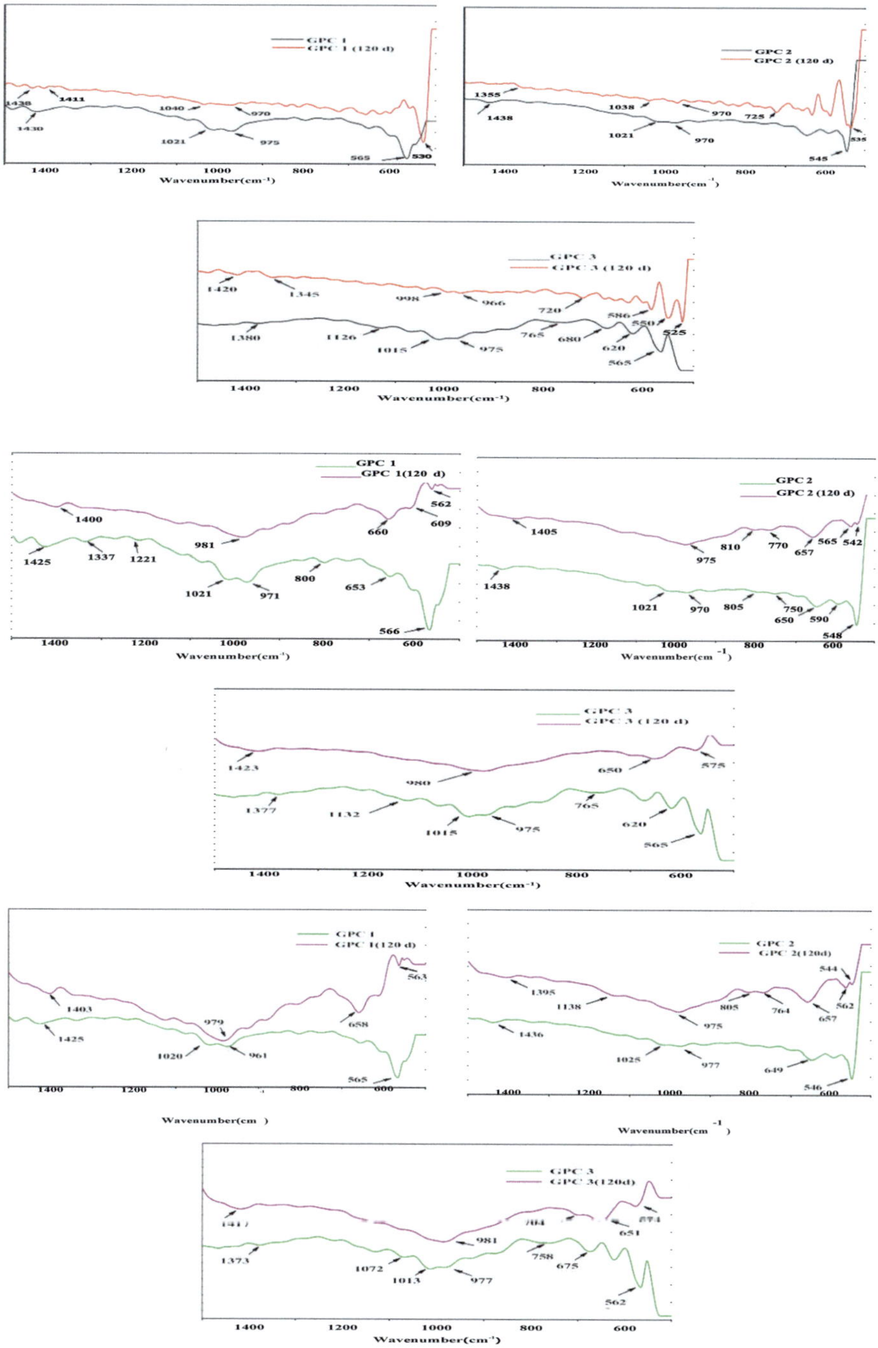

Fig. 9 FT-IR spectrum of the GPC exposed to carbonation, **b** FT-IR spectrum of the GPC exposed to 5% sulphuric acid, **c** FT-IR spectrum of the GPC exposed to 5% sulphate solution

network stretching band has appeared newly at 725 cm^{-1}. This represents the progress of the reaction even with CO_2 exposure. The symmetric stretching vibration present in the composites has experienced the slight shifting to lower region indicating the Al incorporation. The stretching modes are sensitive to the Si-Al composition of the framework and may shift to a lower frequency with the increasing number of tetrahedral aluminium atoms. This indicated that the carbonation has not affected the gel formation and the progress of the reaction. Referring to the FT-IR spectrum after the acid exposure, the finger print region of the geopolymer appeared as a broad range in all the control samples has been changed as narrow with one specific peak around 975 cm^{-1} indicating the presence of geopolymeric gel phase. With the sulphate exposure, the Si- peaks around 800 cm^{-1} have been disappeared from the spectrum for GPC 1 and GPC 3. The broad finger print peak of the geopolymer remains unchanged. This coincides with the enhanced strength results.

4 Conclusion

The behaviour of M40 grade geopolymer concrete designed from ambient-cured FA-GGBS alkali-activated cement has been discussed in detail. The concrete made from GPC mixes exposed to aggressive chemicals for shorter and longer duration. From the results observed, the following conclusions are drawn.

- Among the synthesised GPC mixes in all three exposure conditions, GPC 3 with higher content of GGBS performed better due to the mixed N(C)-SH.
- Physical integrity of samples verified from visual observations and found to be intact with formwork stability
- In terms of weight loss, GPC specimens exhibited a marginal loss upto 2.8% (GPC 3) in acidic medium but slightly gained in sulphate environment at 120 days of exposure. Similar such trend is observed in strength of exposed samples. Concentrated sulphuric acid deteriorate the strength of all samples to an extent of 10–15%.
- Under accelerated carbonation test, from the colour changes it was inferred that most of the specimens exhibited partial carbonation on 120 days of exposure to CO_2 environment.
- Structural examination using XRD, IR revealed that new phases of vaterite and sodium sulphate were identified in sulphate exposure, gypsum along with CSH were present in acid-exposed samples. Carbonated samples showed XRD reflections corresponding to phases like Natron and Nahcolite.
- The geopolymeric reaction product normally evolved from partial substitution of Al in Si tetrahedra [Si-O-T](T = Si / Al) assessed from FT-IR spectra found to be unaltered in all the exposed samples conforming to the network stability.

The study outcomes would encourage the researchers to adopt alkali-activated FA-GGBS geopolymers for structural applications as they performed well in different aggressive chemical environmental exposures in terms of strength as well as mineralogical characteristics leading to sustainable construction practices.

References

1. Singh NB, Middendorf B (2020) Geopolymers as an alternative to Portland cement: an overview. Constr Build Mater 20(237):117455
2. Amran YM, Alyousef R, Alabduljabbar H, El-Zeadani M (2020) Clean production and properties of geopolymer concrete: a review. J Cleaner Prod 1(251):119679
3. Xu G, Shi X (2018) Characteristics and applications of fly ash as a sustainable construction material: A state-of-the-art review. Resour Conserv Recycl 136:95–109
4. Soong LS, Liu MYJ, Yap SP et al (2022) The potential of geopolymer in development of green coating materials: a review. Arab J Sci Eng 47:12289–12299. Davidovits J (1999) Chemistry of geopolymeric systems, terminology. Geopolymer 99(292):9–39
5. Neville AM (1995) Properties of concrete, vol 4, p 1995. London: Longman
6. Mehta PK, Monteiro PJ (2014) Concrete: microstructure, properties, and materials. McGraw-Hill Education
7. Jan A, Pu Z, Khan KA, Ahmad I, Shaukat AJ, Hao Z, Khan I (2022) A review on the effect of silica to alumina ratio, alkaline solution to binder ratio, calcium oxide+ ferric oxide, molar concentration of sodium hydroxide and sodium silicate to sodium hydroxide ratio on the compressive strength of geopolymer concrete. Silicon 14(7):3147–3162
8. Gao K, Lin KL, Wang D, Hwang CL, Shiu HS, Chang YM, Cheng TW (2014) Effects SiO_2/Na_2O molar ratio on mechanical properties and the microstructure of nano-SiO_2 metakaolin-based geopolymers. Constr Build Mater 53:503–510
9. Sivasakthi M, Jeyalakshmi R (2021) Effect of change in the silica modulus of sodium silicate solution on the microstructure of fly ash geopolymers. J Build Eng 44:102939
10. Ryu GS, Lee YB, Koh KT, Chung YS (2013) The mechanical properties of fly ash-based geopolymer concrete with alkaline activators. Constr Build Mater 47:409–418
11. Srividya T, PR KR, Sivasakthi M, Sujitha A, Jeyalakshmi R (2022) A state-of-the-art on development of geopolymer concrete and its field applications. Case Stud Constr Mater 16:e00812
12. Poornima N, Sivasakthi M, Jeyalakshmi R (2022) Microstructure investigation of the Na/Ca aluminosilicate hydrate gels and its thermal compatibility in fly ash–GGBS cementitious binder. J Build Eng 50:104168
13. Thiyagarajan S, Ramalakshmi KRP, Jayapaul BS, Murugesan J, Ramaswamy J (2024) Experimental and analytical studies on bond strength of geopolymer concrete using pull-out specimens. In: AIP conference proceedings, vol 3187, no 1. AIP Publishing
14. Srividya T, Kannan Rajkumar PR (2023) Durability properties of geopolymer concrete from fly ash and GGBS. In: Recent advances in materials, mechanics and structures, pp 601–608. Springer, Singapore
15. Jegan M, Annadurai R, Rajkumar PK (2023) A state of the art on effect of Alkali activator, precursor, and fibers on properties of geopolymer composites. Case Stud Constr Mater e01891
16. Wallah S, Rangan BV (2006) Low-calcium fly ash-based geopolymer concrete: long-term properties
17. Bakharev T (2005) Resistance of geopolymer materials to acid attack. Cem Concr Res 35(4):658–670
18. Baščarević Z, Komljenović M, Miladinović Z et al (2015) Impact of sodium sulfate solution on mechanical properties and structure of fly ash based geopolymers. Mater Struct 48:683–697
19. Thokchom S, Ghosh P, Ghosh S (2009) Resistance of fly ash-based geopolymer mortars in sulfuric acid. ARPN J Eng Appl Sci 4(1):65–70

20. Ismail I, Bernal SA, Provis JL et al (2013) Microstructural changes in alkali activated fly ash/slag geopolymers with sulfate exposure. Mater Struct 46:361–373
21. Fernández-Jiménez A, García-Lodeiro I, Palomo A (2007) Durability of alkali-activated fly ash cementitious materials. J Mater Sci 42:3055–3065
22. Kong DL, Sanjayan JG (2008) Damage behavior of geopolymer composites exposed to elevated temperatures. Cem Concr Compos 30(10):986–991
23. Pradhan P, Dwibedy S, Pradhan M, Panda S, Panigrahi SK (2022) Durability characteristics of geopolymer concrete-progress and perspectives. J Build Eng 105100
24. Chen K, Wu D, Xia L, Cai Q, Zhang Z (2021) Geopolymer concrete durability subjected to aggressive environments–a review of influence factors and comparison with ordinary Portland cement. Constr Build Mater 279:122496
25. Wasim M, Ngo TD, Law D (2021) A state-of-the-art review on the durability of geopolymer concrete for sustainable structures and infrastructure. Constr Build Mater 291:123381
26. Morandeau A, Thiery M, Dangla P (2014) Investigation of the carbonation mechanism of CH and CSH in terms of kinetics, microstructure changes and moisture properties. Cem Concr Res 56:153–170
27. Pasupathy K, Berndt M, Castel A, Sanjayan J, Pathmanathan R (2016) Carbonation of a blended slag-fly ash geopolymer concrete in field conditions after 8 years. Constr Build Mater 125:661–669
28. Chang CF, Chen JW (2006) The experimental investigation of concrete carbonation depth. Cem Concr Res 36(9):1760–1767
29. Nagajothi, S. and Elavenil, S., 2021. Effect of GGBS addition on reactivity and microstructure properties of ambient cured fly ash based geopolymer concrete. Silicon 13:507–516
30. Revathi T, Vanitha N, Jeyalakshmi R, Sundararaj B, Jegan M, Rajkumar PK (2022) Adoption of alkali- activated cement-based binders (geopolymers) from industrial by-products for sustainable construction of utility buildings-A field demonstration. J Build Eng 52:104450
31. Srividya T, Rajkumar PK (2022) Mechanical and durability properties of alkali-activated binders based paver blocks derived from secondary sources. Case Stud Constr Mater 17:e01561
32. Bernal SA, Pacheco-Torgal F, Labrincha J, Leonelli C, Palomo A, Chindaprasit P (2016) The resistance of Alkali- activated cement-based binders to carbonation. Mortars and Concretes, Woodhead Publishing Limited, Cambridge (UK), Handbook of Alkali-Activated Cements, pp 319–332
33. ASTM C618 (2003) Standard specification for coal fly ash and raw or calcined natural pozzolan for use in concrete. American society for testing and materials. West Conshohocken, PA, USA: ASTM International
34. ASTM C989 (2006) Standard specification for ground granulated blast furnace slag for use in concrete and mortars, vol 4(02)
35. Standard specification for concrete aggregates (1999) ASTM C 33. American Society for Testing and Materials, ASTM specification, Philadelphia
36. ASTM, A., 2014. C136/C136M–14. Standard Test Method for Sieve Analysis of Fine and Coarse Aggregates, ASTM International, West Conshohocken, PA.
37. IS: 2386-1 (1963) Methods of test for aggregates for concrete Part I: Particle size and shape. Bureau of Indian Standards, New Delhi
38. IS: 2386 (1963) Methods of test for aggregates for concrete–Part 3: specific gravity, density, voids, absorption and bulking. Bureau of Indian Standards, New Delhi
39. Srividya T, Kannan Rajkumar PR (2023) Mechanical properties and bond behaviour of steel rebar and fly ash–slag based geopolymer concrete in pull-out test. Innov Infrastruct Solut 8(4):117
40. ASTMC1898-20 (2000) Standard test method for determining the chemical resistance of concrete products to acid attack. ASTM international: West Conshohocken, PA, USA
41. ASTM C1012-04 (2004) Standard test method for length change of hydraulic-cement mortars exposed to a sulfate solution. ASTM international: West Conshohocken, PA, USA
42. EN 13295 (2004) Products and systems for the protection and repair of concrete structures—test methods—determination of resistance to carbonation

43. Bakharev T, Sanjayan JG, Cheng YB (2003) Resistance of alkali-activated slag concrete to acid attack. Cem Concr Res 33(10):1607–1611
44. Mehta A, Siddique R (2017) Sulfuric acid resistance of fly ash based geopolymer concrete. Constr Build Mater 146:136–143
45. Rajamane NP, Nataraja MC, Lakshmanan N, Dattatreya JK, Sabitha D (2012) Sulphuric acid resistant ecofriendly concrete from geopolymerisation of blast furnace slag.
46. Alexander AE, Shashikala AP (2022) Studies on the microstructure and durability characteristics of ambient cured FA-GGBS based geopolymer mortar. Constr Build Mater 347:128538
47. Komljenović MM, Baščarević Z, Marjanović N, Nikolić V (2012) Decalcification resistance of alkali- activated slag. J Hazard Mater 233:112–121
48. Bernal SA, Provis JL, Rose V, De Gutierrez RM (2011) Evolution of binder structure in sodium silicate- activated slag-metakaolin blends. Cement Concr Compos 33(1):46–54
49. Puertas F, Fernández-Jiménez A (2003) Mineralogical and microstructural characterisation of alkali-activated fly ash/slag pastes. Cement Concr Compos 25(3):287–292
50. Oh JE, Monteiro PJ, Jun SS, Choi S, Clark SM (2010) The evolution of strength and crystalline phases for alkali-activated ground blast furnace slag and fly ash-based geopolymers. Cem Concr Res 40(2):189–196
51. Haha MB, Lothenbach B, Le Saout GL, Winnefeld F (2011) Influence of slag chemistry on the hydration of alkali-activated blast-furnace slag—Part I: Effect of MgO. Cem Concr Res 41(9):955–963
52. Fernández-Díaz L, Fernández-González Á, Prieto M (2010) The role of sulfate groups in controlling $CaCO_3$ polymorphism. Geochim Cosmochim Acta **74**(21):6064–6076
53. Lee NK, Lee HK (2016) Influence of the slag content on the chloride and sulfuric acid resistances of alkali- activated fly ash/slag paste. Cem Concr Compos 72:168–179
54. Thokchom S (2014) Fly ash geopolymer pastes in sulphuric acid. Int J Eng Innov Res 3(6):943–947
55. Bernal SA, Provis JL, Brice DG, Kilcullen A, Duxson P, van Deventer JSJ (2012) Accelerated carbonation testing of alkali-activated binders significantly underestimate the real service life: the role of the pore solution. Cem Concr Res 42(10):1317–1326
56. Goldsmith JR, Graf DL (1958) Relation between lattice constants and composition of the Ca-Mg carbonates. Am Mineral 43(1):84–101
57. Bernal SA, Provis JL, Walkley B, San Nicolas R, Gehman JD, Brice DG, Kilcullen AR, Duxson P, van Deventer JS (2013) Gel nanostructure in alkali-activated binders based on slag and fly ash, and effects of accelerated carbonation. Cem Concr Res 53:127–144
58. Breck DW (1974) Zeolite molecular sieves: structure, chemistry and use, pp 415–418. Wiley-Interscience, New York

Application of Geographical Information Systems in Rooftop Rainwater Harvesting

R. Suresh Kumar, Peddireddy Sreekanth Reddy, and G. S. Brunda

Abstract Water plays a key role in the earth's ecosystem and is a vital natural resource for sustaining life and the environment within the present cenario of worldwide global warming. The supply of water is a major concern due to rapid urbanization and poor recharge of groundwater due to concrete infrastructure, especially in metropolitan cities. In order to counter water scarcity problems, rainwater harvesting is one of the simplest and most effective practices. In India, Hyderabad is a fast-growing city in terms of population and development for a decade and faces severe water problems during the summer season. To decline these issues, a study was conducted on a particular location, Miyapur. To evaluate the potentiality of rain harvesting, surface areas of roofs are obtained by employing geographical information systems (GIS), and further interpretations were done to obtain a slope map, contour map, and flow direction map. Further, the total water demand for the selected area was calculated considering the prevailing population. The potential rainwater harvesting sites are identified based on the GIS map according to the lowest elevation and drainage direction of the study area. The results of this study provide rainwater harvesting potential and water demand to avoid water shortage in the study area. Rainwater harvesting provides water at the purpose of demand, and it is an effective and eco-friendly solution to meet the various demands.

Keywords Water quality · Compressive strength · Tensile strength

R. S. Kumar
Department of Civil Engineering, Institute of Aeronautical Engineering, Dundigal, Hyderabad, Telangana, India

P. S. Reddy · G. S. Brunda (✉)
Department of Civil Engineering, Acharya Institute of Technology, Bengaluru, Karnataka, India
e-mail: brundags@gmail.com

P. S. Reddy
Department of Civil and Infrastructure Engineering, Indian Institute of Technology Dharwad, Dharwad 580011, India

G. S. Brunda
Department of Civil Engineering, National Institute of Technology Mizoram, Aizawl, Mizoram, India

Y. K. Guruprasad et al. (eds.), *Recent Advances in Infrastructure Development and Management—Volume 1*, Lecture Notes in Civil Engineering 570,
https://doi.org/10.1007/978-981-96-1554-4_25

1 Introduction

With the effect of industrialization, the number of industries has increased for a decade, resulting in the consumption of water in large quantities, especially in Hyderabad. On the other hand, post-utilized water may be composed of added pollutants that can contaminate land and water bodies. It cannot be suggested for household and drinking activities. To avoid such kind of issue, rainwater harvesting is one of the technologies that can collect, deliver, and store rainwater from various types of roofs, rock catchment and land surface. It has gained much significance as it is a modern, water-saving device and it is very simple technology. It is one of the best methods which is used to conservation of rainwater and to increase the water demand in many areas where there is a scarcity of water. It is very simple, eco-friendly, and effective one. It involves diverting and storing rainfall water from various roofs of buildings/ houses. Generally, it is useful in coastal, urban, and hilly areas. If this technique is properly installed and maintained, then the water which is collected and stored can be used for various consumptive needs. Many studies adopted different techniques to address the problems of water scarcity and recharge of ground water, reported different degrees of success. With the increasing awareness on the water demand and their quality, studies pertain to modelling of runoff, geographic information system (GIS), and remote sensing (RS) have gained more importance in the selection of suitable sites for water recharging/harvesting structures [1–4]. Acknowledging it, Singh et al. [5] employed RS and GIS in the analysis of Soankhad Watershed, Punjab. In addition to land use map, soil map, slope map, and digital elevation model (DEM) hydroprocessing used along with satellite image to estimate the capacity of rainwater harvesting.

A very minimal study is in light utilizing the efficiency of GIS in rainwater harvesting, which is having more potency. The main aim of the current study is to evaluate the rooftop rainwater harvesting potential and water demand by developing a geospatial database for the study area. Due to the ever-growing demand for water, the collected water can be used by recharging the water into the ground by simple infiltration techniques. Further, it improves the quality of groundwater by infiltrating the stored rainwater through simple filtration devices.

2 Materials and Methodology

In order to evaluate the rooftop rainwater harvesting and the water demand, we selected the study area as Miyapur. Miyapur is located at 22.5 kms Northwest from Hyderabad. It is located in Hyderabad district of Telangana state in India. Its geographical are 17° 29.808′ N, 78° 21.684′ E. It is located at an elevation of about 536 m above mean sea level. This area receives the highest rainfall during the rainy season in a year the average rainfall is 745 mm. For the selected area, the existing population is 8600. As this area consists of many industries, schools, colleges, shopping malls, and IT sectors, there is an essential of water in huge amounts. So, to provide the water, the water is collected from the various types of roofs to evaluate the total amount of water collected and calculate the water demand based on the existing population.

2.1 Google Earth Pro

Google Earth Pro: Google earth pro is a geospatial desktop application and global computer program, which is used to identify or locate the study area. The data obtained from this will be saved in Kml/Kmz format, which is further converted into shape files in QGIS, in order to calculate the roof area. The typical photograph of Google. Earth Pro Software tool can be seen in Fig. 1.

Fig. 1 Typical photograph of Google Earth Pro used in the current study

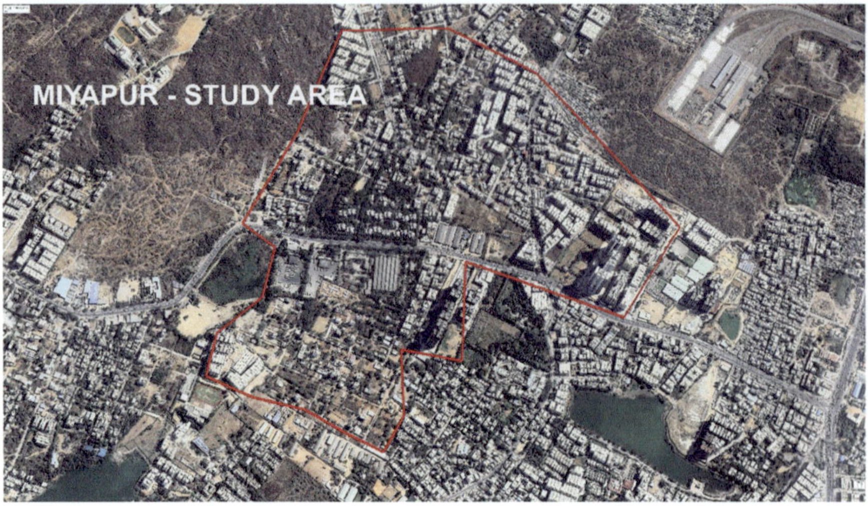

Fig. 2 Base map for the study area by using Google Earth Pro

2.2 *Quantum GIS (QGIS)*

QGIS is a free software platform for geographic information systems (GIS) which supports the analysis of global data. Using QGIS, the roof area for the study map can be digitized. During the process, the Kml/Kmz file from google earth pro is exported to QGIS software. Further, the type of roof material used in the building is identified by using satellite images and google maps. In order to get the roof area, a geometry calculation tool will be available in QGIS software. The typical photograph indicating the digitized map of the study area is shown in Fig. 2.

2.3 *Roof Area Digitization by Using GIS*

From Google Earth Pro, the digitized different rooftops were imported to QGIS. The Kml/Kmz format files are then converted to the shapefiles by QGIS tool. In QGIS, the roof area for different rooftops, boundaries of roofs, boundaries of road lines are digitized. So QGIS tool is used to find the roof area. The procedure to be followed to digitize the roof area is (Figs. 3, 4, 5 and 6).

1. First open QGIS tool and then, Go to layer > Add layer > Add raster layer.
2. Now raster image is visible in QGIS, then select rooftops of different buildings of the study area and digitize roof area.
3. Save roof area as SHP file.
4. Now convert raster layer to vector layer by importing all shape file.

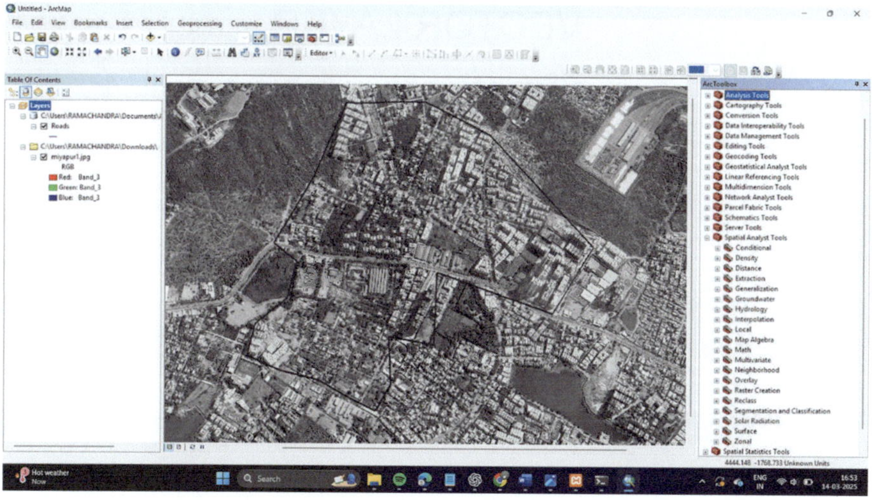

Fig. 3 Add vector layer

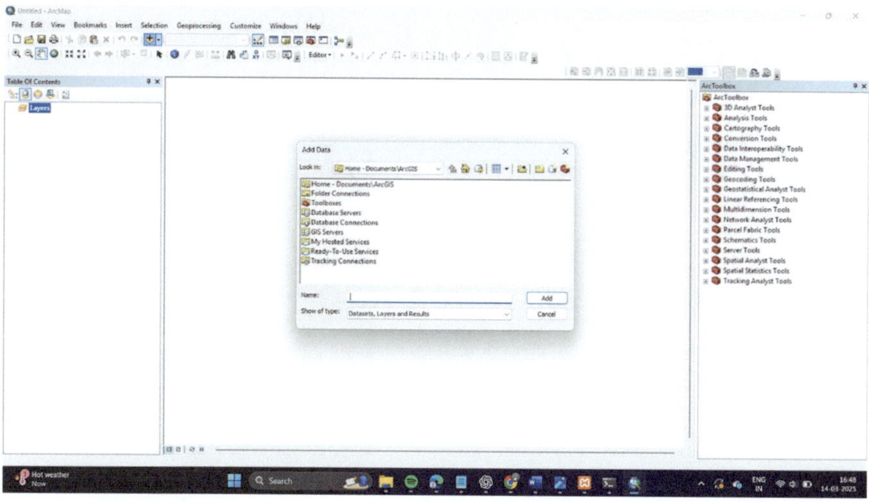

Fig. 4 Attribute pop up for polygon

From this digitization by using QGIS, the total roof area is 302219 m^2, out of which for concrete roofs, the total area is 297177.01 m^2, for galvanized iron roo,f the total roof area is 2941036 m^2, and for asbestos roofs, the total roof area is 2100.63 m^2. By employing the time series model formula, the synthetic or generated values are determined. The future rainfall data for the next 20 years based on statistical parameters of historical data are generated. The synthetic values are generated in order to compare the graphs between historical and generated rainfall data.

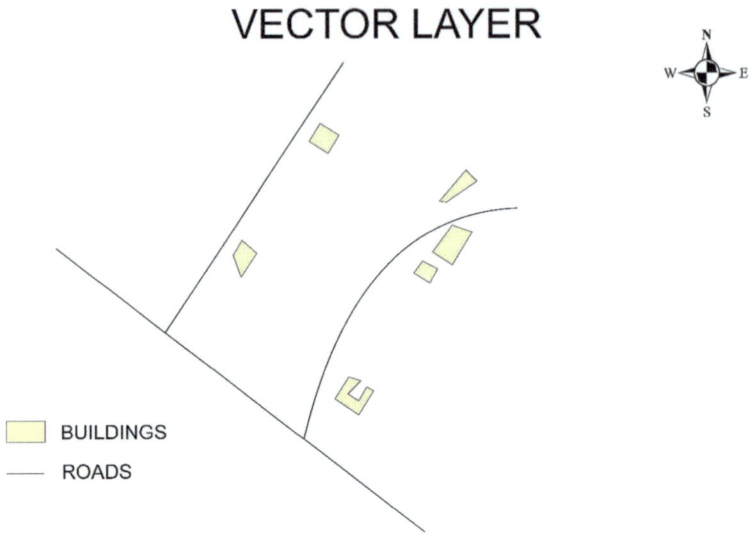

Fig. 5 Vector image

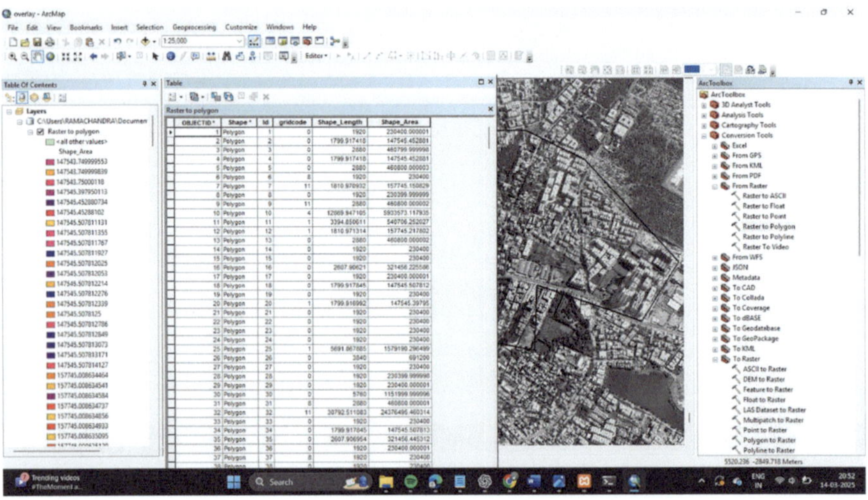

Fig. 6 Areas through field calculator and calculated areas

3 Results and Discussion

By employing QGIS, the total roof area is measured at 302,219 m^2, out of which for concrete roofs the total roof area is 297,177.01 m^2, for galvanized iron roofs the total roof area is 2941.36 m^2 and for asbestos roofs the total roof area is 2100.63 m^2.

From the obtained data, a relationship is established between monthly rainfall data and their corresponding mean values and depicted in Fig. 7. The monthly rainfall data are being taken from the Indian Meteorological Department (IMD), Begumpet, Hyderabad. Further, the comparison is made between the historical and generated rainfall data for mean values, and their measures are listed in Table.1. Interestingly, the above graph revealed that the historical and generated mean values observed are approximately equal. Table 1, it shows the statistical parameters for the standard deviation for historical rainfall data and generated rainfall data. The historical data for the standard deviation are for 20 years (2000–2020), and synthetic or generated rainfall data are forecasted for the next 20 years (2021–2041). Based on these, the comparison line graph is plotted for both the historical and generated rainfall data to know whether the historical and generated standard deviation rainfall that is generated is approximately equal to the historical data.

In addition, the comparison is made between the historical and generated rainfall data for standard deviation values and depicted as Fig. 5 resembling the trends of Figs. 7 and 8 also displays the similar trend portraying that the historical and generated standard deviation values are approximately equal. In both cases of Figs. 3, 6 and 7, historical and generated data are approximately the same, revealing the authenticity of the employed method.

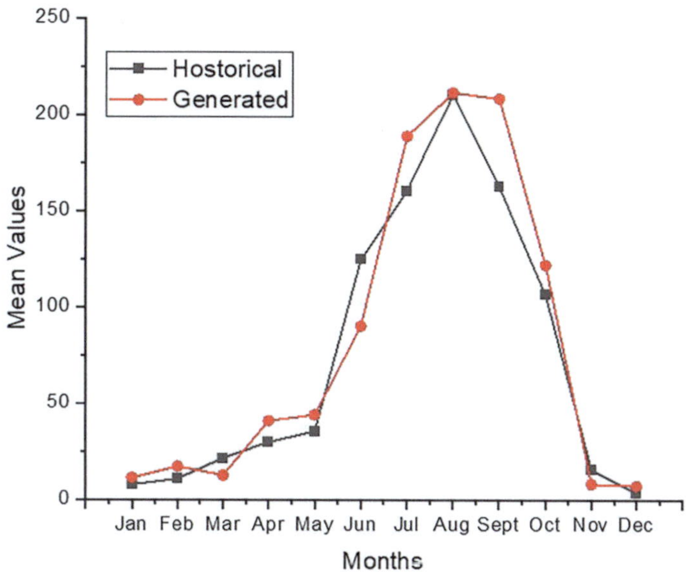

Fig. 7 Line graph comparing the mean historical and mean generated data

Table 1 Historical and generated rainfall data mean and standard deviation values

Months	Mean rainfall data (μ)		Statistical rainfall data (ρ)	
	Historical	Generated	Historical	Generated
January	7.7	11.34	13.51	8.73
February	10.7	17.10	21.91	16.67
March	21.4	12.60	36.81	20.14
April	29.8	40.81	26.37	21.74
May	35.3	43.90	35.05	35.32
June	124.6	90.02	63.12	44.54
July	160.4	188.74	89.61	109.31
August	201.4	211.49	110.31	129.50
September	162.7	208.37	96.42	112.71
October	106.3	121.83	91.57	71.87
November	15.5	8.09	18.68	10.43
December	3.5	7.27	10.48	9.26

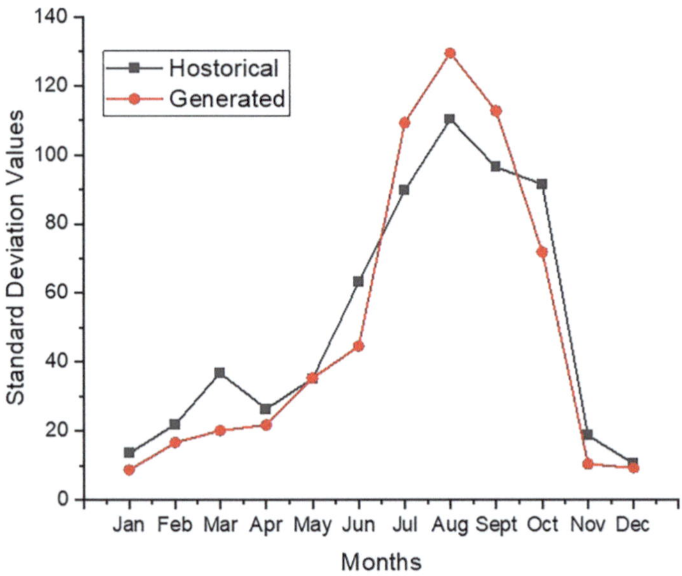

Fig.8 Line graph comparing the standard deviation historical and generated data

3.1 Estimation of Rooftop Rainwater Harvesting Potential

Rooftop rainwater harvesting is the total amount of water that is collected from the various rooftops of the study area. Based on the digitization reports which means the

Table 2 Estimation of rooftop rainwater harvesting potential

Sl. no.	Type of roof	Area (m^2)	Rooftop Coefficients	Average annual rainfall (mm)	$S = R * A * C_r$ (Cumecs)	$S = R * A * C_r$ (L)
1	Concrete	297,177.01	0.70	1009.64	210,029.2575	210,029,257.5
2	Galvanized iron	2941.36	0.80	1009.64	2672.7432	2,672,743.2
3	Asbestos	2100.63	0.90		1009.64	1696.7040
	Total area	3,02,219			2,14,398.70	

Table 3 Estimation of water demand for the study area

No. of households	Total population	No. of days	Per capita demand (LPCD)	Water demand (Cumecs)
700	8600	365	200	627,800

data obtained in the Google Earth Pro and QGIS (based upon the roof area and type of roofs) the total quantity of water that can harvested in the study area is calculated. Rooftop rainwater harvesting potential for the study area is estimated by using Gould and Nissen Formula. Gould and Nissen formula $S = R * A * C_r$ (Table 2).

3.2 *Estimation of Water Demand for the Study Area*

Water demand is calculated based on the existing population of the study area and the per capita demand for various needs. According to IS 1172:1993 Code under ordinary conditions, the minimum per capita demand for any town/city with full flushing system should be taken as 200 LPCD (Table 3).

Water demand = Per capita demand * Total Population of study area * No of days.

4 Conclusions

In this present study, analysis is made to estimate the rooftop rainwater harvesting potential to meet the general demands of water in the study area.

- Geospatial tools Google Earth Pro is used to digitize the rooftops and QGIS which is used to calculate the area of roofs area used. The geospatial applications QGIS and Google Earth Pro are very helpful where extensive field works are involved.

- Comparison of both historical rainfall data and generated rainfall data which is obtained from the Thomas Fiering Model is generated and plotted in the graph. The graph obtained between the historical rainfall data and generated rainfall data is approximately equal.
- To estimate the Rooftop Harvesting Potential Gould, Nissen formula is used. From Table 2, it shows that total amount of water that can be harvested from various rooftops is 214398.70 Cumecs and the total annual water demand are 627,800 Cumecs which is quieter enough to overcome water scarcity in the study area.
- Rainwater harvesting technique is a cost-effective solution for areas such as residential buildings having large roof area. It is an alternative technique for ever-growing water demand issues.
- The current study revealed that the highest amount of water can be collected in the selected area from various roofs and the water demand for the existing population is also high. So, the collected water can meet the demands of the people in the future when the availability of water is less than required and also the water can be directly recharged into the wells by using some infiltration techniques.

References

1. Padmavaty AS, Ganesha Raj K, Yogarajan N, Thangavel P (1993) Check dam site selection using GIS approach. Adv Space Res 13(11):123–127
2. El-Awar FA, Makke MK, Zurayk RA, Mohtar RH (2000) A hydro-spatial hierarchical method for siting water harvesting reservoirs in dry. Appl Engi Agri 16(4):395–404
3. Ravishankar MN, Mohan G (2005) A GIS based hydrogeomorphic approach for identification of site-specific artificial-recharge techniques in the Deccan Volcanic Province. J Earth System Sci 114(5):505–514
4. De Winnaar G, Jewitt GPW, Horan M (2007) A GIS-based approach for identifying potential runoff harvesting sites in the Thukela River basin,South Africa. Phys ChemEarth Parts A/B/C 32(15):1058–1067
5. Singh JP, Singh D, Litoria PK (2009) Selection of suitable sites for water harvesting structures in Soankhad watershed, Punjab using remote sensing and geographical information system (RS&GIS) approach. J Indian Society Remote Sens 37:21–35

Investigation on Partial Replacement of Eggshell Powder with Cement

Ambreshwar, G. S. Brunda, and Peddireddy Sreekanth Reddy

Abstract In the production of concrete, cement holds utmost significance due this production of GHG is increased and it is hazardous to environment. To minimize the burden on the environment in search of substitution material for the cement is consider as eggshell powder. The current experimental study investigates the viability of using eggshell powder (ESP) in concrete manufacturing in place of some of the cement. The motto of this finding is to assess the mechanical and durability properties of the resulting concrete mixtures with varying levels of ESP content. The bulk of waste eggshells are thrown into landfills without the treatment. As a result, suitable alternatives are necessary for handling garbage in an environmentally acceptable manner. The goal of the experiment is to use eggshell powder as a minimum cement addition. Eggshell powder is substituted in percentages of 10, 15, and 20%. Experimental study indicates the strong properties of eggshell-based concrete, such as the compressive strength test. It has been discovered that adding 10% eggshells powder to the concrete decreases 8% of compressive strength.

Keywords Replacement of cement · Compressive strength · Egg cell compressive strength

Ambreshwar (✉)
Department of Civil Engineering, Guru Nanak Dev Engineering College, Bengaluru, Karnataka, India
e-mail: ambreshwarj@gmail.com

Ambreshwar · G. S. Brunda
Department of Civil Engineering, National Institute of Technology Mizoram, Aizawl, Mizoram, India

G. S. Brunda · P. S. Reddy
Department of Civil Engineering, Acharya Institute of Technology, Soladevanahalli, Bengaluru, Karnataka, India

P. S. Reddy
Department of Civil and Infrastructure Engineeirng, Indian Institute of Technology Dharwad, Dharwad, India

Y. K. Guruprasad et al. (eds.), *Recent Advances in Infrastructure Development and Management—Volume 1*, Lecture Notes in Civil Engineering 570,
https://doi.org/10.1007/978-981-96-1554-4_26

1 Introduction

The purpose of this work is to give a study on the partial substitution of eggshell powder for cement. The world's most popular building material, concrete, produces a large quantity of carbon dioxide emissions during manufacture. Eggshell powder is a waste material generated in large quantities by the food industry and has the potential to be used in place of some of the cement. Therefore, reducing the environmental effect of the concrete industry requires the use of other substances in the production process according to Jabbar et al. [7]. Also in the similar context, Vasudevan and Wei [1] has conducted the research by varying 5, 10, 15, and 20% of chicken-dried eggshell powder with cement and stated that minimization of voids is occurred and concrete has become denser in nature and water requirement also reduces. Wei et al. [2] examined the utilization of various percentage of eggshell powder with cement and investigated that the water absorption is lower as compare to the normal concrete and eggshell powder loses its weight when subjected to the acid attack. Hama et al. [9] the researcher considered the different proportion of eggshell powder for cement replacement and studied the internal microstructure by scanning electron microscope (SEM), and identified that the internal microstructure is more denser that the reference one. Khan et al. [4] reported that test has been carried by using machine learning (ML) techniques, incorporated the various percentage of eggshell powder for cement replacement, and examined that for 7.5% percentage of replacement 45.7% of water absorption reduction will takes place. Paruthi et al. [8] reported that usage of the eggshell powder as substitute material for the cement increases the strength parameters. This study aims to examine the impact of using eggshell powder in place in particular cement on the characteristics of concrete. The workability and compressive strength of concrete mixes with different amounts of eggshell powder replacement are evaluated as part of the study. After that, the specimens are cast and cured according to standard method. The specimens are tested for compressive strength and workability. The outcomes are contrasted with the specimens used as controls, which lack of eggshell powder. The impact of several variables, it will effect on the property of concrete mix. This study has the potential to lower the cost of producing concrete, lessen the impact of cement manufacturing on the environment, and offer a long-term solution for eggshell waste management. The building industry may be significantly impacted by the study's findings and help create more environmentally friendly and sustainable building materials.

2 Objectives

1. Evaluate the behaviour analysis of eggshell powder (ESP) when replaced with cement in the concrete.
2. Determine the optimum dosage of ESP in concrete.

3. Study the behaviour of fresh and harden concrete when replacing the eggshell powder with cement.

3 Materials and Methodology

3.1 Steps to be Followed Collection and Preparation of Eggshells

Concrete mixtures can benefit from the use of eggshells as a source of calcium carbonate, which increases the concrete's compressive strength. After procuring eggshells from nearby restaurant chains, they underwent a thorough water wash to get free of any dirt or contaminants. The eggshells were subsequently exposed to direct sunlight for a few days until they started to chip. Eggshells used in concrete should first be carefully cleaned, and the most effective way to dry them is through a hot oven or the sun at a temperature of 45°. The chicken eggshell can be pounded into a finely ground powder with a mortar and crusher once they are totally dry. The dried shells should be added to a powerful food processor or spice grinder. Blend the shells until a fine powder is formed, breaking them down entirely. As a maximum of three months, the ESP can be stored in a sealed container in a cold, dry location. ESP can be mixed into a cement paste or put straight to the concrete mixture when using it in concrete. The required strength and other characteristics of the concrete will determine how much eggshell powder is added to the mixture. Constructing concrete mixes 0, 10, 15, and 20% by weight of cement were substituted with different proportions of eggshell powder for producing concrete mixes. The mix ratios were established with the goal of achieving a 20 MPa peak compressive strength after 28 days (Fig. 1; Table 1).

The dry materials are carefully mixed to prepare the mixture for concrete, and the cement is substituted with varying amounts of ESP. The fresh qualities of the concrete mix are examined and recorded in Table 2, and moulding of Concrete cube was cast in laboratory with standard cube dimension of (150 × 150 × 150 mm) shown in Fig. 2. The specimens were compacted using a vibrating table to eliminate any voids and to ensure proper compaction. The specimens were cured under standard conditions of temperature (27 ± 2 °C) and humidity (95 ± 5%) for 7, 14, 21, and 28 days represented in Fig. 3.

Testing of specimens: Compressive strength and workability were all assessed on the specimens. The compressive capacity was evaluated using an IS: 516:1959 the compression testing equipment. Using the IS: 1199:1959 slump test, the concrete's workability was confirmed. Table 1 contains a list of the specific elements that are utilized in the concrete mix.

Fig. 1 Process of mixing concrete

Table 1 Details of materials and proportions used in present study

Materials used for kg/m^3	W/C ratio	Cement	FA	CA	Eggshell powder
0% ESP	0.5	6.8	10.2	20.4	0
10% ESP	0.5	6.3	10.2	20.4	0.1
15% ESP	0.5	6.05	10.2	20.4	0.15
20% ESP	0.5	5.8	10.2	20.4	2.0

Table 2 Physical properties of materials

Si. no.	Materials	Test conducted	Results	IS codes
1	Cement	Fineness	6%	IS 516–1959
2	Coarse aggregate	Impact value	18.52%	IS: 2386- (Part IV)–1963
3	Coarse aggregate	Flakiness index	7.4%	
4	Coarse aggregate	Elongation index	33.2%	
5	Fine aggregate	Specific gravity	2.663	IS: 2720 (Part 4–1985
6	Coarse aggregate	Specific gravity	2.69	IS: 2386 (Part 3)–1963

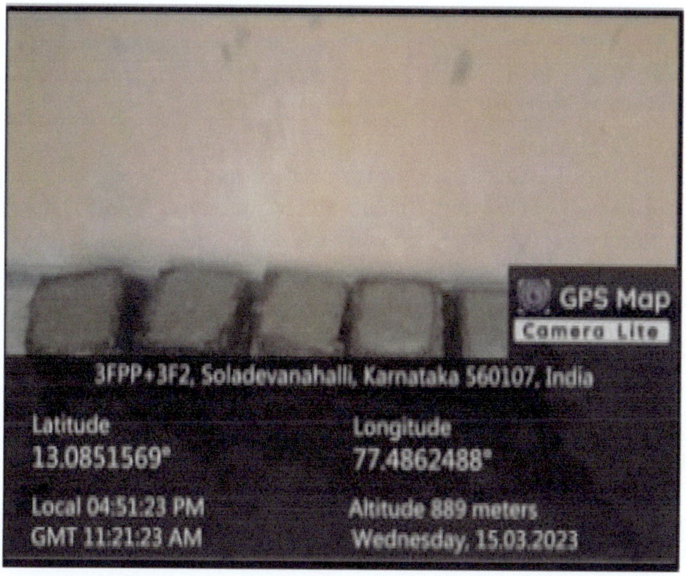

Fig. 2 Casting cubes in laboratory

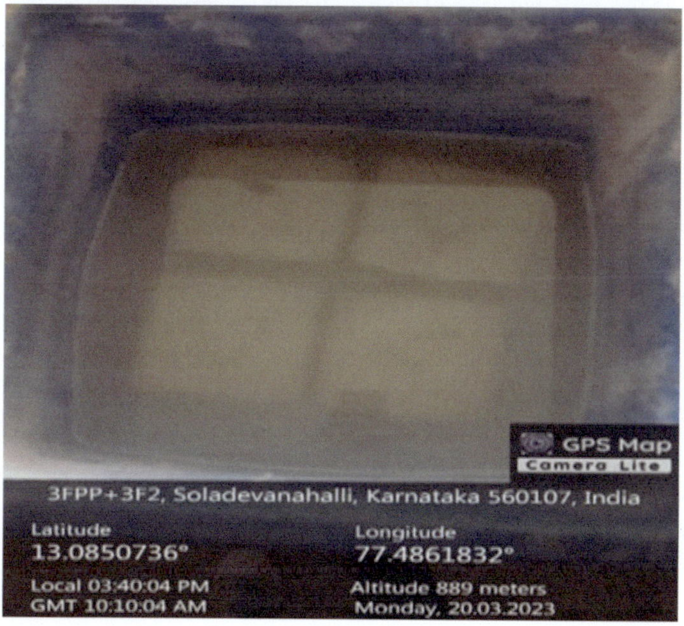

Fig. 3 Curing cubes in laboratory

4 Results and Discussion

4.1 *Physical Properties*

The physical properties are the very important parameters to check the quality of the any materials for concrete mix. In the similar line, same has been determined for different material used in the present study and listed in Table 2.

4.2 *Tests Conducted on Fresh Concrete*

4.2.1 Workability of Concrete Conducted by Using Slump Cone Method

The slump test is used to determine the ease working with concrete, and it is the prime parameter to use the determining behaviour in fresh state. The slump test is conducted, and results are tabulated in Table 3. We can clearly observe that the water–cement ratios are kept as same for all the mixes and the slump value is showing the same for the 0% ESP and 10% of ESP. Later on however the replacement of ESP is increasing the concrete mix. Slump values are decreased due to increasing the powder and calcium content in paste. The segregation and bleeding are reduced by increasing the paste content. For acknowledging this, Vasudevan and Wei [1] reported that the water absorption is decreased up to optimum dosage and workability will increase; later on, it will get decreased. Wei et al. [2] also reported that there is lower the water demand and increases the workability. Ishak et al. [3] reported that there is declination in the slump when increases the ESP. Teara et al. [5] reported the mitigation in the slump when increasing the proportion of ESP. Hamada et al. [6] there is a reduction in the slump value when adding the higher percentage of ESP in the concrete.

Table 3 Slump cone test

Water–cement ratio (%)	Eggshell powder (%)	Slump (cm)
0.5	0	30
0.5	10	30
0.5	15	29
0.5	20	27

Table 4 Concrete's compressive strength for different ages

Type of mix	7th	14th	21st	28th
Egg cell 0%	17.83	18.42	18.98	22.83
Egg cell 10%	18.24	15.55	16.91	20.23
Egg cell 15%	14.67	13.1	14.45	16.47
Egg cell 20%	14	10.85	13.71	15.26

4.3 Mechanical Properties of Concrete

4.3.1 Compressive Strength

Compressive tests are performed on samples for testing at known ages. At each age, at least three specimens, preferably from separate batches, must be prepared for testing. The cubes are put in compressed apparatus for testing such that the load is carried to the cube's opposing sides as cast. The weight is applied unless the specimen being examined breaks. After 7, 14, 21, and 28 days of concrete hardening, objects were compressed. The compressive strength test is performed as per standard guidelines. The results are tabulated in Table 4, and the comparison between for various percentage of replacement is shown in Fig. 5. From the results, we can clearly conclude that along with the age of the concrete the compressive strength also increases and the replacement of 10% eggshell with the cement is recommended because there is smaller reduction in the compressive strength and its achieving the target strength at 28 days of curing. For acknowledging this, Vasudevan and Wei [1] the compressive strength of the 5% replaced ESP with cement has a higher strength when compared to control mix and it is having 26.6 Mpa strength at 28 days of curing, and later on, there is declination in the strength by increasing the ESP content. Wei et al. [2] Eggshell powder is replaced from 0 to 10% with 2.5% of variation. This 2.5% of variation is showing the higher strength than the reference mix at the age of 28 days, i.e. 26.925 Mpa, and by increasing percentage of replacement, the strength is declining. Hamada et al. [6] reported that there is higher the compressive strength at lower replacement; later by increasing the percentage of replacement, the compressive strength is decreased. Ishak et al. [3] noted that by incorporating the 10% of ESP there is a 39% declining the compressive strength at 28 days (Figs. 4, 5).

4.4 Conclusions

The investigation of substituting some of the cement in M20 grade concrete with eggshell powder to ascertain its compressive strength has shown that eggshell powder can be used as a supplementary cementitious material, which can replace cement up to 10% without significantly affecting the compressive strength of concrete.

Fig. 4 Testing of cube under CTM

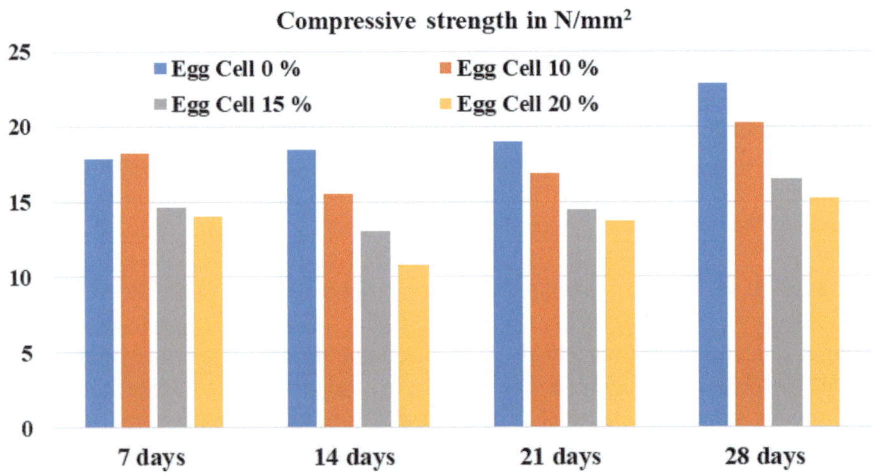

Fig. 5 Concrete's compressive strength for different ages

The compressive strength of concrete with eggshell powder replacement decreased as the strength increased but the decrease was not significant to 10% replacement level. Beyond that, the reduction in compressive strength was significant. However, it is important to note that the study was compressive strength test is conducted for a specific M20 grade of concrete, and the results may not be applicable for other grades of concrete. Additionally, the study was limited to only a few curing periods, and further research is needed to determine the long-term effects of eggshell powder replacement on the durability and performance of concrete. In conclusion, the study has shown that eggshell powder can be used as a partial replacement for cement in M20 grade concrete without significantly affecting its compressive strength. However, research is needed to determine the optimal level of eggshell powder replacement for different grades of concrete and its long-term effects on concrete properties.

References

1. Vasudevan G, Wei SC (2020) Utilisation of eggshell powder (ESP) as partial replacement of cement incorporating superplasticizer. In: IOP conference series: materials science and engineering, vol 840, pp 012016. IOP Publishing https://doi.org/10.1088/1757-899X/840/1/012016
2. Wei CB, Othman R, Ying CY, Jaya RP, Ing DS, Mangi SA (2020) Properties of mortar with fine eggshell powder as partial cement replacement. Mater Today: Proc. https://doi.org/10.1016/j.matpr.2020.07.240
3. Ishak MY, Zamani MN (2019) Eggshell as the partial replacement of Portland cement in the production of concrete. In: IOP conference series: materials science and engineering, vol 849, pp 012034. IOP Publishing. https://doi.org/10.1088/1757-899X/849/1/012034
4. Khan K, Ahmad W, Amin MN, Deifalla AF (2022) Investigating the feasibility of using waste eggshells in cement-based materials for sustainable construction. J Market Res. https://doi.org/10.1016/j.jmrt.2023.02.057
5. Teara A, Doh SI (2020) Mechanical properties of high strength concrete that replace cement partly by using fly ash and eggshell powder (SERIES2020). Phys Chem Earth, Parts A/B/C 120:102942. https://doi.org/10.1016/j.pce.2020.102942
6. Hamada HM, Tayeh BA, AI-attar A, Yahaya FM, Muthusamy K, Humada AM (2020) The present state of the use of eggshell powder in concrete: a review. J Build Eng. https://doi.org/10.1016/j.jobe.2020.101583
7. Abduljabbar S et al (2021) An experimental study on the partial replacement of cement with eggshell powder in concrete. Adv Civ Eng 2021:6656199
8. Paruthi et al (2023) Sustainable cement replacement using a waste eggshell A Review on mechanical properties of eggshell concrete and strength prediction using artificial neural network. Case Studies in construction materials
9. Hama et al (2023) Predicting optimum content of eggshell ash for best mechanical and structural properties of concrete. E3S Web Conf 427:02014

Smart Helmet for Workers Safety in Construction Site Using Internet of Things (IoT)

Hem Chand Sai Muppala and Akshayakumar V. Hanagodimath

Abstract The revolutionary impact of this technology is encapsulated by the use of Smart Helmets in construction, which include a range of sensors including GPS, radio frequency, pulse, accelerometer, body temperature, and an SOS button. By incorporating real-time tracking, communication, health monitoring, and emergency response features, these cutting-edge helmets provide a full safety solution. The integration of various sensors is emphasized in the abstract as a way to improve spatial awareness, enable quick communication, avoid accidents by detecting falls, keep track of worker health, and give an immediate SOS mechanism. The abstract essentially acts as a succinct introduction to the revolutionary potential of Smart Helmets to improve productivity and safety at construction sites.

Keywords GPS · Temperature probe · Pulse · Radio frequency · SOS · Accelerometer · Gyroscope · Smart helmet · Tracking · Buzzer · Room detection

1 Introduction

In the dynamic realm of construction, where innovation meets necessity, the Smart Helmet emerges as a groundbreaking solution, seamlessly integrating the power of the Internet of Things (IoT) to redefine worker safety. This cutting-edge helmet harnesses an array of IoT technologies, including GPS for precise location tracking, radio frequency and room detection for heightened situational awareness, and an SOS button for immediate response to emergencies. But its capabilities extend beyond mere alert systems—with sensors monitoring body temperature, pulse, accelerometer, and gyroscope, the Smart Helmet becomes a guardian of worker health. This convergence of IoT features not only ensures real-time communication and hazard awareness but also proactively manages worker well-being. As the construction

H. C. S. Muppala (✉) · A. V. Hanagodimath
Department of Civil Engineering, M. S. Ramaiah University of Applied Sciences, Bengaluru, Karnataka, India
e-mail: hemchandsai16@gmail.com

Y. K. Guruprasad et al. (eds.), *Recent Advances in Infrastructure Development and Management—Volume 1*, Lecture Notes in Civil Engineering 570,
https://doi.org/10.1007/978-981-96-1554-4_27

industry steps into a new era, the Smart Helmet stands as a beacon of technological synergy, a safeguard against risks, and a testament to our commitment to creating safer, smarter work environments.

2 Literature Review

In response to a 4.5% increase in construction worker fatalities in the Korean industry, a sensing safety helmet, equipped with a three-axis accelerometer sensor, was developed to accurately monitor proper, improper, or no helmet usage, addressing a critical gap in existing studies with a high accuracy of 97.0%. [8]. The study explores the growing interest in Internet of Things (IoT) technology, particularly in the construction industry, aiming to analyze its practical applications through a detailed literature review and investigation of current use cases [1, 2]. An IoT-based autonomous system, integrating radio frequency, directional antennas, and ultrasound waves, to monitor and warn construction workers in danger zones, enhancing safety through a wearable device with a power-efficient design and light-powered wireless nodes for continuous operation [3, 6]. IoT-based Real-Time Locating Systems (RTLS) in construction safety provide comprehensive insights into methodologies and challenges and emphasizing their significant contribution to safety management in areas such as monitoring, accident prevention, and on-site training [9, 13].

3 Objective

The objective of implementing a Smart Helmet for worker safety in construction sites is to establish an unparalleled standard of occupational security through the seamless integration of Internet of Things (IoT) technologies. By incorporating GPS, the Smart Helmet aims to provide precise location tracking, ensuring rapid response and emergency assistance. The utilization of radio frequency and room detection facilitates a heightened level of situational awareness, alerting workers to potential hazards and promoting a proactive safety culture. The inclusion of an SOS button ensures immediate communication in critical situations, reducing response times and enhancing emergency management. Furthermore, the integration of body temperature, pulse, accelerometer, and gyroscope sensors seeks to not only monitor and alert to health-related issues but also to create a comprehensive understanding of workers' physical well-being, thereby fostering an environment where safety is not just a response but an intrinsic part of every construction task. In essence, the objective is to elevate construction site safety to unprecedented levels, mitigating risks and prioritizing the health and security of the workforce through innovative IoT-driven technology.

4 Problem Statement

In the construction industry, despite numerous safety measures, the persistent challenge of ensuring the well-being of workers' demands a paradigm shift. The absence of a comprehensive safety solution has led to a concerning frequency of accidents and delayed emergency responses. Traditional safety gear falls short in providing real-time data crucial for accident prevention and health monitoring. This lacuna is particularly pronounced in environments where hazards are dynamic and unpredictable. The need for a holistic safety solution is underscored by the lack of effective communication tools and immediate response mechanisms during emergencies. The amalgamation of Internet of Things (IoT) technologies, including GPS, radio frequency, room detection, SOS button, body temperature, pulse, accelerometer, and gyroscope in a Smart Helmet, aims to address these critical gaps. The problem at hand is the imperative to revolutionize construction site safety by mitigating the inadequacies of current safety protocols, fostering a proactive approach to risk management, and prioritizing the swift response to emergencies for the welfare of the workforce.

5 Methodology

The methodology for implementing Smart Helmets in the context of worker safety on construction sites encompasses a multi-faceted approach. Firstly, an in-depth analysis of the construction site environment will be conducted to identify specific safety challenges and requirements. Following this, the integration of IoT technologies, including GPS, radio frequency, room detection, SOS button, body temperature, pulse, accelerometer, and gyroscope, will be meticulously designed and engineered into the helmet. The implementation process involves rigorous testing in controlled environments to ensure the seamless functionality and interoperability of each IoT component. Real-world simulations and scenario testing will be conducted to evaluate the helmet's responsiveness in dynamic and hazardous conditions. A comprehensive training program will be developed to familiarize workers with the functionalities of the Smart Helmet and encourage proper usage. To measure the effectiveness of the Smart Helmet in enhancing safety, quantitative data such as incident rates, response times, and health metrics will be collected and analyzed. Additionally, qualitative feedback from workers, supervisors, and safety officers will be gathered through surveys and interviews to gauge user satisfaction and identify areas for potential improvement. The methodology thus adopts a holistic and iterative approach, combining technological implementation, testing, training, and ongoing evaluation to ensure the successful integration of Smart Helmets for optimal worker safety in construction sites.

6 Proposed Solution

6.1 Prototype

The Smart Helmet is designed to enhance worker safety on construction sites through the integration of Internet of Things (IoT) technologies, representing a groundbreaking advancement. The helmet incorporates a compact GPS module for precise location tracking (see Fig. 1), ensuring immediate response during emergencies. Utilizing radio frequency and room detection sensors, the prototype provides real-time hazard alerts, enhancing situational awareness. The inclusion of an SOS button enables instant communication for urgent situations, facilitating swift assistance. Moreover, the helmet integrates body temperature and pulse monitoring features, along with an accelerometer and gyroscope, creating a comprehensive health monitoring system. The prototype's design emphasizes user comfort, durability, and seamless integration, ensuring that workers can perform their tasks efficiently while benefiting from advanced safety functionalities. This prototype, a convergence of cutting-edge technologies, stands as a tangible example of how the Internet of Things can revolutionize worker safety in the challenging and dynamic environment of construction sites (see Fig. 2).

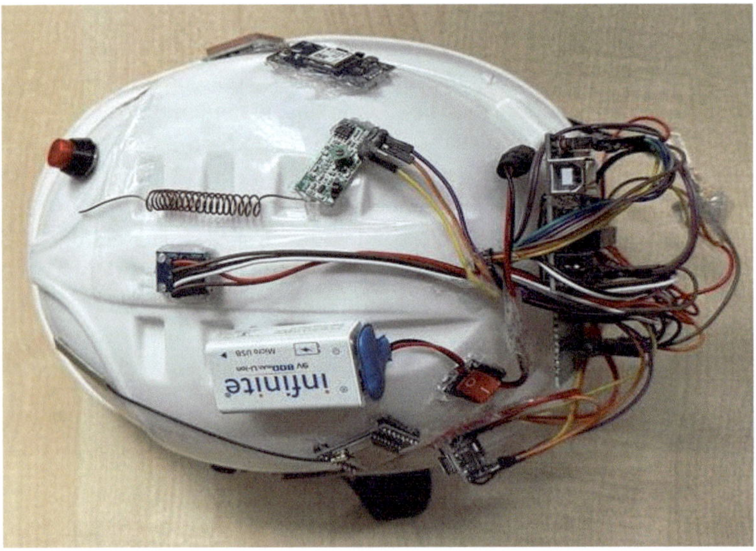

Fig. 1 Smart helmet

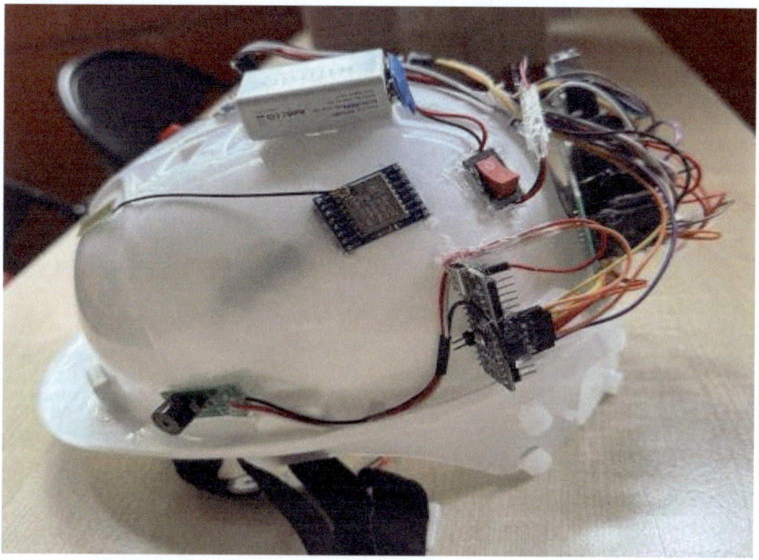

Fig. 2 Developed smart helmet

6.2 *Prototype*

The circuit diagram for a Smart Helmet geared toward worker safety in construction sites involves an intricate integration of various components. At the core of the design is a microcontroller unit (MCU), serving as the central processing hub (see Fig. 3). The GPS module interfaces with the MCU to provide precise location data, enabling real-time tracking. Radio frequency and room detection sensors are linked to the MCU (see Fig. 4), ensuring continuous monitoring of the environment for potential hazards. The SOS button, a critical component, establishes a direct link to emergency services through wireless communication protocols. Health monitoring components, including body temperature and pulse sensors, connect to the MCU to enable constant health assessment. The accelerometer and gyroscope contribute to fall detection and movement analysis, interfacing with the MCU to trigger alerts when necessary. The circuit design should prioritize power efficiency and compactness, ensuring the Smart Helmet remains practical for daily use in the construction setting. This comprehensive integration of IoT components into the circuitry forms the backbone of a Smart Helmet, creating a robust and interconnected safety system for construction workers.

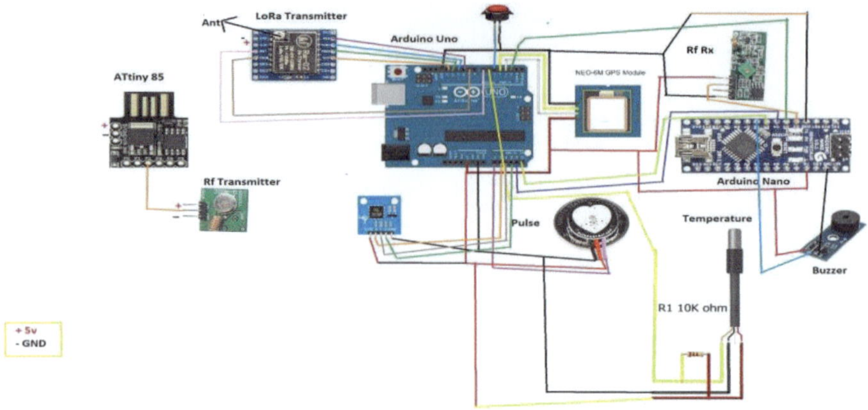

Fig. 3 Circuit diagram

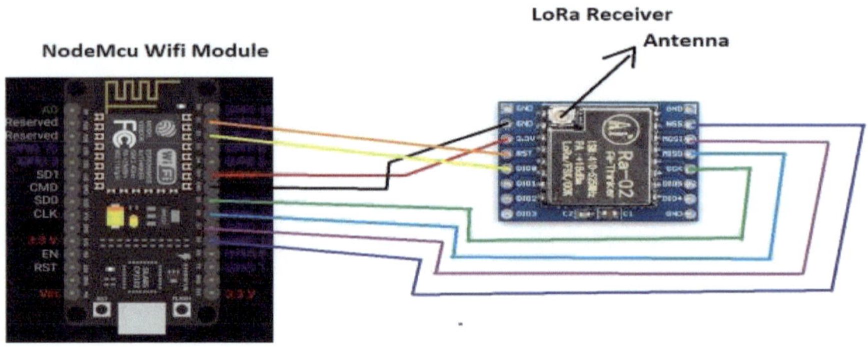

Fig. 4 LoRa Wi-fi module circuit diagram

7 Results and Discussions

Validating the results with existing medical equipment available in the market. There is a little discrepancy in the values since further research in the area is necessary, and new advances in the future will improve accuracy and provide safer working conditions (see Table 1). The potential of Smart Helmets to change safety paradigms has come to light in discussions around the industry of construction. These helmets are equipped with a variety of sensors, including GPS, radio frequency, pulse, accelerometer, body temperature, and an SOS button. Real-time tracking made possible by the integration of these sensors allows for accurate location monitoring and efficient site management. Radio frequency technology is also used to improve communication between managers and employees, promoting quick information interchange. The conversation also emphasized the need of accelerometer sensors in quickly detecting falls and sudden movements, resulting in timely alarms and replies. The conversations

Table 1 Comparison of smart helmet with medical equipment

Person	Medical equipment		Smart helmet	
	Pulse	Body temperature	Pulse	Body temperature
Person 1	87	98.2	82	95
Person 2	93	98.3	89	94
Person 3	97	97.9	91	93
Person 4	93	98.2	89	96
Person 5	85	97.3	83	92

highlighted how Smart Helmets combine sensor technologies to build a comprehensive safety ecosystem, significantly improving both individual well-being and general site safety.

8 Conclusion

The integration of Smart Helmets in construction, utilizing a variety of sensors including GPS, radio frequency, pulse, accelerometer, body temperature, and SOS button, has led to the conclusion that they have the potential to be revolutionary and to provide higher levels of safety. These helmets offer a holistic solution to solving numerous safety concerns, and they represent a substantial advancement in construction site safety. Real-time spatial awareness and communication are made possible by the GPS sensor and radio frequency technology, improving worker safety and site administration. Together with health monitoring sensors, the accelerometers capacity to detect falls and unexpected movements helps to ensure the well-being and general health of workers. The SOS button serves as a vital lifeline in crisis and guarantees quick action. In summary, the combination of these sensors in Smart Helmets not only saves lives but also changes the way construction sites are safe, paving the way further improvements in worker safety.

References

1. Arslan V, Ulubeyli S, Kazaz A (2019) The use of internet of things in the construction industry
2. Campero-Jurado I et al (2020) Smart helmet 5.0 for industrial internet of things using artificial intelligence. MDPI 20(21):6241
3. Chellappa V (2022) Fatal fall incidents in the Indian construction industry: a case study analysis. Inst Civ Eng 175(3):87–92
4. Iniyan EU, PA et al (2020) An embedded smart affordable helmet for construction workers to reduce fatality rate using IoT device. Elementary Educ Online 19(2):1768–1777
5. Jeevana V, Kulkarni G (2018) Internet of things (IOT) to prevent delays of construction industry. Int J Pure Appl Math 118:1037–1041

6. Kanan R, Elhassan O, Bensalem R (2018) An IoT-based autonomous system for workers' safety in construction sites with real-time alarming, monitoring, and positioning strategies, pp 73–86. Elsevier
7. Katiyar IA, Kumar P (2021) A review of internet of things (IoT) in construction industry: building a better future. Int J Adv Comput Sci Eng 3(2):65–72
8. Kim SH, Wang C, Min SD, Lee SH (2018) Safety helmet wearing management system for construction workers using three-axis accelerometer sensor. MDPI 8(12):2400
9. Kishore AN, Suman AV (2021) Accidents at construction sites—a legal perspective. Int Res J Mod Eng Technol Sci 3(6)
10. Nandhini M, Priya GP, Nadhini S, Dinesh K (2018) IoT based smart helmet for ensuring safety in industries. Int J Eng Res Technol (IJERT) 6(4):1–4
11. Patil N et al (2022) IOT based smart helmet for construction workers. Int J Adv Res Sci, Commun Technol (IJARSCT) 2(8):466–485
12. Patni S (2022) Statistically analyzing fatal accidents pattern of construction activities in India. Asian Rev Civ Eng 11(2):10–14
13. Sabu J, Kumar DDR (2022) A review of IoT based construction site and labour safety monitoring systems. Int J Eng Res Technol (IJERT) 10(6)
14. Shewale N et al (2022) Smart helmet for industrial workers safety. Int Res J Eng Technol (IRJET) 9(4)
15. Zhong RY et al (2017) Prefabricated construction enabled by the internet of things. Autom Constr 76:59–70

Industrial Wastes and Their Utilization in the Manufacturing of Interlocking Blocks

Peddireddy Sreekanth Reddy and K. Anand Goud

Abstract Conventional fire bricks contaminate the environment as well as induce land subsidies due to illegal mining activity, especially in hilly terrains with rapid growth in construction. Though researchers employed industrial waste by-products like fly ash in the production of blocks as a replacement, it is confined to a certain extent by inducing ordinary Portland cement (OPC). The current chapter investigates the influence of gypsum on compressive strength characteristics of solid interlocking blocks prepared from waste by-products FA and quarry dust. The test specimens were subjected to compressive strength with a curing period of 28 days. From the test results, the maximum compressive strength of 8.5 MPa is obtained at a curing period of 28 days with optimum content of 6% cement and 4% gypsum along with 50% FA, 30% quarry dust, and 10% of 6 mm aggregates (i.e., CG-2), which is comparatively greater than the conventional bricks. The obtained compressive strength is complying the requirement stated by IS1077 (i.e., 3.5 MPa) and IS-3495 (Part1) (i.e., 7.5 MPa). These blocks can be used as an alternative for red brick masonry in construction. The cost comparison with conventional red bricks revealed that the interlocking blocks are cheaper than the convention red bricks and reduced the construction cost by 33%. It does not require any cement mortar for bonding and plastering, which reduces cement consumption by 20–30%.

Keywords Red bricks · Environmental issue · Cement · Industrial waste · Interlocking blocks · Masonry blocks

---

P. S. Reddy (✉)
Department of Civil Engineering, Acharya Institute of Technology, Soladevanahalli, Bangalore, Karnataka, India
e-mail: sreekanth2648@acharya.ac.in

K. Anand Goud
Department of Civil Engineering, Institute of Aeronautical Engineering, Dundigal, Hyderabad, Telangana, India

P. S. Reddy
Department of Civil and Infrastructure Engineering, Indian Institute of Technology Dharwad, Chikkamalligwad, Dharwad, Karnataka, India

Y. K. Guruprasad et al. (eds.), *Recent Advances in Infrastructure Development and Management—Volume 1*, Lecture Notes in Civil Engineering 570,
https://doi.org/10.1007/978-981-96-1554-4_28

1 Introduction

Construction sector plays a predominant role in the country's infrastructural development. Red bricks made from clay/soil have been the most commonly used building material for many decades and to date. Globally, Asian countries contribute to two-thirds of clay brick production which is approximately 87% [1] and rounds to 1500 billion per annum [2]. It leads to rapid mining activity of red clay/soil, which influences the environment and causes land subsidence, especially in hilly terrains. In order to mitigate the environmental effects due to conventional bricks and largely generating industrial waste, researchers paid attention to the development of alternative material, i.e., cement blocks/bricks. In this context, Islam et al. [3] evaluated the compressive strength of cement and sand-based block and reported it as 5.28 MPa with the optimum contents of 5% cement, 13.33% fine, and 26.67% coarse sand. Similarly, several researchers adopted different proportions of cement and reported different degrees of success. On the flip side, the utilization of ordinary Portland cement (OPC) in the manufacturing of cement-based blocks triggered its consumption. In addition, the production of OPC is a very energy-intensive process resulting in the liberation of greenhouse gases in an enormous way and consumes large quantities of raw materials. These gases decline the quality of air and lead to serious health hazards [4].

Increasing awareness of environmental aspects, and the development of alternative eco-friendly materials in construction, especially in masonry, enhanced the utilization of industrial end products. Fly ash is one of the by-products generated in large quantities after the combustion of coal in thermal power plants. Globally, its generation rounds to approximately 2800 MMT per annum [5]. Among these, its utilization is confined to 53% only and the rest is left to dumping or stockpiling in lagoons in turn contaminates the environment. To reduce the impacts of cement and FA, many researchers replaced cement with FA and produced FA-based cement blocks for masonry purposes. In this context, Dayananda and Gowda [6] employed 90% of granulated blast furnace slag (GBFS) and 10% of cement and reported a compressive strength of 10.87 MPa. Similarly, Islam et al. [7] employed 10% cement and 25% FA and reported a compressive strength of 8.10 MPa. Very recently, Soharu et al. [8] reported a compressive strength of 8.53 MPa, with optimum contents of 60% FA, 30% M-sand, and 10% of Portland slag cement. To further decline the utilization of cement, research is in progress to employ other waste materials along with FA to replace the cement content. Dayananda and Gowda [6] reported a compressive strength of 5.94 MPa for a curing period of 28 days with optimum contents of 30% FA, 60% GBFS, and 10% Portland slag cement. Very limited studies are in light of delineating the influence of other waste materials along with FA and cement, which reduces the environmental footprints. To address the same, the current study evaluates the compressive strength behavior of interlocking blocks by varying the cement and gypsum contents along with fixed contents of FA, quarry dust, and coarse aggregate of 6 mm. It can be employed as an alternative material to red clay bricks, which declines cement consumption by 20–30%.

Table 1 Physical properties of different materials used in the current study

Properties	Cement	Fly ash	Aggregates		Gypsum
			Coarse	Fine	
Fineness (mm)	8				
Specific gravity	3.15	3.1	2.74	2.68	2.2
Density (kg/m^3)		2700	1600	1790	

2 Material and Methods

Ordinary Portland cement (OPC) of 53-grade gypsum was used in the study, which was collected from the local suppliers. Quarry dust and coarse aggregate of 6 mm were collected from Guru Raghavendra crushers, Nellore, which was used for the study purpose. Fly ash of class F from Reliance thermal power station, Nellore, was used. The physical properties of the materials were determined in accordance with the respective standards and enlisted in Table 1. The compressive strength of the prepared specimens was determined by following the guidelines of ASTM C39 [9].

2.1 Sample Preparation

The sample preparation involves the variation of cement (i.e., 0, 2, 4, 6, 8, and 10%) and gypsum (i.e., 0, 2, 4, 6, 8, and 10%) contents with fixed FA of 50%, 30% quarry dust, and 10% of 6 mm aggregates [10, 11] which are locally called as jelly chips. These fixed contents have been finalized by relying on the literature studies. These materials were mixed thoroughly by adding sufficient water content in a pan mixer until they reached homogeneity. Later, the mix is transferred to the hydraulic-operated mold, in which compression is applied. The dimensions of the blocks are maintained as per the guidelines of IS 12894 [12]. The typical photograph of the sample prepared and used in the study can be found in Fig.1. The prepared samples were cured for a duration of 3, 7, and 28 days before testing. Three number of samples were prepared for each mix and curing period in order to obtain accuracy. From the figure, the interlocking grooves can be observed in all directions.

3 Results and Discussion

The physical properties of interlocking bricks prepared in the current study are listed in Table 2. In addition, the physical properties of burnt bricks are also included in the same table for better comparison. From the table, it can be observed that the physical properties of the interlocking block prepared in the current study are in line with the

Fig. 1 Typical photograph of a specimen prepared in the current study

requirements of masonry material properties specified by IS 1077 [13]. The designations and their corresponding mixes are listed in Table 3. The compressive strength test is the basic test to judge the suitability of any building blocks in the construction activity. Several adopted the same for evaluating the strength characteristics of masonry bricks and concrete mixes. The same has been followed in the current study to evaluate the strength behavior of interlocking blocks. Fig.2 depicts the influence of cement and gypsum contents on the compressive strength of interlocking blocks with curing periods of 3, 7, and 28 days. The measured compressive strength for different mix proportions is listed in Table 3. From the figure, it can be noticed that the compressive strength decreases with the increment in gypsum contents.

On the flip side, the compressive strength increased with the curing period. The maximum compressive strength of 9.5 MPa is obtained with the optimum contents of 10% of cement for a curing period of 28 days. However, as per IS 1077 [13], the burnt clay brick should possess a minimum compressive strength of 3.5 MPa. Similarly, IS-3495 (Part 1) [14] also stated that the compressive strength of brick should not be

Table 2 Physical properties of burnt brick and interlocking block

S. no.	Properties	Burnt brick [13]	Interlocking block (Current study)
1	Dimension (cm)	19 * 9 * 9	30 * 23 * 13
2	Density (kg/m^3)	1900	2500
3	Compressive strength (N/mm^2)	3.5–35	8.5
4	Water absorption (%)	Not more than 20	18
5	Efflorescence	Moderate	Moderate

Table 3 Mix designation and strength characteristics of samples tested in the current study

S. no.	Mix ID	Cement, C (%)	Gypsum, G (%)	Compressive strength (MPa)		
				3 days	7 days	28 days
1	CG-0	10	0	6.7	7.4	9.5
2	CG-1	8	2	5.2	6.8	8.8
3	CG-2	6	4	5	6.2	8.5
4	CG-3	4	6	3	4.8	6.2
5	CG-4	2	8	2.8	3.6	5.7
6	CG-5	0	10	1.5	2.4	3.1

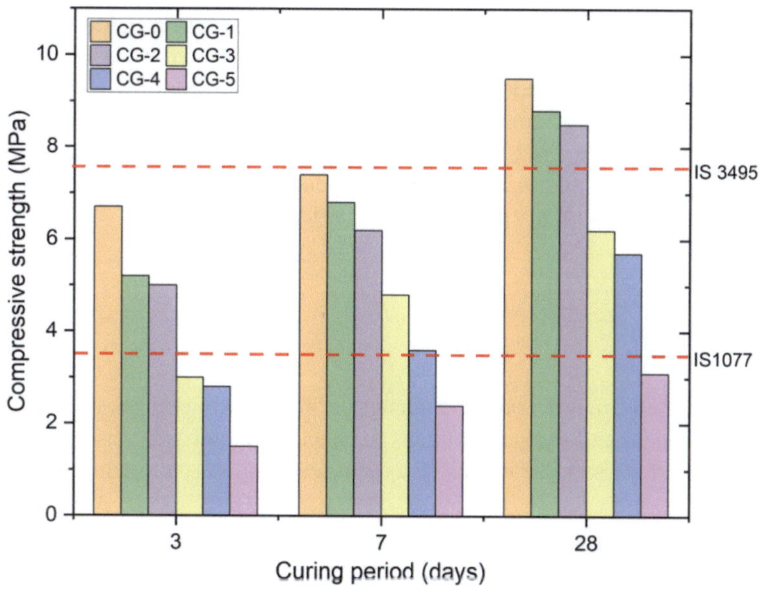

Fig. 2 Influence of cement and gypsum on compressive strength of specimen used in the current study

Table 4 Detailed cost comparison between the conventional red bricks and interlocking blocks prepared in the current study

S. no.	Wall dimensions	Materials	
		Bricks	Interlocking Blocks
1	10 m * 10 m	No. 8450 * Rs. 9 = Rs. 76050	No. 1450 * Rs. 35 = Rs.50750

< 7.5 MPa. At this juncture, it can be noticed from the figure that the compressive strength obtained at optimum contents of CG-2, i.e., 6% cement and 4% gypsum complies with the requirements of both codes. The interlocking blocks prepared from these contents can be employed for masonry activity in construction, which declines the consumption of cement. As these blocks comprise interlocking grooves, which do not require any cement mortar for their bonding. Further, it possesses a very smooth surface replicating the plastering texture. The utility of these blocks in the construction activity declines the overall consumption of cement by 20–30%.

Cost Perspective

In addition to other technical parameters, the cost associated with the materials that are going to be employed has a significant effect on the construction activity. In this context, a cost comparison is made in assessing the feasibility of employing the conventional red bricks and blocks prepared in the current study. For better understanding on the variation of cost with respect to materials, a 10 m * 10 m wall dimensions has been selected and listed in Table 4. From the table, it can be observed that the cost of conventional red bricks are becoming costlier than the interlocking blocks prepared from industrial waste by-products. Considering the economic and strength parameters, it is recommended to employ interlocking blocks as a replacement material to conventional red bricks.

4 Conclusion

In the current study, the compressive strength of interlocking blocks is evaluated by varying the cement and gypsum contents to check their suitability as an alternative in masonry works. From the extensive experimentation, the following conclusions are derived.

- The maximum compressive strength of 9.5 MPa is obtained with the optimum contents of 10% of cement alone for a curing period of 28 days, which is comparatively higher side than the conventional red bricks.
- The maximum compressive strength of 8.5 MPa is obtained at an optimum mix of CG-2, i.e., 6% cement and 4% gypsum along with fixed contents of 50% FA, 30% quarry dust, and 10% of 6 mm aggregates 50%.

- Though cement alone offers a higher strength than the CG-2, it leads to huge consumption of cement which increases the levels of greenhouse gases in the atmosphere.
- The compressive strength obtained in the case of CG-2 is in good agreement with the standards of IS 1077 (i.e., 3.5 MPa) and IS-3495 (Part1) (i.e., 7.5 MPa), which can be employed as building blocks in the construction activity. Further, it declines the overall consumption of cement by 20–30%.
- The interlocking block prepared in the current study can be a good alternative to conventional bricks, which can be used in the construction of partition walls in any kind of structure and compound walls for large areas.
- The cost comparison revealed that the interlocking blocks are cheaper than the convention red bricks and reduced the construction cost by 33%.

References

1. Weyant C, Athalye V, Ragavan S, Rajarathnam U, Lalchandani D, Maithel S, Baum E, Bond TC (2014) Emissions from South Asian brick production. Environ Sci Technol 48(11):6477–6483
2. Fudge C, Fouad F, Klingner R (2019) Autoclaved aerated concrete. In: Developments in the formulation and reinforcement of concrete, pp 345–363. Woodhead Publishing
3. Islam MS, Tausif-E-Elahi Shahriar AR, Nahar K, Hossain TR (2020) Strength and durability characteristics of cement-sand stabilized earth blocks. J Mater Civ Eng 32(5):04020087 (2020)
4. Campbell-Lendrum D, Prüss-Ustün A (2019) Climate change, air pollution and noncommunicable diseases. Bull World Health Organ 97(2):160
5. Assi A, Bilo F, Zanoletti A, Ponti J, Valsesia A, La Spina R, Zacco A, Bontempi E (2020) Zero-waste approach in municipal solid waste incineration: Reuse of bottom ash to stabilize fly ash. J Clean Prod 245:118779
6. Dayananda N, Keerthi Gowda BS (2017) Prediction of properties of fly ash and cement mixed GBFS compressed bricks. Mater Today: Proc 4(8):7573–7578
7. Islam MS, Elahi TE, Shahriar AR, Mumtaz N (2020) Effectiveness of fly ash and cement for compressed stabilized earth block construction. Constr Build Mater 255:119392
8. Soharu A, Naveen BP, Sil A (2022) Fly ash bricks development using concrete waste debris and self-healing bacteria. J Mater Cycles Waste Manage 24(3):1037–1046
9. ASTM C39 (2015) Standard test method for compressive strength of cylindrical concrete specimens, Annual Book of ASTM Standards. ASTM International, West Conshohocken, PA
10. Rai B, Kumar S, Satish K (2014) Effect of fly ash on mortar mixes with quarry dust as fine aggregate. Adv Mater Sci Eng 2014:1–7
11. Makaka G (2014) Influence of fly ash on brick properties and the impact of fly ash brick walls on the indoor thermal comfort for passive solar energy efficient house. Comput Water, Energy, Environ Eng
12. Indian Standard: IS 12894 (2002) Pulverized fuel ash-lime bricks–specifications. Bureau of Indian Standards, New Delhi
13. Indian Standard: IS 1077 (1992) Common burnt clay building brick specifications. Bureau of Indian Standards, New Delhi
14. Indian Standard: IS 3495 (Part 1) (1976). Methods of tests of burnt clay building bricks. Bureau of Indian Standards, New Delhi

Shear Behaviour of Concrete Beams Without Transverse Reinforcement

Thoukeerahmed A. Aralimatti and Manjunath Ishwaragol

Abstract Shear resistance is vital for reinforced concrete beams' safety. This study assesses shear strength without stirrups and examines concrete properties—compressive and split tensile strength. Concrete properties follow M30 ratio, tested after 28 days. Shear strength of beams, using same mix, is tested with varying depths. Shear values, without shear reinforcement, consistently 9.07–9.14% lower. Shear strength increases by 6.82% with deeper beams. Results are compared with formulas and codes; European code shows 0.34% difference due to similar provisions. Formulas differ significantly: JSCE (1986) 32.96% higher, attributed to unique limitations.

Keywords Compressive strength · Split tensile strength · Shear strength · Longitudinal reinforcement · Transverse reinforcement

1 Introduction

Shear resistance is vital in structural design for reinforced concrete beams. Often, stirrups provide transverse reinforcement, but in some cases, this is impractical or costly. This study explores shear behaviour without such reinforcement. Shear failure in reinforced concrete beams is a concern, so understanding behaviour without transverse reinforcement is important. This study enhances knowledge about shear behaviour in these scenarios, aiding safer design and assessments. Investigating reduced reinforcement assesses its impact on shear capacity and failure mechanisms. Even after 70 years of research into comprehending the shear behaviour of reinforced concrete elements, this topic is resurging in importance for structural engineers. This renewed focus comes as a result of the recent updates in shear provisions within the Australian concrete design standard, AS 3600, and the introduction of a new one-way shear design equation by ACI 318 [1].

T. A. Aralimatti (✉) · M. Ishwaragol
Depertment of Civil Engineering, BLDEA's V. P. Dr. P. G. Halakatti College of Engineering and Technology, Vijayapura, Affiliated to VTU, Belagavi, Karnataka, India
e-mail: thoukeerahmed25@gmail.com

Y. K. Guruprasad et al. (eds.), *Recent Advances in Infrastructure Development and Management—Volume 1*, Lecture Notes in Civil Engineering 570,
https://doi.org/10.1007/978-981-96-1554-4_29

The study's significance lies in advancing understanding of shear behaviour without transverse reinforcement, informing better design practices. The project evaluates reduced reinforcement's effects on shear capacity and failure modes. Identifying shear transfer mechanisms in concrete beams is challenging due to the intricate stress redistributions that occur post-cracking, making the process complex and unclear [2]. The findings provide valuable insights for engineers dealing with limited or impractical transverse reinforcement. Based on the load–deflection curves, it can be inferred that the test parameters positively influence the ductility of RC deep beams [3].

The study also assesses the effectiveness of IS code provisions in predicting shear behaviour. Understanding compressive strength is crucial for concrete structure safety. It ensures durability and performance. Design codes stipulate minimum compressive strength to ensure structural integrity. It is important to note that design code provisions, including factors like load factors, strength reduction factors, and material strength reduction factors, are not considered in this evaluation. This omission removes the influence of conservatism from these factors, allowing for a more comprehensive assessment of code approaches based on their design philosophy [2].

Understanding concrete behaviour under different loads is vital for structural integrity. Shear forces are critical, especially in elements without transverse reinforcement. This project delves into shear behaviour, providing insights into mechanisms, influencing factors, and empirical formula effectiveness.

Tensile strength is important for concrete's structural behaviour. Split tensile strength tests show how concrete responds to tension, crucial for structures subjected to bending or tension. This project aims to enhance safety, efficiency, and cost-effectiveness in reinforced concrete beam design by addressing shear behaviour without transverse reinforcement. It provides practical guidance for projects where transverse reinforcement is not viable with an increase in the shear span ratio, the shear capacity, and cracking load of the beam decrease, while its deflection tends to increase [4].

While omitting shear reinforcement has advantages, it's important to consider trade-offs and limitations. Simplicity, cost-effectiveness, and design flexibility are advantages. However, there's reduced shear capacity, increased risk of failure, and potential non-compliance with codes. Calculating shear strength involves evaluating experimental outcomes against various codes and empirical equations [2]. This comparison validates the reliability of formulas in predicting shear capacity. Discrepancies between formulas and data highlight potential limitations, driving code enhancement for precise structural guidelines.

2 Materials

The materials and methodology section offers a comprehensive overview of the study's foundation, encompassing the materials and methodologies employed. This section is a crucial reference for comprehending the components and techniques of the experimental programme. The study delves into the utilization of minerals, casting processes, and testing procedures, with a focus on cement, aggregates, water, and steel reinforcement.

Concrete mix design is vital in determining ingredient proportions tailored to project-specific needs. The preliminary material test data phase involves essential assessments of cement, fine aggregates, coarse aggregates, and water. The cement testing aligns with IS 1489-1991 specifications to ensure its quality. Fine aggregate evaluations cover specific gravity, setting times, fineness, and more. Likewise, coarse aggregate examinations encompass specific gravity, fineness modulus, and water absorption. Water's suitability for construction is verified through pH tests.

The outcomes of these tests confirm material quality and enhance the understanding of their behaviour within the concrete mix. Crucial material property information is summarized in tables, further boosting the study's reliability and comprehension.

The provided values are essential for the concrete mix design of M30 grade concrete according to the updated IS 10262:2019 standard, considering a volume of 1 m^3. This designed mix is employed to create three cubes, three cylinders, and nine prismatic beams for evaluating compression, split tensile, and shear strength, respectively. The concrete mix ratio in this scenario is 1 part cement, 1.4 parts fine aggregates, and 2.5 parts coarse aggregates. This ratio dictates the proportions of these components to achieve the desired strength and performance characteristics in the resulting concrete structures.

3 Reinforcement

The bottom reinforcement of a beam is crucial for upholding its structural integrity and load- carrying capability. This reinforcement is essential to counteract the tensile stresses induced by bending moments at the base of the beam, thereby preventing tensile failure. It effectively controls and restricts the formation and propagation of cracks, contributing to the beam's durability and serviceability.

In the context of reducing reinforcement, a deliberate effort was made to adjust the reinforcement in a manner that shifted the failure mode of the beam from flexural to shear. This involved determining the necessary flexural reinforcement based on the beam's design parameters and load conditions. Various factors, including span length, material characteristics, and safety considerations, were considered to establish the minimum required reinforcement for flexural strength. Subsequently, the flexural reinforcement was reduced by around 10–15% with the aim of emphasizing shear failure. It's important to note that this reduction was an initial estimation and could have required further fine-tuning based on additional analysis.

The moment of resistance (M_u) for a reinforced concrete (RC) beam is given by the equation Eq. 1

$$M_u = 0.138 f_{\text{ck}}\, b(d)^2 \tag{1}$$

The equation, Eq. 1 represents the relationship between the moment of resistance (M_u) of a reinforced concrete (RC) beam and various factors that influence its strength.

$$M_u = 0.87 f_y A_{st} d \left[1 - \left(\frac{A_{st}}{b_d}\right)\left(\frac{f_y}{f_{ck}}\right)\right] \quad (2)$$

The equation, Eq. 2 is from the Indian standard code IS-456 which helps to find the reinforcement with the help of the moment of resistance.

The reinforcement provided for the project pertains for three distinct beam specimens labelled as *A*, *B*, and *C* according to their depth. Specimen *A* exhibits an initial moment of resistance of 19.018×10^6 N-m and utilizes an initial steel reinforcement area of 301.153 mm^2, which is subsequently reduced by 10–15% to 271.037 mm^2. This specimen is reinforced with two bars of 12 mm diameter. Specimen *B* displays an initial moment of resistance of 31.438×10^6 N-m and an initial steel area of 381.981 mm^2, reduced to 343.782 mm^2 after a 10–15% reduction. It is reinforced with three bars of 12 mm diameter. Specimen *C*, with an initial moment of resistance of 47×10^6 N-m, employs an initial steel reinforcement area of 473.525 mm^2, reduced to 420.496 mm^2 (10–15% reduction). This specimen is reinforced with two bars of 16 mm diameter. The variations in moment of resistance, steel area, and reinforcement configuration across the specimens illustrate distinct design considerations and potential modifications in beam structures (Figs. 1 and 2).

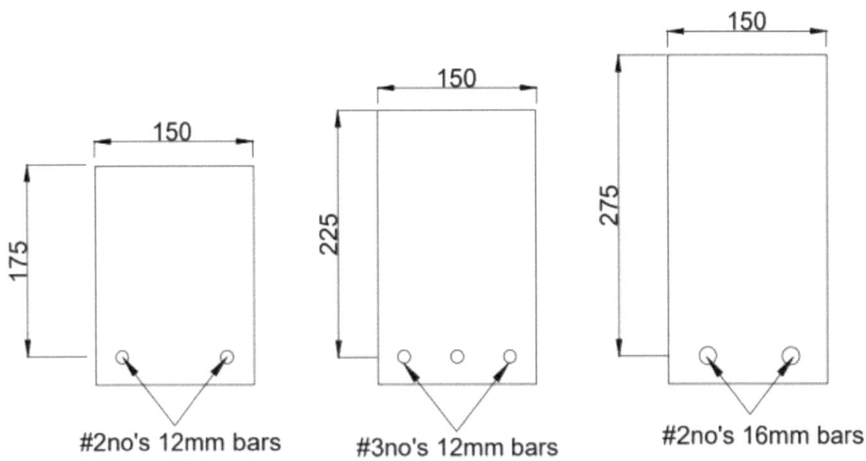

Fig. 1 Cross-section of beams specimens

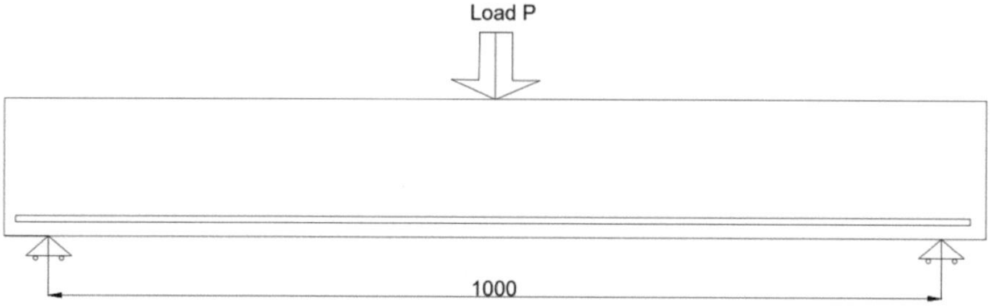

Fig. 2 Longitudinal section of the beam

4 Experimental Work

The experimental work in this project involves several stages, including concrete mixing and casting, compressive strength testing, split tensile strength testing, and shear strength evaluation. Concrete specimens are prepared and subjected to axial and diametrical loading to determine their mechanical properties. Shear behaviour is assessed on concrete beams of varying depths, and the project aims to compare empirical shear strength predictions with experimental findings.

The process encompasses meticulous specimen preparation using precise moulds, accurate material measurement, and careful compaction. Testing involves employing a Universal Testing Machine for shear tests, as well as separate tests for compressive and split tensile strength. Shear strength calculations consider concrete compressive strength and different beam sizes. The casting procedure is detailed for beams of varying dimensions, using M30 grade concrete and specific steel reinforcement. The process includes reinforcement placement, concrete casting, demoulding, and subsequent controlled curing. The project aims to enhance structural design practices and advance concrete engineering knowledge through systematic data collection, statistical analysis, and result interpretation (Fig. 3).

5 Results and Discussion:

The presentation and analysis of the results obtained from the experimental project, offering valuable insights into its outcomes. The subsequent discussion delves into the implications of these findings, comparing them with different empirical formulas and Codal provisions.

Compressive strength, a crucial mechanical property, reflects a material's ability to withstand axial loads without permanent deformation or failure. The compressive strength tests were conducted on concrete specimens of dimensions 150 $\times$ 150 $\times$ 150 mm, treated for 7 and 28 days. The results show a significant increase in strength over the curing period. The compressive strength at 28 days demonstrates substantial

Fig. 3 The shear strength testing process conducted using a universal testing machine (UTM)

enhancement compared to the 7-day strength, indicating continuous concrete development and improved load-bearing capacity with longer curing times. The average compressive strength for the test concrete came 34.91 N/mm^2 which is satisfactory to the mix design ratio 1:1.4:2.5, which was mentioned above.

Split tensile strength, another important mechanical property, characterizes a material's resistance to tensile forces applied perpendicular to the loading axis. The split tensile tests were performed on concrete specimens 150 mm in diameter and 300 mm in height, cured for 7 and 28 days. The results illustrate progressive growth in strength over time, i.e. 3.6 N/mm^2 on an average. Greater tensile strength at 28 days suggests improved structural integrity and load- bearing capability, emphasizing the significance of proper curing duration for optimal concrete performance.

Shear strength in beams refers to a beam's capacity to resist forces acting perpendicular to its longitudinal axis, causing internal layers to slide past each other. This property is critical for maintaining beam stability and structural integrity under various loading conditions, contributing significantly to overall safety and performance. Shear strength in beams is a fundamental mechanical property that measures a material's resistance to forces acting parallel to its cross-sectional area, often leading to deformations or failure along shear planes, the outcomes of the tests can be observed in Table 1 (Figs. 4, 5 and 6)).

The use of empirical equations to calculate shear strength in relation to compressive strength is crucial for a specific type of concrete, especially for beams with only bottom reinforcement. The fact that all beams failed due to shear forces after 28 days highlights the importance of accurately predicting these shear forces. This emphasizes the need for appropriate shear reinforcement design to prevent such failures. The situation underscores the critical role of thorough shear reinforcement detailing in beam design, ensuring overall structural integrity.

In the project, the absence of transverse reinforcement complicates the shear behaviour of concrete beams. The shear failure of the beams underscores the significance of precisely assessing shear strength, influenced by factors like concrete

Table 1 The investigational outcomes for shear strength gained from the conducted tests

Specimen no	Dimensions (mm)	Load at first crack (kn)	Ultimate load (kn)	Applied shear force V_u (kn)	Experimental shear strength (kn)
A1	150 × 200	119	241	120.5	22
A2		124.6	247.3	123.64	22.5
A3		114.7	234.4	117.2	21.39
B1	150 × 250	189.0	306.8	153.4	28.01
B2		191.2	321.49	160.74	29.34
B3		163.4	297.55	148.77	27.16
C1	150 × 300	294	482.36	241.18	44.03
C2		207.5	425.56	212.78	38.84
C3		218.6	465.6	232.8	42.5

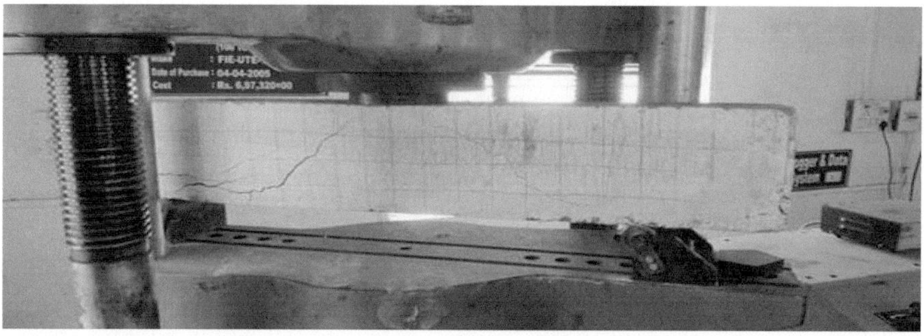

Fig. 4 The initial appearance of a crack in the concrete specimen

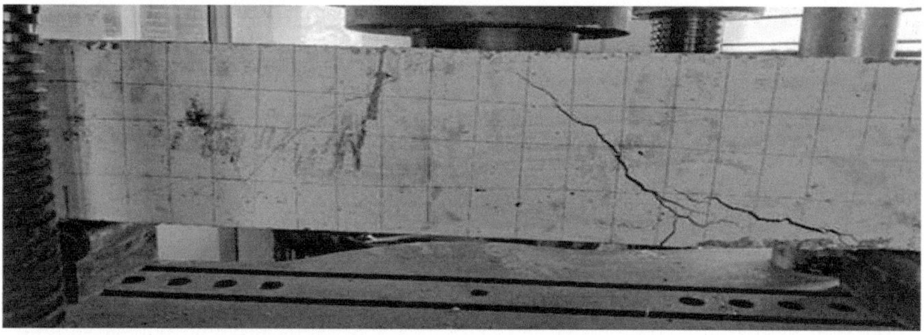

Fig. 5 The shear failure of the concrete beam

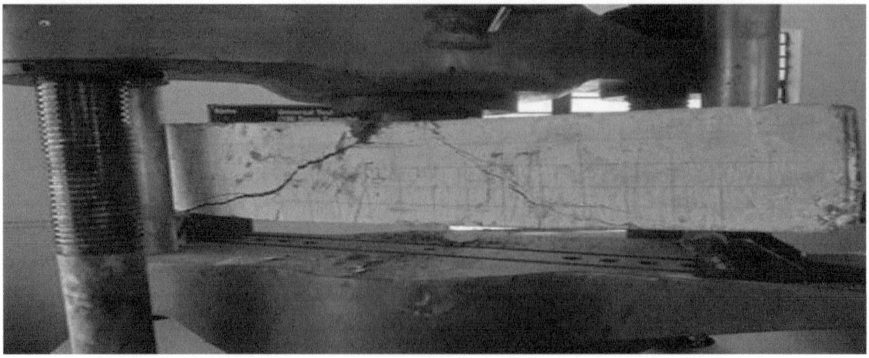

Fig. 6 The shear failure of the concrete beam

strength, beam dimensions, and loading conditions. Empirical equations linking shear strength to compressive strength provide a realistic design method to simplify this prediction.

The lack of transverse reinforcement likely led to shear failures in your study. Transverse reinforcement, like stirrups, is essential to distribute shear forces and prevent brittle shear failure planes. This highlights the necessity for careful shear reinforcement detailing. Well- planned and properly positioned shear reinforcement significantly improves the beams' ability to withstand shear forces and avoid premature failure.

The calculated percentage discrepancies between the experimental shear strength values and the ultimate load values for each specimen fall within a narrow range, ranging from 9.14 to 9.07%. This indicates a robust correlation between the experimental shear strength values and the ultimate loads. Each specimen consistently exhibits a variation in this range (9.14–9.07%). Additionally, with an increase in depth, there is a corresponding 6.82% increase in shear strength.

6 Comparison of the Shear Strength with Different Empirical Formulas and Codal Provisions.

Various design methods and formulas are available for calculating the shear capacity of concrete beams, and the choice depends on the specific design code or standard in use. Common methods include the ACI (American Concrete Institute) method (ACI318-02), Zsutty's equation, JSCE (1986), Canadian Standard Association (CSA), and Eurocode 2.

The ultimate shear strength is computed to enable the analysis and comparison of experimental outcomes across different codes and empirical equations. The comparison of shear strength results using various empirical formulas and code provisions holds significant importance in structural engineering. This practice validates design assumptions, examines parameter sensitivity, ensures compliance with safety regulations, boosts design confidence, optimizes solutions, accommodates diverse material properties, educates engineers, and adapts to evolving design standards.

This systematic process elevates the overall quality and dependability of structural designs while nurturing a deeper comprehension of shear strength principles among engineers. In Table 2, we can observe the evaluation of shear strength outcomes obtained from different tests and methodologies.

Figure 7 illustrates a comparative analysis of shear strength concerning the depth of the beam. The findings are summarized in bullet points as follows:

The Zsutty method consistently yielded shear strength values surpassing the experimental values, exhibiting an average percentage difference of 6.14% less than the experimental results.

In contrast, the JSCE (1986) approach demonstrated a tendency to generate higher shear strength readings, with an average percentage difference of 32.96% higher than the experimental values. Notably, the shear strength readings obtained through the CSA method frequently exceeded the experimental values by around 3.40%.

Meanwhile, the Eurocode 2 method exhibited a pattern of providing higher shear strength values, with an average percentage difference of approximately 0.36% less than the experimental values. These comparative findings shed light on the performance of different methods in predicting shear strength based on beam depth, contributing valuable insights to the realm of structural engineering.

The shear strength data obtained from our investigation exhibited noteworthy trends when compared to findings from other studies. Specifically, whereas those from the Zsutty et al. method tended to be higher for most beams, barring the first series. An interesting observation was that the EC-2 (Eurocode 2) predictions closely aligned with our results from the first series of beams, but demonstrated

Table 2 Evaluation of shear strength outcomes obtained from different tests and methodologies

Beam no	A_{sl}	ρratio $= A_{sl}/b_w d$	Shear strength (kN)					
			Experimental value	ACI31 8-02	Zsutty	JSCE (1986)	Canadian standard association (CSA)	Eurocode 2
A1	271.03	0.0103	22	23.96	27.51	23.63	28.03	30.32
A2	271.03	0.0103	22.5	23.96	27.51	23.63	28.03	30.32
A3	271.03	0.0103	21.39	23.96	27.51	23.63	28.03	30.32
B1	343.78	0.0102	28.01	30.80	38.21	29.28	34.56	36.48
B2	343.78	0.0102	29.34	30.80	38.21	29.28	34.56	36.48
B3	343.78	0.0102	27.16	30.80	38.21	29.28	34.56	36.48
C1	402.49	0.0097	44.03	37.65	49.27	31.25	40.66	41.82
C2	402.49	0.0097	38.84	37.65	49.27	31.25	40.66	41.82
C3	402.49	0.0097	42.5	37.65	49.27	31.25	40.66	41.82

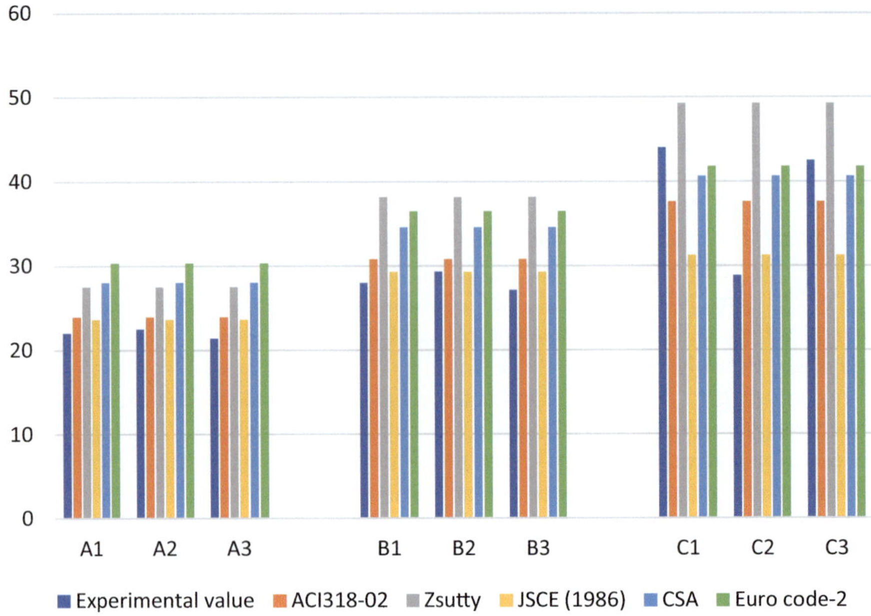

Fig. 7 The evaluation of shear strength outcomes obtained from various empirical equations

closer agreement with the ACI 18–02 and JSCE (1986) methods for the subsequent series.

The influence of stirrup reinforcement on enhancing shear strength was apparent across all beam depths, underscoring the pivotal role of precise design. While our experimental values increasingly approached the predictions of the EC-2 method as beam depth increased, variations persisted.

This highlighted the inherent diversity of findings across different studies, even when considering the same code or standard. These comparative insights emphasize the complexity of shear strength behaviour and underscore the need for meticulous design approaches in ensuring structural integrity.

7 Summary

The objective of the paper was to evaluate shear strength in concrete beams without transverse reinforcement. The concrete used was of M30 grade with mix proportion ratio 1:1.4:2.5, and properties like compressive strength and split tensile strength were examined. The study found that the M30 concrete exceeded 30 N/mm^2 compressive strength after 28 days, aligning with the designed mix ratio. Split tensile strength averaged at 3.6 N/mm^2. Experimental shear strength results for prismatic beams consistently fell within a narrow range of 9.14–9.07%, increasing by 6.82% as beam depth grew. Eurocode 2 predictions closely matched experimental values with a 0.36% average difference due to similar reinforcement calculation methods. When compared to other methods, Zsutty et al.'s showed a 6.17% similarity and JSCE (1986) differed by 32.96%. This emphasized the impact of different methods on shear strength predictions. The experiment provided insights into shear behaviour and highlighted the importance of accurate design and method selection.

References

1. Jayasinghe T, Gunawardena T, Mendis P (2022) Assessment of shear strength of reinforced concrete beams without shear reinforcement: a comparative study between codes of practice and artificial neural network. Case Stud Constr Mater 16:e01102
2. Junga S, Kim KS The University of Seoul, 90 Jeonnong-dong, Dongdaemun-gu, Seoul 130–743. Republic of Korea
3. Akkaya HC, Aydemir C, Arslan G (2022) The Authors. Published by Elsevier Ltd
4. Huang' Z, Miao X, Shen J, Li H, Zhang J (2021) Institution of Structural Engineers. Published by Elsevier Ltd
5. Ortiz Navas F, Juan NG, Leiva Herdocia G, Serna P, Cuenca E (2021) Institution of Structural Engineers. Published by Elsevier Ltd
6. Ivan SI, Snezˇana BM, Nikola T (2017) Elsevier Ltd. All rights reserved
7. Zende AA, Khadiranaikar RB, Momin AIA (2023) Inderscience Publishers (IEL)
8. IS 456 (2000) Plain and reinforced concrete—code of practice
9. IS 1489–1 (1991) Specification for Portland pozzolana cement, Part 1: flyash based [CED 2: Cement and Concrete]
10. IS 10262 (2019) Concrete mix proportioning—guidelines
11. IS 383 (1970) Specification for coarse and fine aggregates from natural sources for concrete

Effect of Stratified Soil Deposit on Seismic Response of Elevated Water Tanks

K. S. Jayadeep and L. Govindaraju

Abstract Functionality of elevated water tanks post-earthquakes has been a major issue, and seismic resilience is of utmost concern. The behaviour of elevated water tanks is different when compared with conventional buildings, considering they have a huge mass of water supported on a tall staging and designing such structures requires specialized analysis. Previous studies have shown the failure pattern of such elevated tanks where improper staging selection and overestimating seismic responses have been the major causes. The conventional approach of fixed base analysis assumes the foundation to be rigid which is contradictory in actual site conditions. In actuality, the soil–structure interaction amplifies the response of the structure resting especially in soft soils and stratified soil deposits compared to stiff soil or rock, and disregarding such a complex interaction has proven to be detrimental to the performance of structures leading to premature collapse during intense ground shaking. The aseismic design of such tanks demands to account for all these effects and in addition to it, the dynamic effect from sloshing waves. The present study examines the dynamic behaviour of elevated water tanks through a comprehensive numerical analysis considering the effect of sloshing and soil– structure interaction.

Keywords Soil–structure interaction · Stratified soil · Sloshing wave

1 Introduction

Earthquakes are the most devastating of all the natural disasters and the destruction they cause have created a tremendous impact and concern about the need for earthquake-resistant design by posing significant challenges. Every single day,

K. S. Jayadeep (✉)
Department of Civil Engineering, R N S Institute of Technology, Channasandra, Bangalore, India
e-mail: jayadeepks77@gmail.com

L. Govindaraju
Department of Civil Engineering, University Visvesvaraya College of Engineering, Bangalore, India

Y. K. Guruprasad et al. (eds.), *Recent Advances in Infrastructure Development and Management—Volume 1*, Lecture Notes in Civil Engineering 570,
https://doi.org/10.1007/978-981-96-1554-4_30

elevated water tanks ensure that communities have access to clean water and the tank's elevated positions make it easier to use natural gravity to supply water in an efficient and effective manner, and to ensure such functionality they must be resilient to all kinds of forces including seismic activity. In contrast to regular structures, elevated water tanks exhibit unique seismic behaviour since they have a huge water mass supported on a tall staging due to which they are extremely vulnerable to lateral forces. The damages observed during past major earthquakes such as Bhuj (2001) have illustrated the failure pattern of such elevated tanks where improper staging selection and overestimating seismic responses have been the most significant causes [1, 2]. During earthquakes, the behaviour of a structure not only depends on superstructure but is greatly influenced by the response and interaction between super structure and the underlying soil which is the effect of soil–structure interaction (SSI). These effects are usually ignored by adopting fixed base analysis while estimating a structure's seismic response. The conventional approach of fixed base analysis assumes the foundation to be rigid which is contradictory in actual site conditions. In actuality, the soil–structure interaction amplifies the response of the structure resting especially in soft soils and stratified soil deposits compared to stiff soil or rock, and disregarding such a complex interaction has proven to be detrimental to the performance of structures leading to premature collapse during intense ground shaking [3]. The aseismic design of such tanks demands to account for all these effects and in addition to it, the dynamic effect from sloshing waves. The effect of sloshing waves alters the motion of tanks during ground shaking which was the major cause of failure during the Chilean earthquake of 1960. Considering the effects of sloshing and soil–structure interaction, this paper presents the dynamic behaviour of elevated water tanks through a comprehensive numerical analysis (Fig. 1).

Scope and Objectives

Scope

To study the effect of stratified soil deposits on seismic response of elevated water tanks

Objectives

To evaluate the dynamic behaviour of elevated water tank in stratified soil conditions.

To study the behaviour of stratified soil deposits and the layer effect of soil mass.

To evaluate seismic response considering the effect of sloshing water mass.

To evaluate the frequency variation of stratified soil mass and to correlate it with seismic responses.

Fig. 1 Elevated water tank

2 Idealization of Elevated Water Tanks

The concept of idealization of elevated water tanks plays a major role in evaluating seismic responses and designing such structures require specialized analysis. Elevated water tanks are never fully filled, there is always some free space for movement of water. Sloshing is a phenomenon that occurs when there is a free surface for water movement, and has a significant effect. Housner in his work [4] clearly explained the effect of sloshing and the concept of idealization of elevated water tanks in accounting the dynamic forces from sloshing considering the relation between the motion of water in tank with respect to tank and motion of whole structure with respect to ground. During an earthquake when an elevated water tank containing liquid vibrates, there is a relative movement of water inside the tank container exerting pressures on walls and slabs of the water tank, and to evaluate these pressures exerted, the idealization plays a key role.

As per IS 1893 (Part 2):2014 [5] elevated water tanks can be idealized as.

(a) Single lumped mass model
(b) Two-mass model.

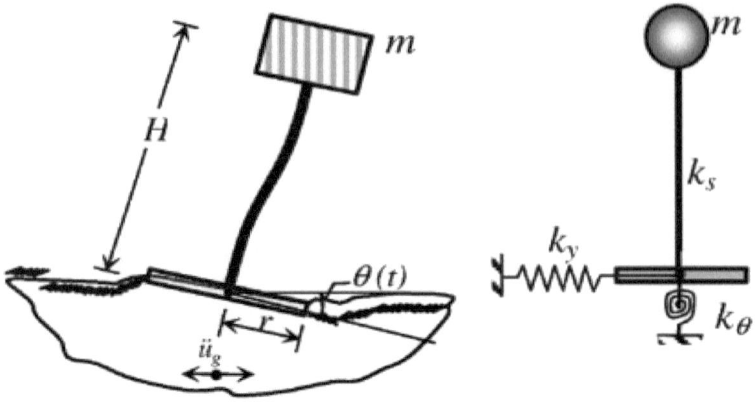

Fig. 2 Single lumped mass model of elevated water tanks [6]

For a water tank to be considered as a single lumped mass model (one-mass model), there must be no free surface for water movement, i.e. either the tank must be fully filled with no free surface, or it must be empty without water, but when there is free surface this single mass concept falls short of establishing the idea of sloshing. Figure 2 shows the single lumped mass model of elevated water tanks.

In case of partially filled water tanks, there is a free surface available for water movement and during earthquake ground shaking, the liquid exerts hydrodynamic pressure. The water in the tank gets divided into two different masses. The liquid in the lower region of the tank behaves like a mass that is rigidly connected to tank and is termed as impulsive mass. The liquid in the upper region of the tank moves in the direction opposite to the movement to tank and is termed as convective mass. The convective mass is the sloshing mass or sloshing wave which alters the dynamic behaviour of container. Figure 3 shows two-mass model of elevated water tanks. This two-mass idealization is also known as spring mass model which is illustrated in seismic resistant design code IS1893 (Part 2): 2014 [5].

2.1 Design Specifications and Structural Configuration

In this study, an RC-reinforced concrete elevated square water tank is considered to observe the behaviour for both fixed base and flexible base (SSI) resting on stratified soil profile. An open square water tank container of dimension 3 m × 3 m × 3m (Freeboard of 0.3 m) is supported on a staging of height 9 m above ground level. The geometric and material properties of the structure are illustrated in Tables 1 and 2. The RC tank is designed in accordance with IS 3370: 1965 [7], guidelines for seismic design of liquid storage tanks [8] and IS 456:2000 [9]. The dynamic analysis is performed as per IS 1893 (Part 1): 2016 [10] and IS 1893 (Part 2): 2014 [5] (Table 3).

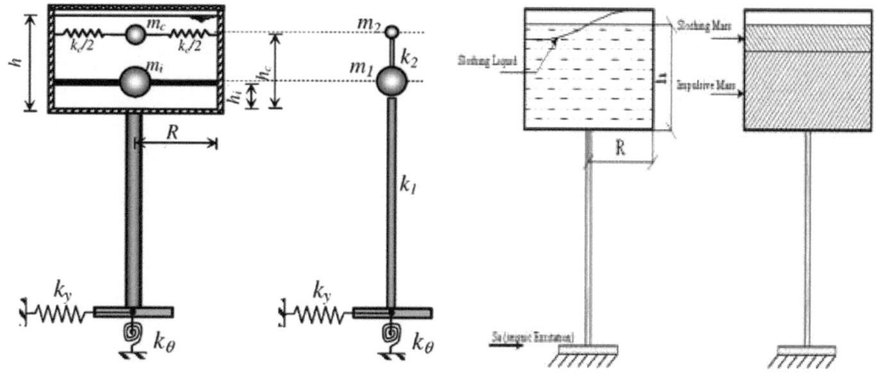

Fig. 3 Two-mass model of elevated water tank [6]

Table 1 Details of geometric properties of the structure

Sl. no.	Structural elements	Dimensions
1	Column	300 mm × 300 mm
2	Beam	300 mm × 400 mm
3	Bracings	300 mm × 300 mm
4	Slab	180 mm
5	Walls	180 mm

Table 2 Details of material properties of the structure

Sl. no.	Material type	Grade
1	Concrete	M-25
2	Steel	Fe-500

Table 3 Description of models

Sl. no.	Model	Description
1	Empty tank model	It refers to the water tank without any water mass and the results evaluated are referred as empty tank condition. The structure is analysed considering its dead load. (Single lumped mass model)
2	Full tank model	It refers to the water tank with water mass and free surface and the results evaluated are referred as full tank condition. The structure is analysed considering its dead load along with water mass (Two-mass model)

The three-dimensional view of water tank is shown in Fig. 4.

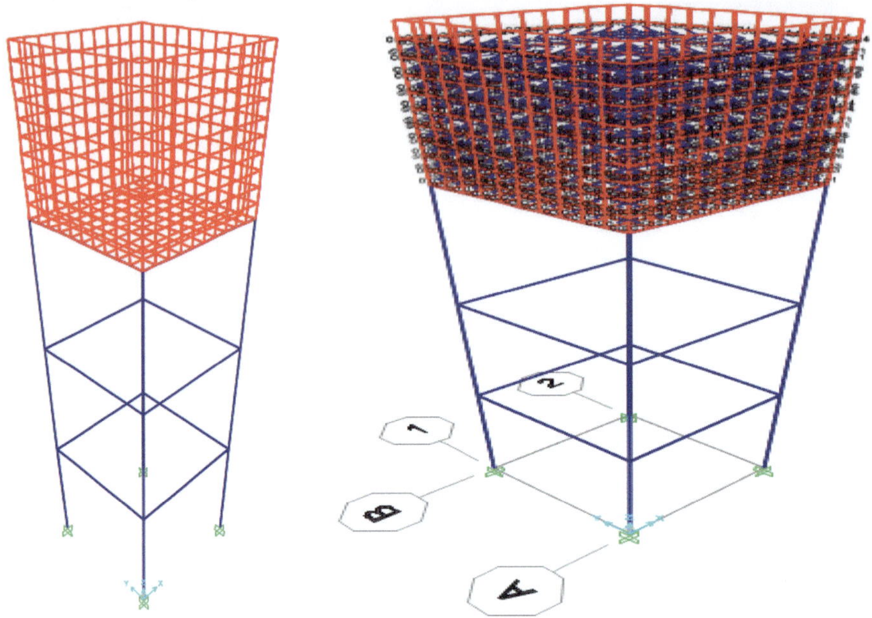

Fig. 4 3D View of elevated water tank

2.2 *Numerical Modelling and Analysis*

The finite element modelling (FEM)-based software tool SAP2000 V18.2 was employed for numerical simulations. The analysis is performed considering special moment resisting frame with the dynamic properties as shown in Table 4and using Bhuj (2001) earthquake acceleration data with peak ground acceleration of 0.1 g. The description of models and its idealization are discussed in Table 3. The hydrodynamic water pressure is applied as impulsive and convective pressures and details are shown in Table 5 (Fig. 5).

In Fig. 6, the impulsive and convective water pressure distribution is shown, where q_i and q_c are the resultant forces which acts at height h_i and h_c from the base of the tank container, and the variation of these pressures along the height of the wall is calculated using relations as given in IS 1893 (Part 2): 2014 [5].

Table 4 Dynamic characteristics

Sl. no.	Parameters	Description
1	Seismic zone	II
2	Soil type	III
3	Importance factor	1.5
4	Response reduction factor	4

Table 5 Details of water pressure considered

Sl. no.	Type of load	Intensity (kN/m^2)
1	Impulsive hydrodynamic pressure	1.44
2	Convective hydrodynamic pressure	0.33
3	Hydrostatic pressure	26.49

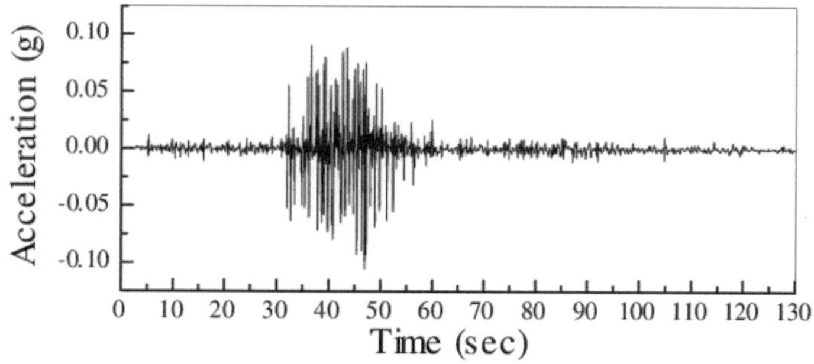

Fig. 5 Bhuj (2001) earthquake acceleration

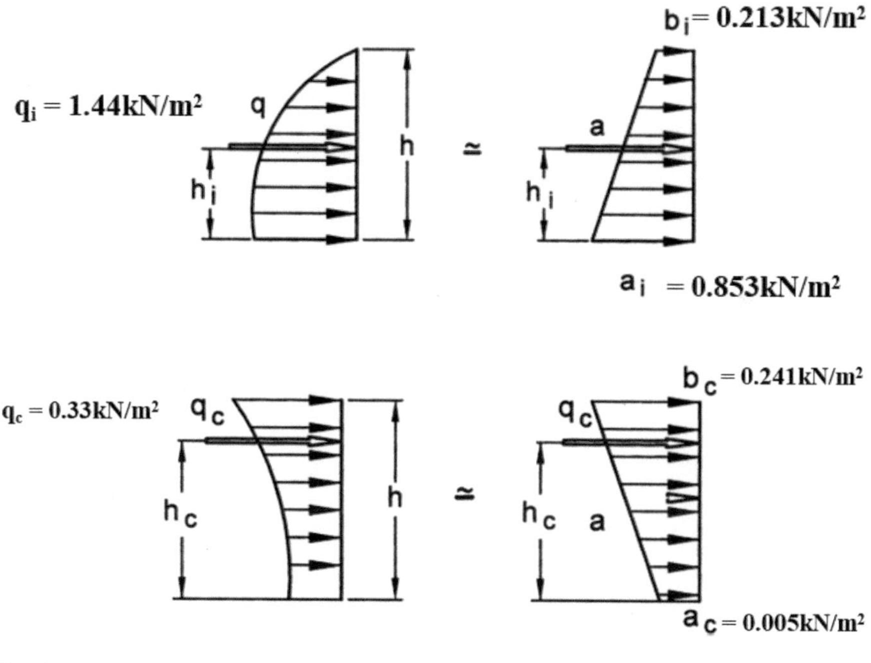

Fig. 6 Impulsive and convective water pressure distribution [5]

Table 6 Soil properties [11]

Soil type	Unit weight (kN/m^3)	Modulus of elasticity (kN/m^2)	Poisson's ratio
Very soft clay	18	15,000	0.4
Loose sand	18.5	24,000	0.3
Hard soil	18	95,000	0.3

Table 7 Details of stratified soil profiles [12]

Soil profile	Description
C1	10 m soft clay + 10 m loose sand + 10 m soft clay
C2	10 m loose sand + 10 m soft clay + 10 m loose sand
C3	15 m loose sand + 15 m soft clay
C4	15 m soft clay + 15 m loose sand
C5	10 m hard soil + 10 m soft clay + 10 m hard soil
C6	30 m hard soil

2.3 Soil–Structure Interaction (SSI)

The effect of the SSI is evaluated considering the Elastic Continuum Approach (FEM Model). Isolated square footing is designed considering the maximum load on individual columns. The footing has dimension of 1.5 m × 1.5 m and is at a depth of 1.5 m below ground level, and the thickness of the footing is 500 mm. The soil and sub- structure are modelled as a continuum; the soil mass has dimensions of 50 m in breadth and 30 m in depth and free field motion is considered, and at the base, fixity is assigned to the soil mass considering the soil below the 30 m depth is a hard stratum. The soil properties are adopted as per Bowles [11], and stratified layers are considered as per [12] and is shown in Tables 6 and 7, respectively.

2.4 Behaviour of Stratified Soil:

To study the behaviour of stratified soil deposits and the layer effect of soil mass, modal analysis is performed to determine the natural period and frequency of the soil mass and the acceleration of the stratified soil layers without considering the water tank (see Fig. 7) and the results obtained are shown below in Tables 8 and 9. The variation of frequency and acceleration shows the effect of stratified soil layers during ground shaking. (see Figs. 8 and 9).

Figure 8 depicts the frequency variation of stratified soil layers. It is evident that the C3 soil profile exhibits lower frequency and the C6 soil profile with higher frequency, with displacement being the lower frequency parameter, this aids in the

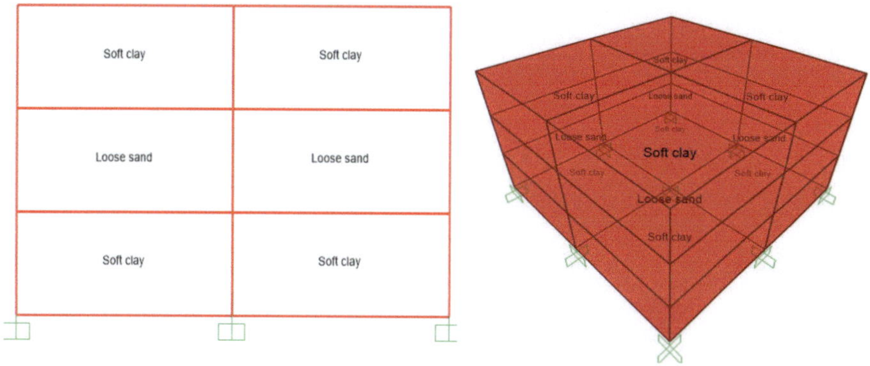

Fig. 7 Stratified soil mass model without super structure

Table 8 Natural period and frequency stratified layers

Modal parameters		C1	C2	C3	C4	C5	C6
Time period (s)	Mode:01	2.571	2.367	2.737	2.255	1.892	1.068
	Mode:02	2.440	2.298	2.572	2.241	1.710	1.057
	Mode:03	2.440	2.298	2.572	2.241	1.710	1.057
Frequency (Cyc/s)	Mode:01	0.388	0.422	0.365	0.443	0.528	0.935
	Mode:02	0.409	0.434	0.388	0.446	0.584	0.945
	Mode:03	0.409	0.434	0.388	0.446	0.584	0.945

estimation of displacements indicating soil mass with lower frequency to exhibit larger displacements.

Figure 9 shows the variation of acceleration in stratified soil layers and it can be observed that C6 soil profile exhibits higher acceleration and C5 soil profile with lower acceleration.

The acceleration variation for the stratified soil mass of depth 30 m is tabulated above, for soil profiles C1, C2, C5 and C6, the soil mass has three stratified layers and hence the acceleration is evaluated at every 10 m depth, and for C3 and C4 soils, two stratified soil layers are considered and acceleration is evaluated at every 15 m depth and the values obtained are indicated in Table 9 (Fig. 10).

3 Results and Discussions

Dynamic time history analysis was performed on elevated water tank using Bhuj earthquake acceleration and their responses in terms of displacement and modal parameters, such as natural period and frequencies are studied for both fixed and flexible base conditions.

Table 9 Acceleration variation in stratified layers

Soil profile	Depth (m)	Acceleration (m/s^2)
C1	0 (G.L)	0.943
	– 10	0.525
	– 20	0.417
	– 30	0.884
C2	0 (G.L)	0.782
	– 10	0.566
	– 20	0.474
	– 30	0.884
C3	0 (G.L)	0.687
	– 15	0.446
	– 30	0.884
C4	0 (G.L)	0.774
	– 15	0.434
	– 30	0.884
C5	0 (G.L)	0.641
	– 10	0.518
	– 20	0.792
	– 30	0.884
C6	0 (G.L)	1.740
	– 10	1.212
	– 20	0.924
	– 30	0.884

3.1 Fixed Base Condition

Tables 10 and 11 illustrate the results from fixed bas analysis, it is evident that the responses are maximum when the tank is fully filled, which is caused by the water's mass and hydrodynamic pressures, which include both convective and impulsive pressures.

The natural period of the water tank is 0.718 s which is 35% more than the natural period of empty tank and the displacement has increased by 16% which is due to the huge mass of water.

3.2 Flexible Base Condition (SSI)

The results obtained from numerical simulations considering stratified soils are tabulated below from Tables 12, 13, 14, and 15. In the study, gross displacements are

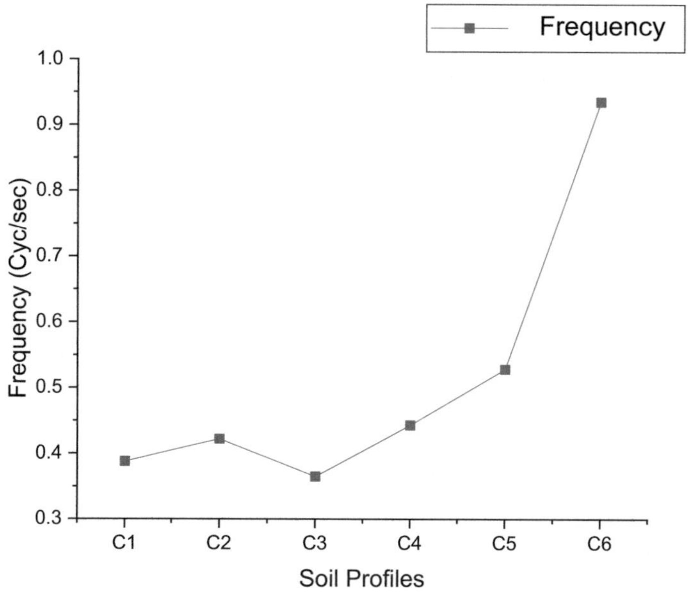

Fig. 8 Variation of frequency in stratified layers

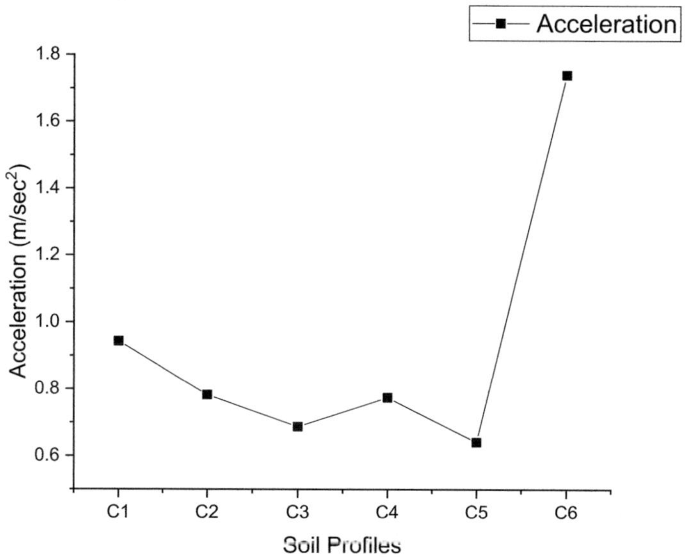

Fig. 9 Acceleration variation in stratified layers

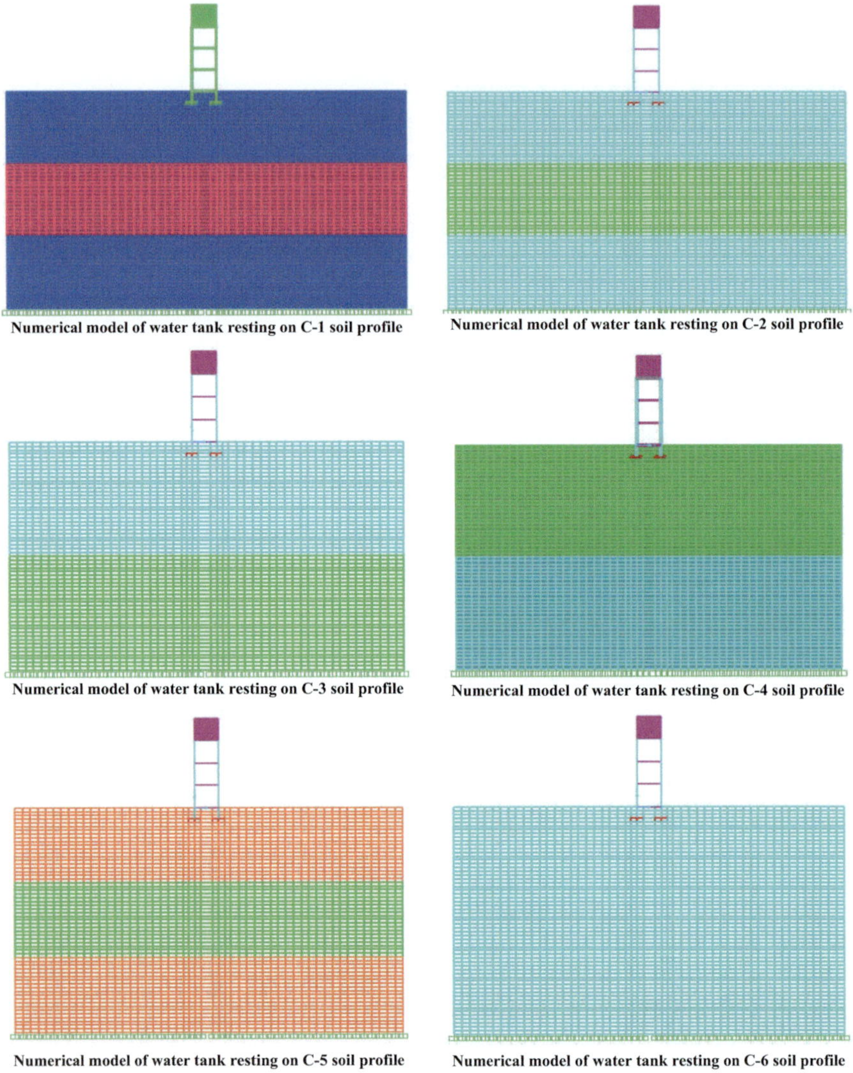

Fig. 10 SSI FEM models

considered and assessed. The net displacement can be computed by making zero displacement at ground level.

Table 13 presents the displacements that were obtained using numerical SSI while taking stratified soil profiles into consideration. It is evident that water tanks resting on stratified soil with lower frequency exhibit larger displacements, which suggests that the tanks will eventually fail. In C5 and C6, the displacement at the base of the tank is significantly less than in other soil profiles, but at the top of the tank, there are

Table 10 Displacement from fixed base analysis

Height (m)	Displacement (mm)	
	Empty tank condition	Full tank condition
12	22.81	26.67
9	22.2	26
6	15.82	18.15
3	6.53	7.42
0	0	0

Table 11 Modal parameters from fixed base analysis

Mode	Time period (s)		Frequencies (Cyc/s)	
	Empty tank condition	Full tank condition	Empty tank condition	Full tank condition
Mode-01	0.528	0.718	1.893	1.392
Mode-02	0.528	0.718	1.893	1.392
Mode-03	0.362	0.433	2.762	2.309

Table 12 Displacement of water tank (Empty tank Condition)

Height (m)	Displacement (mm)					
	C1	C2	C3	C4	C5	C6
12	188.6	153.4	187.2	135.2	94.6	112.8
9	178.5	145.4	177.1	130.5	91.1	108.5
6	164.8	133.9	163.3	132.1	85.2	92.1
3	149	120.5	147.6	114.1	78.3	72.2
0	133.8	107.9	132.5	105.3	72.6	59.9

Table 13 Displacement of water tank (Full tank Condition)

Height (m)	Displacement (mm)					
	C1	C2	C3	C4	C5	C6
12	217.1	188.7	221.9	210.9	174.1	209.2
9	204	177.8	208.77	198.2	166.2	200.2
6	182.7	157.1	185.9	171.7	131.8	157
3	158.2	132	158.6	138.9	85.5	98.4
0	135.5	109.5	134.1	103	72.3	61.1

higher displacements. This is due to resonance that occurs in hard soil layers because of higher ground-level stiffness.

Figure 11 shows the plot of displacement of water tank for different soil profiles, and it is evident that water tank resting on stratified soil profile with lower frequency

Table 14 Modal parameters for empty tank condition

Modal parameters		C1	C2	C3	C4	C5	C6
Time period (s)	Mode:01	2.476	2.305	2.573	2.207	1.714	1.042
	Mode:02	2.476	2.305	2.573	2.207	1.714	1.042
	Mode:03	2.203	2.034	2.287	1.940	1.571	0.907
Frequency (Cyc/s)	Mode:01	0.403	0.433	0.388	0.453	0.583	0.959
	Mode:02	0.403	0.433	0.388	0.453	0.583	0.959
	Mode:03	0.453	0.491	0.437	0.515	0.636	1.101

Table 15 Modal parameters for full tank condition

Modal parameters		C1	C2	C3	C4	C5	C6
Time period (s)	Mode:01	2.476	2.306	2.572	2.208	1.715	1.04
	Mode:02	2.476	2.306	2.572	2.208	1.715	1.043
	Mode:03	2.203	2.034	2.287	1.940	1.571	0.907
Frequency (Cyc/s)	Mode:01	0.403	0.433	0.388	0.453	0.582	0.958
	Mode:02	0.403	0.433	0.388	0.453	0.582	0.958
	Mode:03	0.453	0.491	0.437	0.515	0.636	1.101

exhibiting larger displacement value compared to other soil profiles and maximum displacement is observed in C3 soil profile with frequency of 0.388.

Figure 12 shows the variation of fundamental frequency obtained for soil mass without water tank and is compared with fundamental frequency obtained for SSI with water tank, and the results shows good correlation indicating the influence of stratified soils and layer effect on response of structure.

3.3 *Sloshing Wave Height*

The heights of sloshing waves for various zones with varying soil conditions are computed and it is observed that as the seismic zone changes from II to V, the height of the sloshing waves increases.

Figure 13 shows the variation of sloshing wave with zones, and it can be observed that in higher seismic zones the provided freeboard height of 0.3 m is insufficient due to which additional hydrodynamic forces will act on tank walls and roof slab (Table 16).

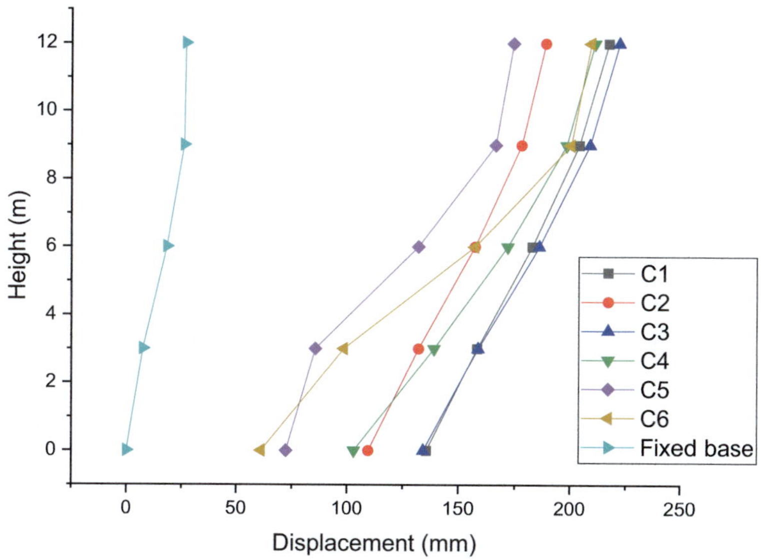

Fig. 11 Displacement versus height for full tank condition

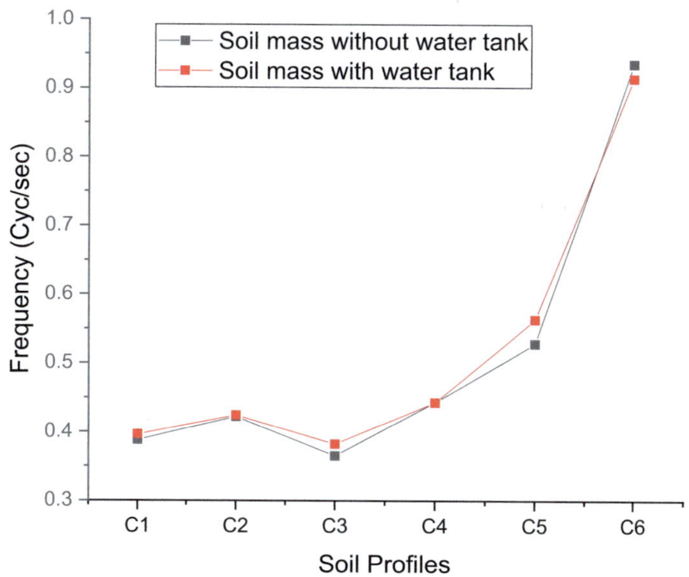

Fig. 12 Frequency variation in soil profiles

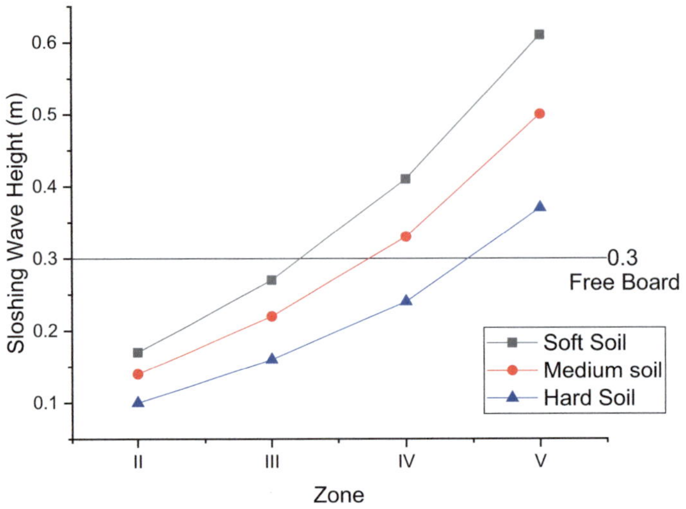

Fig. 13 Variation of sloshing wave with zones

Table 16 Sloshing wave height

Zone	Soil type	Sloshing wave height (m)
II	Soft soil	0.17
	Medium soil	0.14
	Hard soil	0.10
III	Soft soil	0.27
	Medium soil	0.22
	Hard soil	0.16
IV	Soft soil	0.41
	Medium soil	0.33
	Hard soil	0.24
V	Soft soil	0.61
	Medium soil	0.50
	Hard soil	0.37

4 Conclusion

Numerical simulation was performed on an elevated water tank resting on stratified soil deposits and was subjected to an earthquake acceleration of 0.1 g, and the results were evaluated in terms of displacement, natural period, and frequency.

The following conclusions were drawn from the study:

1. The displacement in full tank condition is higher than the empty tank condition and hence the design will be governed by full tank condition.

2. The maximum displacement is observed for C3 soil profile compared to all the other stratified layers because C3 soil profile has lower frequency.
3. The effect of stratified soil layers alters the frequency of soil mass due to which the displacement variation of the structure can be observed in stratified soil layers.
4. Whenever a structure is resting on stratified soil deposits, frequency of soil mass must be determined so that if soil mass has lower frequency, it will lead to larger displacements in structures resting on them.
5. The fundamental frequency of soil mass without water tank and soil mass with water tank are in good correlation (see Fig. 12), and this shows the effect of stratified soil on seismic response.
6. In case of C5 and C6 soil profile, it can be observed from (see Fig. 11) that the displacement at base is very less compared to other soil profiles but at top of the tank higher displacements are observed only in these (C5 and C6) soil profile, this is because of resonance that occurs in hard soil layers because of higher stiffness at ground level.
7. Figs. 7 and 8 show the variation of natural frequency and acceleration of stratified soil deposits and the layer effect of soil and the responses obtained correlates with the responses from SSI with tank.
8. The stratified soil layers increase the natural period of the structure by 3 times in (C1, C2, C3, C4) soil profiles indicating the effect of amplification.
9. In C5 and C6, the soil profile consists of hard soil and due to which the soil mass shows higher frequency.
10. The seismic response of water tank resting on stratified soil deposits depends on the natural frequency of the soil mass, with displacement being inversely proportional to the frequency.
11. The sloshing wave height is calculated for different zones considering different soil types (see Fig. 13), and maximum sloshing wave is observed in Zone V for soft soils.
12. The displacement from the sloshing waves alone cannot be distinguished since the displacement obtained from the analysis considers the overall water mass including the mass of structure, the displacement due to sloshing waves can be computed through an experimental study.

References

1. Rai D (2003) Performance of elevated tanks in M_w 7.7 Bhuj earthquake of January 26th, 2001. J Earth Syst Sci 112:421–429. https://doi.org/10.1007/BF02709269
2. Soroushnia S, Tavousi S, Omidinasab F, Beheshtian N, Soroushnia S (2011) Seismic performance of RC elevated water tanks with frame staging and exhibition damage pattern. Procedia Eng 14:3076–3087. https://doi.org/10.1016/j.proeng.2011.07.387
3. Mylonakis G, Gazetas G (2000) seismic soil-structure interaction: beneficial or detrimental? J Earthq Eng 4(3):277

4. Housner GW (1963) The dynamic behavior of water tanks. Bull Seism Soc Am 53(2):381–387. https://doi.org/10.1785/BSSA0530020381
5. IS 1893 (Part 2) (2014) Criteria for Earthquake Resistant Design of Structures Part 2 Liquid Retaining Tanks. Bureau of Indian Standards, New Delhi
6. Livaoğlu R, Doğangün A (2006) simplified seismic analysis procedures for elevated tanks considering fluid–structure–soil interaction. J Fluids Struct 22(3):421–439
7. IS3370 (PART1) (1965) Concrete structures for the storage of liquids (General Requirements)
8. IITK-GSDMA (2007)"Guidelines for seismic design of liquid storage tanks. National Information Centre for Earthquake Engineering, IIT Kanpur
9. IS-456 (2000) Plain and reinforced concrete-code of practice. Bureau of Indian Standards, New Delhi
10. IS 1893 (Part 1) (2016) Criteria for earthquake resistant design of structures. Bureau of Indian Standards, New Delhi
11. Bowles JE (1997) Foundation analysis and design, 5th edn. McGraw-Hill, Inc
12. Basavangowda GM, Govindaraju L (2017) Influence of stratified soil on seismic response of pile supported building. Int J Adv Res, Ideas, Innov Technol 3(6)

Micro-Shell of Paraffin@SiO_2 Phase Change Material Embedded in Fly Ash–GGBS Geopolymer for Energy Conservation in Buildings

Rithikaa Thanigaiselvan, A. Mullaimalar, N. Vanitha, and R. Jeyalakshmi

Abstract Microencapsulated phase change materials (MPCM) enclose the core material inside the shell, providing higher thermal regulation performance by expanding the heat transfer surface and preventing the leakage of core materials. This study adopted sol–gel methodology to synthesize Paraffin@SiO_2 to enhance the latent heat storage capacity. The same is implemented in different doses to study their thermal gradient performance in ambient cured fly ash–Ground Granulated Blast Furnace slag (4:1) alkali geopolymer (Gp). The liquid migration test results proved the outstanding tightness of paraffin in the MPCM capsule. Though adding MPCM has slightly reduced the compressive strength, no significant characteristic changes are observed in IR reflection. This confirmed the integration of MPCM in a geopolymeric binder. The heat flow and internal temperature were measured from the thermal test conducted on a geopolymer panel to evaluate the effectiveness of PCM. Compared to the neat PCM, the composites demonstrated high thermal performance with a decrease in indoor temperature up to 3–4 °C. It is attributed to the latent heat capacity and thermal stability of the MPCM, as derived from the DSC test results. To conclude, this finding demonstrates that eco-friendly geopolymers effectively increase the thermal inertia and reduce energy consumption in buildings.

Keywords Geopolymer · Phase change material · Microencapsulation · Paraffin · Thermal energy storage capacity

---

R. Thanigaiselvan · A. Mullaimalar · N. Vanitha · R. Jeyalakshmi (✉)
Department of Chemistry, College of Engineering and Technology, SRM Institute of Science and Technology, Kattankulathur, Tamil Nadu 603203, India
e-mail: jeyalakr@srmist.edu.in

Y. K. Guruprasad et al. (eds.), *Recent Advances in Infrastructure Development and Management—Volume 1*, Lecture Notes in Civil Engineering 570,
https://doi.org/10.1007/978-981-96-1554-4_31

1 Introduction

Phase change materials (PCM) can store and release significant amounts of thermal energy during the phase transition. Buildings with PCM integration have better thermal energy storage capabilities which lowers energy usage, owing to their excellent thermal and chemical stability, low vapour pressure, minimal to no supercooling, high latent heat storage capacity and self-nucleating behaviour [1–3]. PCM can be incorporated into different parts of buildings such as walls, ceilings, windows and floors [4]. This is an effective method for maintaining the indoor temperature appropriately while consuming less energy. Building heat regulation accounts for about 40% of all energy use in cities, which can help to minimize the energy consumption significantly. The most common classifications of PCM's are organic, inorganic and eutectic [5, 6]. Paraffin is a straight-chain alkane combination with a melting point of 35 °C. The melting point and latent heat of fusion rise with increasing chain length. They are considered safe, dependable and economical because they are not corrosive or subcooling [7].

In addition to the efficiency of thermal energy exchange, the direct use of phase change materials in building materials can cause PCM to seep from the surface and low-viscosity liquids of melted hydrocarbons to diffuse throughout the building material [8, 9]. To avoid the above-mentioned challenges, the microencapsulation method is employed in the preparation of PCM. Microencapsulation is the process of enclosing solid particles or liquid droplets in a coating or embedding them in a homogeneous or heterogeneous matrix to create tiny capsules with various beneficial characteristics. The inorganic shell materials have several desirable properties, including superior mechanical durability, chemical inertness, great thermal stability and relatively high thermal conductivity which makes them advantageous instead of organic ones [10]. Due to its surface modification susceptibility to different functional groups, silica (SiO_2) is utilized for microencapsulating organic PCM [11]. Inorganic shell materials have additional advantages than organic shell materials, with such as toxicity, flammability, poor heat transfer performance and poor thermal stability. As a result, inorganic shell materials have received greater attention recently [12].

Reports on using paraffin as a phase change material by direct incorporation or encapsulation in a polymer matrix shell in geopolymer concrete have been studied by different researchers. Sukontasukkul et al. investigated PCM in lightweight geopolymer panels using paraffin combined with crushed aerated concrete as coarse aggregate. Since the PCM-incorporated geopolymer panels took longer to reach their peak temperature, they provided superior thermal insulation and storage [13]. Organic shell materials containing paraffin as the core material, such as polystyrene crosslinked with divinylbenzene and melamine–formaldehyde, were studied by Cao et al. The geopolymer's MPCM improved performance and the inertness of the polymer [14]. The integration of PCM with expanded clay aggregate in geopolymer mortar was investigated by Frattini et al. According to the results, coated aggregate with a paraffin concentration of between 30 and 40% by weight exhibited good mechanical properties, high fire resistance, and no detectable CO_2 or smoke emissions [15]. According to Rasoul et al., the inclusion of PCM drastically reduced the compressive strength of geopolymer mortar, mostly due to the reduced unit weight and stiffness of the material. Even then, the geopolymer mortar containing up to 20% PCM was still sufficiently strong for construction [16].

This study focuses on preparing paraffin as a core phase change material encapsulated by SiO_2 shell material, by sol–gel synthesis, which has a melting point in the range of 35–40 °C. The physical, mechanical, thermal and durability of geopolymer composites incorporating PCM and MPCM were investigated. Various analytical and thermal techniques such as functional group analysis by FT-IR, morphology by SEM analysis and latent heat from DSC are conducted. A building model was fabricated to find the temperature measurements between the material composite and the surroundings. These findings pave the way for offering sustainable and green buildings that consume less energy.

2 Materials and Methods

GGBS from the steel industry and FA from coal-based thermal power plants (ASTM C618a standard) were utilized as binder materials. The chemical composition of these is listed in Table 1. The alkali solution used was a mixture of NaOH and Na_2SiO_3. The commercial sodium silicate procured from Kuttuva Silicates Private Ltd (Madurai, India) has a silica modulus (SiO_2/Na_2O Ms) of 2.4 and sodium hydroxide at a concentration of 49%. Alkali activator solution adopted in the study has a molar ratio of 1.1 for Na_2O to SiO_2. The following materials were used to synthesize phase change materials: Paraffin, tetraethyl ortho silicate, sodium dodecyl sulphate (SDS), ethanol, hydrochloric acid and distilled water. Microencapsulated Paraffin@SiO_2 is prepared by sol–gel synthesis.

2.1 *Preparation of Microencapsulated P@SiO_2.*

About 0.2 g of sodium dodecyl sulphate was dissolved in 100 ml distilled water. To this solution, 10 g of paraffin is added and continuously stirred at 600 rpm for 2 h at

Table 1 Chemical composition of FA and GGBS by XRF analysis

Chemical composition (wt%)	CaO	SiO_2	Al_2O_3	K_2O	Fe_2O_3	SO_3	Loss of Ignition (%)
Fly ash	1.56	60.94	27.53	2.18	5.97	0.13	1.4
GGBS	55.25	21.58	14.88	0.48	1.78	–	1.7

70 °C. This is oil in water (O/W) emulsion. A total of 20 g of tetraethyl silicate, 20 g of anhydrous ethanol, and 40 g of distilled water were added separately to create the sol solution. Drops of hydrochloric acid were added to the mixture to bring its pH down to 2–3. A magnetic stirrer was used to agitate the liquid for an hour at 500 rpm while maintaining a temperature of 60 °C. The sol solution was added to the emulsion, which was vigorously stirred for 2 h at a rate of 800 rpm and 75 °C. Finally, white powders were collected in filter paper and washed with distilled water. The microencapsulated composites were dried in an oven at 50 °C overnight.

Preparation of Geopolymer Specimens. As a homogenization process, fly ash (80 wt%) and GGBS (20 wt%) were dry mixed in a Hobart mixer for 5 min. Then, sodium silicate solution was slowly added in a liquid-to-solid ratio of 0.40. Samples were prepared by the addition of 10% of paraffin and P@SiO_2 (5, 10, and 20%) to the geopolymer binder. The slurry mixes cast in the mould of dimension 50 mm^3 are kept at the standard age of 28 days. The PCM and MPCM was added as the last step of homogenization to avoid physical damage to the microcapsules. The designated sample IDs are: control is Gp, Paraffin admixed geopolymer is Gp-P and P@SiO_2 integrated geopolymer is Gp-P@SiO_2.

Characterization Studies and Fabrication of Experimental Setup. Compressive strength tests were conducted using an Automatic Compression Testing Machine (AIMIL COMPTEST 2000, India) for 28 days. A Bruker IFS 66v/S FT-IR spectrometer fitted with an ATR accessory set and a KRS-5 lens was used to acquire FT-IR spectra. With 32 scans, the absorbance spectra were obtained at a resolution of 2 cm^{-1} and a scanning speed of 5 kHz, covering the 4000–400 cm^{-1} range. Field emission scanning electron microscopy (FESEM) was performed to determine the sample's morphology using a JEOL JSM 6300 microscope outfitted with a tungsten filament electron source and a 20 kV accelerating voltage. The *X*-ray spectrometer Energy Dispersive Spectroscopy (EDS) was employed to determine the chemical compositions of the phases found. A Pyris Diamond differential scanning calorimeter (DSC) was employed to determine latent heat and melting temperature. Pre-weighed samples were subjected to repeated heating and cooling cycles for physical stability.

The thermal energy storage (TES) capacity was determined using a fabricated setup, as shown in Fig. 1. The setup comprised a computer, a data acquisition system, K-type thermocouples, a heating lamp and a building model. The test specimen was placed on top of a box composed of rectangular insulating material. A building model, aluminium hood with dimensions of 0.25 m * 0.20 m * 0.19 m, lined with glass wool further covered in fireproof insulation. The thermocouples were installed to record the temperatures at the exterior of the sample (TC 1), inside of the sample (TC 2), the centre of the sample holder, which is indoor (TC 3), and outside environment to record temperature fluctuations. The ambient temperature was measured for the test at 25 ± 2 °C. The specimen was exposed to a radiation lamp for 80 min before being cooled to room temperature. A data logger system automatically recorded the samples change in temperature.

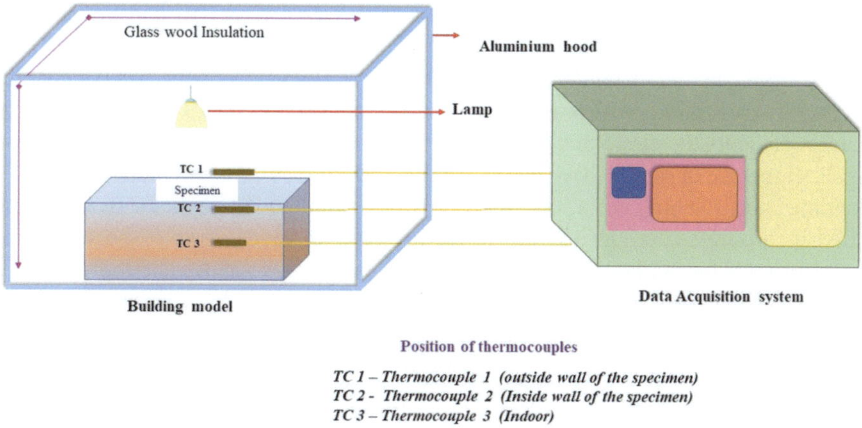

Fig. 1 Experimental setup for TES

3 Results and Discussion

3.1 Compressive Strength

The 28-day ambient cured geopolymer matrices are tested for compressive strength. The compressive strength of Gp, Gp- P@SiO_2 $^{(5\%)}$, Gp- P@SiO_2 $^{(10\%)}$ and Gp-P@SiO_2 $^{(20\%)}$ is 43.4 MPa, 38.8 MPa, 36.2 MPa and 33.73 MPa, respectively. The compressive strength of geopolymer decreases, although a smaller amount of MPCM is added. This is because MPCM is easily broken with a little force due to its low mechanical strength and hardness, making the matrix more porous by inducing void spaces [17]. Additionally, they establish weak links with the matrix and do not participate in geopolymerization reactions.

Leakage Test. Leakage test results of the samples paraffin, P@SiO_2, Gp- P and Gp- P@SiO_2 are represented in Fig. 2. The thermal cycle is continued till 100 times. The samples were heated for 30 min at 40 °C by keeping weighed quantity on the filter paper. At this heating phase, the PCM began to melt. We can see the liquid traces on the filter paper bearing Paraffin and Gp-P samples from visual observation. In contrast, there was no leakage from the prepared P@SiO_2 and Gp-P@SiO_2. Significantly, the capsules are intact as it was observed even after 100 heat cycles.

Microstructure Analysis. The FT-IR spectra of the P@SiO_2 composite and paraffin are shown in Fig. 3a. The peaks at 2919 and 2849 cm^{-1} indicate the stretching vibration of C–H, and the peaks at 1464 cm^{-1} represent the deformation vibration of $-CH_2$ and $-CH_3$. The in-plane rocking vibration of $-CH_2$ is at 724 cm^{-1}. In P@SiO_2, the bending vibration of Si–O–Si is seen from the peaks 1074, 799 and 450 cm^{-1}, along with all the functional groups of paraffin [18]. Since no other functional group assignments are present, it can be concluded that the P@SiO_2 composite is solely the outcome of the physical interaction between SiO_2 and paraffin.

Figure 3b shows the FT–IR spectra of raw materials FA, GGBS and geopolymer composites. The asymmetric stretching vibrations of the tetrahedral unit of Si -O are seen at 1093 cm^{-1} in FA and 969 cm^{-1} in GGBS. The band at 950 cm^{-1} in GP denotes the gel cross-linking, which is situated between FA and GGBS geopolymer. The band represents the stretching vibration of O–C–O at 1415 cm^{-1} from carbonation of alkali surface of matrix. Two bands at 1650 and 3306 cm^{-1} are related to H–O–H's bending and stretching vibrations in geopolymer samples [19]. In the FTIR spectra of PCM-incorporated composites, GP-P and GP-P@SiO_2 seen to feature the distinctive peaks of paraffin, without significant changes in the Si–O–T tetrahedral unit. The effectiveness of the physical encapsulation is demonstrated by the absence of any changes in geopolymeric functional groups when PCM is present.

SEM pictures and the EDAX pattern were used to observe the morphology and microstructure of the paraffin and P@SiO_2 composite, as shown in Fig. 4. An SEM image shows the flaky structures of paraffin and quasi-spherical particles when microencapsulated. The microcapsule agglomeration has been seen with coarse and

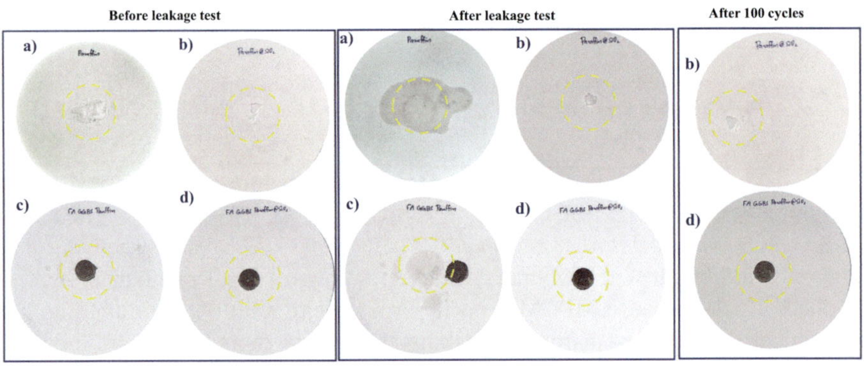

Fig. 2 Leakage test of **a** paraffin, **b** P@SiO_2, **c** Gp- P and **d** Gp- P@SiO_2

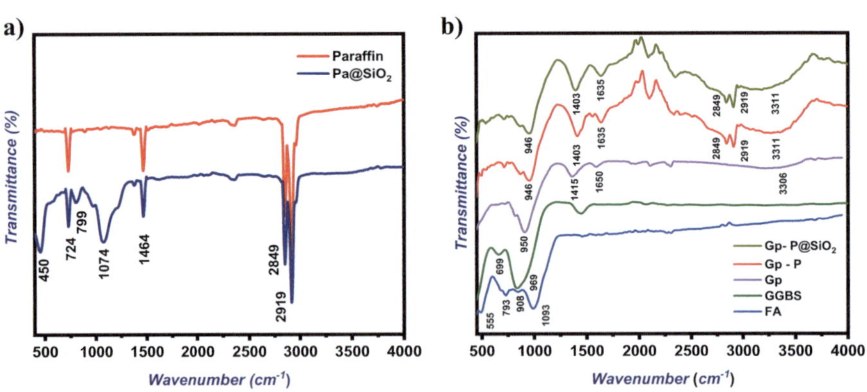

Fig. 3 FTIR analysis of PCM and geopolymer composites

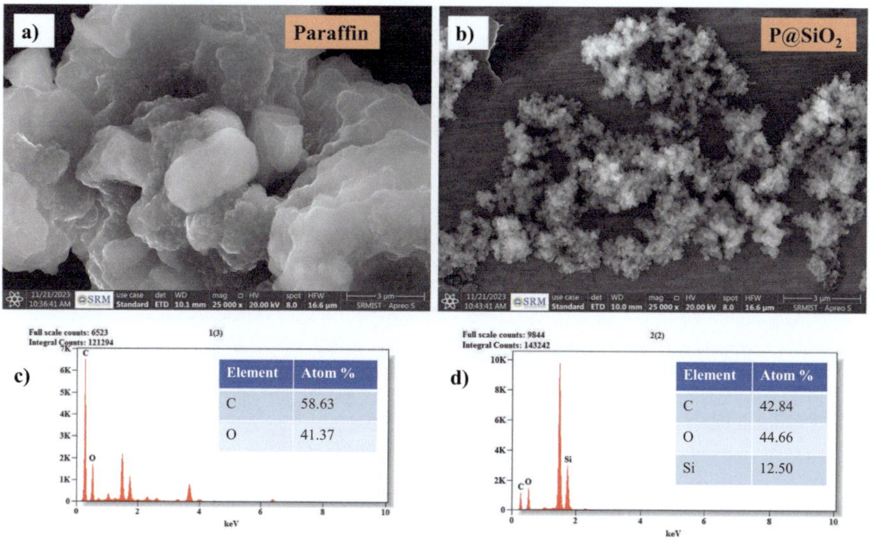

Fig. 4 SEM microstructure and EDAX pattern of Paraffin and P@SiO_2

rigid surfaces. This may be due to the acceleration of the reaction mechanism towards the surface of paraffin droplets [20]. SEM images represented in Fig. 5a clarified the phases of unreacted fly ash, GGBS, geopolymeric gel, and PCM. The highlighted regions in Fig. 5b and 5c, represent the morphology of Paraffin and P@SiO_2 present in the geopolymer matrix. Since geopolymer composites are highly heterogeneous, adding PCM and MPCM has no significant effect in predicting the morphology.

4 Thermal Analysis

Figure 6 represents the endothermic curve of Gp, Gp-P and Gp-P@SiO_2 composites, from differential scanning calorimetry. For the control geopolymeric samples, Gp, an endotherm peak, is observed at 92.1 °C, representing the dehydration of occluded water. These endotherms are common at this temperature [21]. The composites containing paraffin showed entirely different thermograms. The integrated paraffin evolved while heating the sample as an endothermic peak at 43.6 °C, 57.2 °C and 42.6 °C, 56.7 °C and 69.6 °C, respectively. The calculated latent energy is 18.116 J/g and 15.695 J/g for Gp-P and Gp-P@SiO_2, respectively. The impact of shell SiO_2 can explain the difference in latent energy among them.

Fig. 5 SEM microstructure **a** Gp, **b** Gp-P, and **c** Gp-P@SiO_2

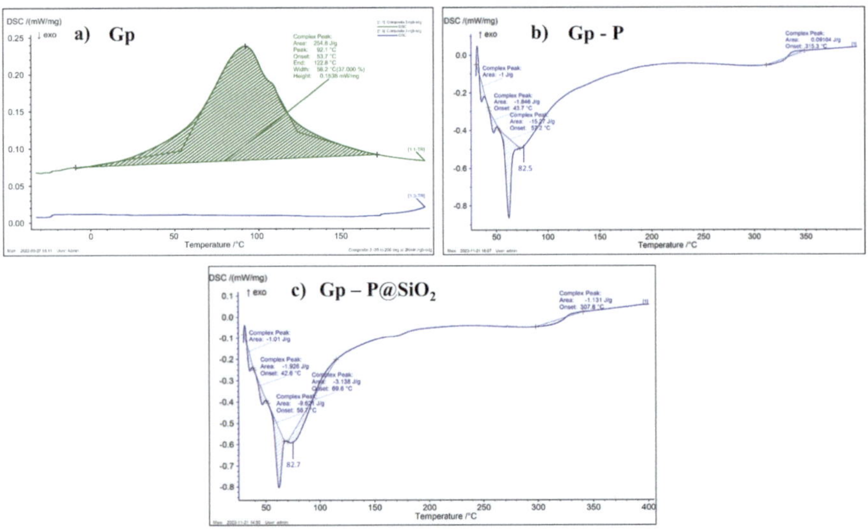

Fig. 6 DSC curves of geopolymer composites

5 Thermal Energy Storage Capacity

The PCM-impregnated FA–GGBS geopolymer binder, Gp-P and Gp-P@SiO_2, with different weight percentages of Paraffin (5%) and P@SiO_2 (5, 10 and 20%) and control geopolymer binder, Gp, was thermally measured. The temperature on the outside surface is continuously monitored on heating and cooling. The indoor temperature measured is shown in Fig. 7.

Upon heating, when paraffin starts to melt, the indoor temperature curves of the samples showed a maximum peak temperature at the time of 80 min for Gp, Gp-P $^{5\%}$, Gp-P@SiO_2 $^{5\%}$, Gp = P@SiO_2 $^{10\%}$, Gp-P@SiO_2 $^{20\%}$ is 36 °C, 34 °C, 33 °C, 31 °C and 30 °C, respectively. The control without PCM attained the peak temperature of 36 °C and slowly fell with the integration of PCM. Among different doses, the one with 20% microcapsule addition showed the lowest temperature. The test room featuring a top board containing 20% of P@SiO_2 has an indoor peak temperature of 4 °C lower than the test room featuring a control geopolymer binder. Therefore, 20% P@SiO_2 in a geopolymer binder lowers energy consumption by reducing indoor temperature variation, and the function is improved as the mass ratio of P@SiO_2 increases.

It is noted that the indoor temperature of all the samples has increased and the attained temperature at the end of 140 min, the temperature increase was found to be 25 °C, 22 °C, 22 °C, 25 °C and 28 °C for Gp, Gp-P $^{5\%}$, Gp-P@SiO_2 $^{5\%}$, Gp-P@SiO_2 $^{10\%}$, Gp – P@SiO_2 $^{20\%}$, respectively. The obtained indoor temperature and a corresponding drop in temp calculated from TC 2 and TC 3 values and the

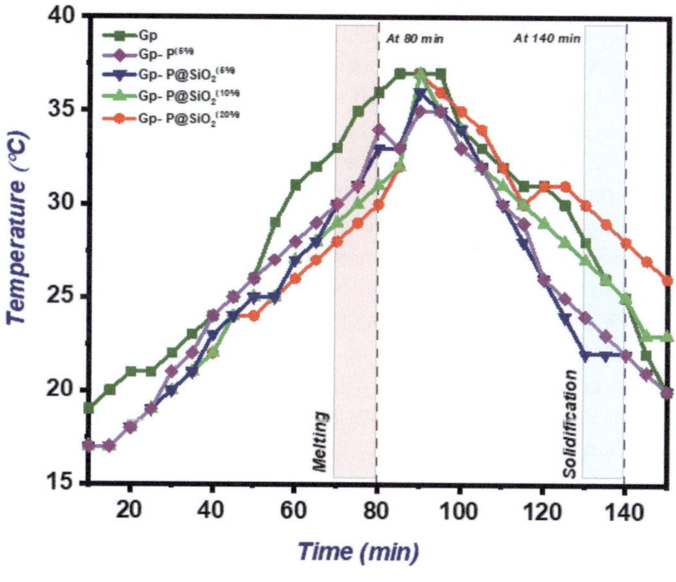

Fig. 7 Indoor temperature, TC 3

Table 2 Temperature measured at TC 2 and TC 3 (heating and cooling)

Temperature (°C)	Geopolymer composites				
	Gp	Gp-P (5%)	Gp-P@SiO_2 (5%)	Gp-P@SiO_2 (10%)	Gp-P@SiO_2 (20%)
Heating process (End of 80 min)					
Inside (TC 2)	37	36	36	34	34
Indoor (TC 3)	36	34	33	31	30
Decrease in indoor temperature (TC 2-TC 3)	1	2	3	3	4
Cooling process (End of 140 min)					
Inside (TC 2)	26	21	21	24	24
Indoor (TC 3)	25	22	22	25	28
Increase in indoor temperature (TC 2-TC 3)	1	− 1	− 1	− 1	− 4

corresponding drop in indoor temperature for the different specimens are included in Table 2.

6 Conclusion

The present study summarizes the preparation and properties of microencapsulated paraffin and SiO_2 shell and its implementation in ambient cured FA–GGBS. The integration of MPCM in the geopolymer matrix slightly declines the compressive strength of the matrix from 43 to 33 MPa, which still satisfies the desirable characteristic strength of structural elements. No liquid traces of PCM were seen on filter paper during the leakage test till 100 cycles for the samples, P@SiO_2 and Gp-P@SiO_2, depicting that microencapsulation has been done successfully. FTIR spectrum of Paraffin and P@SiO_2 replicated the bands of paraffin without altering the asymmetric stretching of Si–O–T, fingerprint region of geopolymer. The thermal energy storage performance (TES) was found to be significant at the optimal dose of 10–20% of P@SiO_2 in the geopolymer matrix. It greatly helps lower the indoor temperature by 3–4 °C, during the heating and increasing the indoor temperature by 4 °C, during the cooling. The results of this study demonstrate the application of microcapsules of PCM in green alternate geopolymers for meeting the heat storage demand for building materials.

Acknowledgements This work was supported by the Department of Science and Technology, Government of India (GOI), under the grant DST/TDT/WMT/2017 14/03/18, GOI. Authors greatly acknowledge the utilization of facilities provided by NRC, the Department of Chemistry, SRM Institute of Science and Technology. The authors acknowledge Technical assistants, the Strength of Material Laboratory, the Department of Mechanical Engineering, SRM Institute of Science and Technology for the compression testing facility.

Credit Authorship and Contribution Statement Rithikaa Thanigaiselvan: Investigation, data curation, original draft.

Mullai Malar. A and Vanitha. N: Investigation, data curation.

R. Jeyalakshmi: Conceptualization, project administration and review as well as editing.

Declaration of Competing Interest The authors declare that they have no known competing financial interests or personal relationships that could have appeared to influence the work reported in this paper.

References

1. Cemil Alkan; Ahmet Sari (2008) Fatty acid/poly(methyl methacrylate) (PMMA) blends as form-stable phase change materials for latent heat thermal energy storage. Sol Energy 82(2):118–124
2. Suppes GJ, Goff MJ (2003) Shailesh lopes: latent heat characteristics of fatty acid derivatives pursuant phase change material applications. Chem Eng Sci 58(9):1751–1763
3. Sari A (2003) Thermal characteristics of a eutectic mixture of myristic and palmitic acids as phase change material for heating applications. Appl Therm Eng 23(8):1005–1017
4. Abderrahman M, Taoufiq M, Azeddine M (2018) Phase change materials for building applications: a thorough review and new perspectives. Buildings 8(5):63
5. Anisur MR, Mahfuz MH, Kibria MA, Saidur R, Metselaar IHSC, Mahlia TMI (2013) Curbing global warming with phase change materials for energy storage. Renew Sustain Energy Rev 18:23–30
6. Cao L, Su D, Tang Y, Fang G, Tang F (2015) Properties evaluation and applications of thermal energy storage materials in buildings. Renew Sustain Energy Rev 48:500–522
7. Ahmet S, Cemil A, Cahit B (2014) Micro/nano encapsulation of some paraffin eutectic mixtures with poly(methyl methacrylate) shell: preparation, characterization and latent heat thermal energy storage properties. Appl Energy 136:217–227
8. Shi X, Memon SA, Tang W, Cui H, Xing F (2014) Experimental assessment of position of macro encapsulated phase change material in concrete walls on indoor temperatures and humidity levels. Energy Build 71:80–87
9. Abhat A (1983) Low temperature latent heat thermal energy storage: heat storage materials. Solar Energy 30(4):313–332
10. Sivanathan A, Dou Q, Wang Y, Li Y, Corker J, Zhou Y, Fan M (2020) Phase change materials for building construction: an overview of nano-/micro-encapsulation. Nanotechnol Rev 9(1):896–921
11. Mosstafa K, Massoud G, Ali M (2017) Cobalt ferrite nanoparticles ($CoFe_2O_4$ MNPs) as catalyst and support: magnetically recoverable nano-catalysts in organic synthesis. Nanotechnol Rev 7(1):43–68
12. Guruprasad A, Yaxue L, Lingkun L, Guiyin F (2017) Synthesis, characterization and applications of microencapsulated phase change materials in thermal energy storage: a review. Energy Build 144:276–294
13. Sukontasukkul P, Nontiyutsirikul N, Songpiriyakij S, Sakai K, Chindaprasirt P (2016) Use of phase change material to improve thermal properties of lightweight geopolymer panel. Mater Struct 49(11):4637–4645
14. Cao VD, Pilehvar S, Salas-Bringas C, Szczotok AM, Bui TQ, Carmona M, Rodriguez JF, Kjøniksen AL (2019) Thermal analysis of geopolymer concrete walls containing microencapsulated phase change materials for building applications. Solar Energy 178:295–307

15. Frattini D, Roviello G, Ferone C, Ricciotti L, Cioffi R (2015) Geopolymer-based composite materials containing PCM for thermal energy storage. In: Ceramic for energy, Italy
16. Shadnia R, Zhang L, Li P (2015) Experimental study of geopolymer mortar with incorporated PCM. Constr Build Mater 84:95–102
17. Haily E, Ait Ousaleh H, Zari N, Faik A, Bouhfid R, Qaiss A (2023) Use of a form-stable phase change material to improve thermal properties of phosphate sludge-based geopolymer mortar. Constr Build Mater 386:131570
18. Li B, Liu T, Hu L, Wang Y, Gao L (2013) Fabrication and properties of microencapsulated Paraffin@SiO_2, Phase change composite for thermal energy storage. ACS Sustain Chem Eng 1(3):374–380
19. Poornima N, Sivasakthi M, Jeyalakshmi R (2022) Microstructure investigation of the Na/Ca aluminosilicate hydrate gels and its thermal compatibility in fly ash–GGBS cementitious binder. J Build Eng 50:104168
20. Zhang H, Wang X (2009) Synthesis and properties of microencapsulated n-octadecane with polyurea shells containing different soft segments for heat energy storage and thermal regulation. Solar energy materials and solar cells 93(8):1366–1376
21. Ismail N, El-Hassan H (2018) Development and characterization of fly ash–slag blended geopolymer mortar and lightweight concrete. J Mater Civ Eng 30(4):04018029

MIX
Papier aus verantwortungsvollen Quellen
Paper from responsible sources
FSC® C105338

If you have any concerns about our products,
you can contact us on
ProductSafety@springernature.com

In case Publisher is established outside the EU,
the EU authorized representative is:
Springer Nature Customer Service Center GmbH
Europaplatz 3, 69115 Heidelberg, Germany

Printed by Libri Plureos GmbH
in Hamburg, Germany